Topics in Fluorescence Spectroscopy

Volume 10
Advanced Concepts in
Fluorescence Sensing
Part B: Macromolecular Sensing

Topics in Fluorescence Spectroscopy

Edited by JOSEPH R. LAKOWICZ and CHRIS D. GEDDES

Topics in Fluorescence Spectroscopy

Volume 10
Advanced Concepts in
Fluorescence Sensing
Part B: Macromolecular Sensing

Edited by

CHRIS D. GEDDES

The Institute of Fluorescence
Medical Biotechnology Center
University of Maryland Biotechnology Institute
Baltimore, Maryland

and

JOSEPH R. LAKOWICZ

Center for Fluorescence Spectroscopy and
Department of Biochemistry and Molecular Biology
University of Maryland School of Medicine
Baltimore, Maryland

 Springer

The Library of Congress cataloged the first volume of this title as follows:

Topics in fluorescence spectroscopy/edited by Chris D. Geddes and Joseph R. Lakowicz.
  p. cm.
  Includes bibliographical references and index.
  Contents: v. 1. Techniques
  1. Fluorescence spectroscopy. I. Geddes, Chris D. II. Lakowicz, Joseph R.

QD96.F56T66 1991   91-32671
543′.0858—dc20   CIP

ISSN: 1574-1036
ISBN 0-387-23644-9   Printed on acid-free paper

Printed in the United States of America

9 8 7 6 5 4 3 2 1   SPIN 11333586

springeronline.com

CONTRIBUTORS

Caleb Behrend. Department of Chemistry, University of Michigan, Ann Arbor, Michigan, 48109-1055.

Murphy Brasuel. Department of Environmental Health Sciences, University of Michigan, Ann Arbor, Michigan, 48109-1055.

Sarah M. Buck. Raoul Kopelman. Department of Chemistry, University of Michigan, Ann Arbor, Michigan, 48109-1055.

Eun Jeong Cho. Department of Chemistry and Biochemistry, Institute for Cellular and Molecular Biology, The University of Texas at Austin, Austin, TX 78712.

Sonja Draxler. Institut für Experimentalphysik, Karl-Franzens-Universität Graz, A-8010 Graz Austria.

Andrew D. Ellington. Department of Chemistry and Biochemistry, Institute for Cellular and Molecular Biology, The University of Texas at Austin, Austin, TX 78712.

Peter M. Haggie. Departments of Medicine and Physiology, Cardiovascular Research Institute, University of California, San Francisco, CA, 94143-0521.

Tony D James. Department of Chemistry, University of Bath, Bath BA2 7AY UK.

Hui Jiang. Boston University, Chemistry Department and Photonics Center, Boston, MA 02215.

Guilford Jones. II. Boston University, Chemistry Department and Photonics Center, Boston, MA 02215.

Yong-Eun Lee Koo. Department of Chemistry, University of Michigan, Ann Arbor, Michigan, 48109-1055.

Robert Massé. Applied Research and Development, MDS Pharma Services, 2350 Cohen Street, Montreal, QC, Canada.

Eric Monson. Department of Chemistry, University of Michigan, Ann Arbor, Michigan, 48109-1055.

Martin A. Philbert. Department of Environmental Health Sciences, University of Michigan, Ann Arbor, Michigan, 48109-1055.

William S. Powell. Meakins-Christie Laboratories, McGill University, 3626 St. Urbain Street, Montreal, QC, Canada.

Manjula Rajendran. Department of Chemistry and Biochemistry, Institute for Cellular and Molecular Biology, The University of Texas at Austin, Austin, TX 78712.

Alnawaz Rehemtulla.Molecular Therapeutics Inc., Ann Arbor, Michigan, 48109.

Brian Ross. Molecular Therapeutics Inc., Ann Arbor, Michigan, 48109

Seiji Shinkai. Department of Chemistry and Biochemistry, Graduate School of Engineering, Kyushu University, Fukuoka 812-8581 JAPAN.

Richard B. Thompson. Department of Biochemistry and Molecular Biology, School of Medicine and Center for Fluorescence Spectroscopy, University of Maryland, Baltimore, Maryland 21201.

Petra Turkewitsch. Applied Research and Development, MDS Pharma Services, 2350 Cohen Street, Montreal, QC, Canada.

A.S. Verkman. Departments of Medicine and Physiology, Cardiovascular Research Institute, University of California, San Francisco, CA, 94143-0521.

Valentine I. Vullev. Boston University, Chemistry Department and Photonics Center, Boston, MA 02215.

Hao Xu. Department of Chemistry, University of Michigan, Ann Arbor, Michigan, 48109-1055.

PREFACE

Over the last decade fluorescence has become the dominant tool in biotechnology and medical imaging. These exciting advances have been underpinned by the advances in time-resolved techniques and instrumentation, probe design, chemical / biochemical sensing, coupled with our furthered knowledge in biology.

Ten years ago Volume 4 of the Topics in Fluorescence Spectroscopy series outlined the emerging trends in time resolved fluorescence in analytical and clinical chemistry. These emerging applications of fluorescence were the result of continued advances in both laser and computer technology and a drive to develop red/near-infrared fluorophores. Based on the advancements in these technologies, it was envisaged that small portable devices would find future common place in a doctor's office or for home health care.

Today, these past emerging trends in fluorescence sensing are now widely used as either standard practices in clinical assessment or commercialized health care products. Miniature lasers in the form of laser diodes and even light emitting diodes are widely used in applications of time-resolved fluorescence. Computer clock-speed is now not considered a hurdle in data analysis. Even our choice of fluorophores has changed dramatically in the last decade, the traditional fluorophore finding continued competition by fluorescent proteins and semi-conductor quantum dots, to name but just a few.

This volume "Advanced Concepts in Fluorescence Sensing: Macromolecular Sensing" aims to summarize the current state of the art in fluorescence sensing. For this reason we have invited chapters, encompassing a board range of macromolecular fluorescence sensing techniques. Chapters in this volume deal with macromolecular sensing, such as using GFP, Aptamers and fluorescent pebble nano-sensors. This volume directly compliments volume 9 of the Topics in Fluorescence Spectroscopy series, which deals with advanced concepts in small molecule fluorescence sensing.

While many of the changes in recent fluorescence have been well received, its continued growth in the world has created a challenge in trying to archive and document its use. Subsequently Chris D. Geddes has now become co-series editor of the Topics in Fluorescence Spectroscopy series. We have also recently launched the Reviews in Fluorescence series, which co-edited also by Dr's Geddes and Lakowicz and published annually, is meant to directly compliment the Topics in Fluorescence Spectroscopy series, with small chapters summarizing the yearly progress in fluorescence.

Finally we would like to thank all the authors for their excellent contributions, Mary Rosenfeld for administrative support and Kadir Aslan for help in typesetting both volumes 9 and 10.

Chris D. Geddes
Joseph R. Lakowicz
Baltimore,Maryland, US.
August 2004

CONTENTS

7. EXCIMER SENSING

Valentine I. Vullev, Hui Jiang, and Guilford Jones, II

8. LIFETIME BASED SENSORS / SENSING

Sonja Draxler

PROTEIN-BASED BIOSENSORS WITH POLARIZATION TRANSDUCTION

Richard B. Thompson[1]

1.1. INTRODUCTION

Fluorescence-based biosensors employing biological recognition molecules such as proteins offer unmatched selectivity and sensitivity for real time determination and imaging of analytes such as metal ions. In some cases the analyte can be quantitated by changes in fluorescence anisotropy (polarization) which offers all the advantages of classical ratiometric techniques, as well as certain optical advantages, especially for microscopy. This chapter introduces the principles of such sensors and displays some examples of the results which may be obtained by their use.

1.2. PRINCIPLES OF OPERATION

The operating principles of anisotropy-based sensors are fairly straightforward, which is perhaps unsurprising given the relatively simple physics of fluorescence polarization and its measurement (Joseph R. Lakowicz, 1999). Fluorescence emission may be polarized, the degree of which can be described quantitatively by the polarization, p, or the anisotropy, r. The polarization and anisotropy are calculated from observed fluorescence intensities through polarizer(s) oriented parallel ($I_{\parallel}$) and perpendicular ($I_{\perp}$) to the plane of polarization of the excitation **Figure 1.1**):

[1] Department of Biochemistry and Molecular Biology, School of Medicine and Center for Fluorescence Spectroscopy, University of Maryland, Baltimore, Maryland 21201

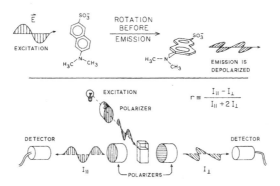

Figure 1.1. Principle of fluorescence anisotropy / polarization.

Anisotropy and polarization are calculated:

$$r = (I_\parallel - I_\perp) / (I_\parallel + 2\,I_\perp) = [(I_\parallel / I_\perp) - 1] / [(I_\parallel / I_\perp) + 2] \tag{1}$$

$$p = (I_\parallel - I_\perp) / (I_\parallel + I_\perp) = [(I_\parallel / I_\perp) - 1] / [(I_\parallel / I_\perp) + 1] \tag{2}$$

The terms "polarization" and "anisotropy" are synonymous in this context and algebraically interconvertible, but the theory of polarized fluorescence is much simpler in terms of anisotropy than polarization, thus we shall employ the former term herein. Note that the term polarization appears in much of the older literature and currently in the high throughput screening and clinical immunoassay literature. Polarization of fluorescence is observed when fluorophores excited with plane polarized light do not rotationally diffuse significantly before emitting. Quantitatively, the degree of polarization may be predicted by the Perrin-Weber equation:

$$r_0 / r = 1 + (\tau / \theta_c) \tag{3}$$

where r_0 is the (excitation wavelength-dependent) limiting anisotropy in the absence of diffusion, r is the observed anisotropy, τ is the fluorescence lifetime, and θ_c is the rotational correlation time of the fluorophore, the inverse of the rotational rate. Rigorous treatments of the time-resolved decay of anisotropy, particularly of non-symmetrical fluorophores, are beyond the scope of this article and have been disclosed elsewhere (Joseph R. Lakowicz, 1999; Steiner, 1991). The rotational correlation time can be estimated:

$$\theta_c = \eta V / RT \tag{4}$$

where T is temperature, η is viscosity, R is the gas constant, and V is the volume of the rotating unit. For more or less spherical macromolecules this can be approximated by the dimensions of the macromolecule, corrected for hydration. Thus a reasonably spherical 45,000 Dalton protein might exhibit roughly a 21 nsec rotational correlation time. For most of the assays it is unnecessary to know θ_c, rather one only need know it approximately to choose a fluorophore with the appropriate lifetime (see below).

To a first approximation, one can thus imagine assaying or detecting an analyte by changes in fluorescence anisotropy if the analyte somehow changes the rotational rate or lifetime of the fluorophore. While the shape and size of the fluorophore would appear to be properties of the molecule and therefore immutable, it is in fact commonplace to attach the fluorophore to a macromolecule and perturb the macromolecule's shape or size to change the anisotropy. Similarly, although the intrinsic lifetime of the fluorophore is a property of the molecule, many means are known for reducing the lifetime by quenching, and they result in measurable changes in anisotropy as well. The basic theory of the effects of changing either rotational or apparent lifetimes is presented below. Although fluorescence anisotropy has been used to look at "microviscosity" of lipid bilayers and micelles as a means of understanding their dynamics (Shinitzky & Barenholz, 1978), as well as assessing fetal lung maturity by the microviscosity of the surfactant, these essentially are studies of the bulk properties of the fluorophore's surroundings, and not of analytes *per se*, and will not be considered further. Similarly, there are many well-known enzyme assays based on fluorescence polarization, which usually rely on a change in size between substrate and product (e.g., protease assays) to provide signal. For the most part, these cannot provide a continuous readout of the level of the analyte, and also will not be considered further.

1.3. ADVANTAGES OF ANISOTROPY-BASED SENSING

Among fluorescence-based sensing approaches, anisotropy-based sensing has a unique portfolio of advantages. First, it is a steady-state (as opposed to a time-resolved) measurement, which requires only simple instrumentation. It can be readily configured in the "T-format"as a true ratiometric measurement (**see Figure 1.1**) as described by Weber nearly fifty years ago to avoid the noisy excitation sources then available (Weber, 1956)(Equations 1 and 2). The advantages of fluorescence ratiometric measurements are widely appreciated based on the use of ratiometric calcium and pH indicators. They include (to a first approximation) immunity from fluctuations in the light source, variations in indicator amount due to bleaching or washout (but see Dinely, et al., for an important caveat regarding indicator amounts) (Dinely, Malaiyandi, & Reynolds, 2002), variations in cell thickness or indicator distribution, and potentially facile calibration *in situ* (Nuccitelli, 1994). Several groups have constructed fluorescence polarization microscopes (Axelrod, 1989; Dix & Verkman, 1990)including confocal microscopes (Bigelow, Conover, & Foster, 2003)and demonstrated images with contrast based on variations in fluorescence polarization. While polarization can be measured merely by changing polarizers in a filter wheel apparatus, electrooptic devices offer the prospect of changing polarization at faster than video frame rates. An important potential advantage is that polarization requires no change of excitation or emission wavelength, making it particularly well-suited for laser-excited microscopies such as confocal microscopy and multiphoton excitation microscopy. While few reports have appeared describing multiphoton excited fluorescence anisotropy sensing (Thompson, Maliwal, & Zeng, 2000), this technique could potentially be incorporated into an imaging application as well. While filtration to avoid scattered light is important, it is a further advantage that one may use the entire emission band for anisotropy measurements instead of a narrow band for wavelength ratiometric measurements. Finally, anisotropy-based sensing offers the possibility (in suitable cases) of an expanded dynamic range (see below)(Thompson, Maliwal, & Fierke, 1998).

One of the strengths of anisotropy-based sensing is its simplicity, since relatively few phenomena perturb the rotational diffusion rate of a molecule: changes in medium viscosity or temperature, as well as changes in molecular size or shape. It is usually trivial to control the temperature and bulk composition of the assay medium, or at least assure they do not change significantly during the short time of the assay. Similarly, few things can influence the lifetime of fluorescence, as this is an inherent property of the molecule. The important exception , as we will see, is fluorescence quenching, but under ordinary conditions most quenchers must be present at near millimolar concentrations to perturb the anisotropy of typical fluorophores with nanosecond lifetimes.

1.4. FLUORESCENCE POLARIZATION IMMUNOASSAY

The earliest example of determination of a chemical analyte by fluorescence anisotropy is the fluorescence polarization immunoassay introduced by Dandliker for antigens that are not macromolecules (Dandliker, Kelly, Dandliker, Farquhar, & Levin, 1973) In its simplest form this is configured as a competition assay wherein a sample containing an unknown amount of analyte is mixed with known amounts of antibody and a fluorescent-labeled analog of the analyte. The fluorescent-labeled analyte competes with the unlabeled for the binding site on the antibody; the bound fluorophores (molecular weight < 1000 Daltons) have their rotational diffusion substantially reduced by tight association with the relatively massive antibody (molecular weight of an IgG is c. 150,000 Daltons) **Figure 1.2**.

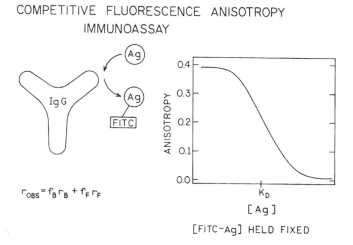

Figure 1.2. Principle of fluorescence polarization immunoassay. Antigen (Ag) in the sample competes with a fixed amount of fluorescent-labeled antigen (FITC-Ag) to bind to a fixed amount of antibody IgG.

For a label like fluorescein with roughly a four nanosecond lifetime, one can calculate using Equations 1 and 2 that the free labeled analyte should exhibit an anisotropy of roughly 0.05 and the bound form, 0.36. As the analyte concentration in the sample increases the proportion of bound fluorescent-labeled analyte decreases, and with it the anisotropy, which is just the arithmetic sum of the anisotropies of the labeled antigen in each form:

$$r_{obs} = r_f f_f + r_b f_b \qquad\qquad (5)$$

where r_{obs} is the observed anisotropy; r_f and r_b are the anisotropies of the free and bound forms, respectively; and f_f and f_b are the fractions of fluorophore free and bound. If the apparent quantum yield changes upon binding, a small correction is necessary (Joseph R. Lakowicz, 1999).

The results of such an immunoassay are depicted in **Figure 1.3**. Evidently such analytes can be quantified by fluorescence polarization immunoassay at the micromolar level and below with good accuracy, if a fluorescent analog of the antigen can be made with similar affinity for the antibody. This technique is still in wide clinical use, particularly for analysis of drugs of abuse in urine specimens. The advantages of such immunoassays are that they are much faster than ELISA'a or RIA's because washing is unnecessary, that the same instrument can be used for a variety of analytes (if the fluorophores are spectrally similar), calibration is straightforward, and for analytes such as drugs of abuse it is difficult to adulterate specimens to produce a false negative. An important advantage is that no modification or purification of the antibody is necessary, such that even antisera can be used, with significant cost savings.

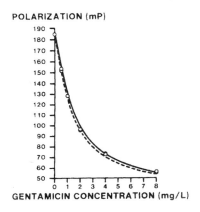

Figure 1.3. Fluorescence polarization immunoassay results for gentamicin; reproduced with permission from (Jolley et al., 1981).

Until recently, such immunoassays were unsuited for macromolecule antigens because the ratios of rotational correlation time to lifetime for free vs. bound antigen were not significantly different for typical nanosecond fluorophores **see Figure 1.4.** The figure depicts the expected anisotropies for a macromolecule antigen (serum albumin in this example, molecular weight 65,000) in the free form and bound to a high molecular weight antibody such as an IgM (molecular weight about one million) for two different label lifetimes. If the label lifetime is relatively short (4 nsec), the long rotational correlation time of the labeled antigen differs little in the free and antibody-bound states (0.02); however, if the label lifetime is longer (400 nsec), the anisotropy difference between free and bound is large enough (0.165) to be really useful. Evidently the anisotropy change will be significant only if the label lifetime is upwards of 100 nanoseconds. Fortunately, the Lakowicz group (Guo, Castellano, Li, & Lakowicz, 1998; Terpetschnig, Szmacinski, & Lakowicz, 1995) introduced the use of long-lived metal to ligand charge transfer (MLCT) probes to immunoassay, many of which have lifetimes approaching one microsecond. With such labels determination of macromolecule antigens can be done by fluorescence anisotropy immunoassay; the results of such an assay are depicted **in Figure 1.5**

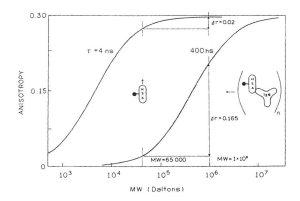

Figure 1.4. Simulated anisotropy as a function of molecular weight for two different fluorescence lifetimes: with a 4 nsec lifetime HSA exhibits a 0.02 anisotropy change upon binding to IgM; with a 400 nsec lifetime it exhibits a 0.165 change. Reproduced with permission from (Terpetschnig et al., 1995).

In this assay a solution of serum albumin labeled with a long lifetime (average lifetime in air = 2.75 microseconds) Re (I) complex is mixed with an anti-serum albumin IgG (mol. wt 150,000 Daltons), and challenged with samples containing varying concentrations of unlabeled serum albumin. The unlabeled antigen competes with the labeled antigen, effectively displacing it from the IgG, and resulting in an increasing proportion of the labeled antigen exhibiting the reduced rotational correlation time of the free form. The longer lifetime of this label and the smaller size of the IgG compared with the IgM in the simulation (**Figure 1.4**) results in reduced anisotropies of both forms, and more particularly a reduced difference (0.04) in the anisotropy between free and bound. Nevertheless, the difference in anisotropy is certainly usable, and these results indicate what the technique is capable of. .

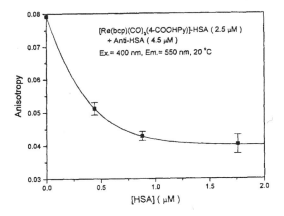

Figure 1.5. Fluorescence polarization immunoassay using a long-lived fluorescent label. Reproduced with permission from (Guo et al., 1998)

For applications requiring continuous measurements such immunoassays are not very useful, for two reasons. First, the fluorescent labeled analyte must be free to diffuse away from its binding site and diffuse rotationally. This makes it difficult to use in a continuous mode because of the risk of the small molecule diffusing away when the probe is exposed to the analyte. Also troublesome are the very slow dissociation rate constants of typical antibodies for their cognate antigens. The very slow off rate means that the binding reaction is often effectively irreversible. For instance, an antibody might have an affinity (K_d) for its cognate antigen of one picomolar (10^{-12} molar); the K_d is merely the ratio of association and dissociation rate constants for the antigen binding reaction. For a small molecule antigen (hapten) the association rate constant might be 10^7 to 10^8 M^{-1} sec^{-1} (nearly diffusion controlled) whereas for a macromolecule antigen, it might be 100-fold slower (10^5 M^{-1} sec^{-1}). Thus the dissociation rate constant in the former case would be 10^{-4} to 10^{-5} sec^{-1}, and for the macromolecule 10^{-7} sec^{-1}. The half-time of these antibody:antigen complexes is then just $0.693/k_{dissoc}$, or 1.9 to 19 hours for the hapten, and more than eleven weeks for the macromolecule antigen. One example exists of a recognition protein in a biosensor (carbonic anhydrase) being engineered to obtain more rapid association and dissociation constants to improve response times (Huang, Lesburg, Kiefer, Fierke, & Christianson, 1996), but to date we know of no antibody which has been molecularly engineered to improve binding kinetics.

1.5. ANISOTROPY-BASED METAL ION BIOSENSING

In general, it would appear difficult to determine the concentrations of most metal ions in solution by fluorescence anisotropy, since most metal ions of interest are not even photoluminescent in solution, and it clearly is difficult to raise antibodies to, for instance, free zinc or copper ions (but see (Darwish & Blake, 2002)). Fortunately, many metal ions can serve as quenchers by various mechanisms, and consequently can be made to report their presence by changes in fluorescence intensity (Fernandez-Gutierrez & Munoz de la Pena, 1985; White & Argauer, 1970), and in some cases, lifetime and anisotropy . For many classical metallofluorescent indicators such as hydroxyquinolines, interaction with metal ions may result in increases in quantum yield and lifetime (Szmacinski & Lakowicz, 1994). However, the lifetime of the indicator with either metal bound or in the absence of metal will typically be in the nanosecond range whereas its rotational correlation time is likely to be less than 100 psec, such that there is a negligible difference in anisotropy between the free and bound forms. Similarly, metallofluorescent indicators that detect analyte by collisional quenching exhibit decreased fluorescence lifetimes as well, but again the decline results in a negligible change in anisotropy for the same reason. Note that using an indicator with a very short lifetime confers no advantage because only very high concentrations of quencher are able to diffuse to and quench the indicator during its brief lifetime. Indicators which form a complex with the analyte are much more sensitive (Grynkiewicz, Poenie, & Tsien, 1985), but indicators of this sort which are quenched typically are completely quenched by the ligand, and consequently the bound form does not contribute to the emission.

We found that the issue of the rapid rotation of the fluorophore compared to its lifetime could be dealt with by attaching it to a macromolecule; moreover, the macromolecule could be chosen to bind particular metal ions with unmatched affinity and selectivity (Fierke & Thompson, 2001), and by judicious choice of the fluorophore, to provide high sensitivity, and even an enhanced dynamic range.

ANISOTROPY – BASED METAL ION SENSING

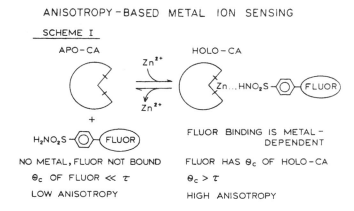

Figure 1.6. Schematic of zinc binding assay by fluorescence anisotropy. Binding of zinc promotes binding of a fluorescent aryl sulfonamide to the protein, increasing its rotational correlation time and anisotropy.

We have devised two methods for determining metal ions by fluorescence anisotropy. The first is based on metal-dependent binding of a fluorescent ligand to a protein, changing its rotational correlation time, the second by having the metal ion change the fluorescence lifetime of a fluorescent labeled protein. One can imagine elaborating either method to ligands other than metal ions. The first method takes advantage of the fact that aryl sulfonamide inhibitors of carbonic anhydrase exhibit much higher affinity for the protein in the zinc-containing holo form than the metal-free apo form. This is unsurprising inasmuch as much of the binding free energy comes from the (weakly acidic) aryl sulfonamide binding to the active site zinc ion as a fourth ligand, in the form of the sulfonamide anion (Maren, 1977) . Thus in the presence of zinc dansylamide binds relatively tightly (Kd ~0.8 uM) to holo carbonic anhydrase, but does not bind measurably to the apoprotein (Chen & Kernohan, 1967; Thompson & Jones, 1993). By binding to the protein the rotational diffusion of the fluorescent aryl sulfonamide is reduced and its anisotropy increases **Figure 1.6**.

Because the increase in rotational correlation time can be substantial (e.g., from perhaps 100 psec for a 500 Dalton inhibitor to 15 nsec for its complex with 30,000 Dalton protein), the increase in anisotropy can be dramatic: an example is shown for the sulfonamide BTCS **Figure 1.7**. In this case the fluorophore exhibits nearly a five-fold increase in its anisotropy upon binding, resulting in an anisotropy increase of 0.20. In terms of polarization this is 0.27, or 270 mP (a nomenclature used in describing polarization immunoassays). This is a substantial change, indicating that the assay can be very accurate. The detection limit

achieved for free zinc concentration in this example is a few picomolar, much better than has
been achieved with small molecule indicators (Kimura & Aoki, 2001; Thompson et al., 2002;
Walkup, Burdette, Lippard, & Tsien, 2000).

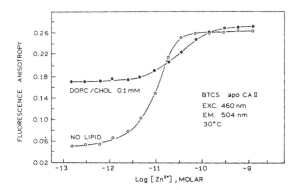

Figure 1.7. Zinc-dependent anisotropy of BTCS (benzothiazolyl coumarin sulfonamide) in the presence of
apocarbonic anhydrase without (open circles) and with DOPC:Cholesterol unilammelar vesicles. Reproduced from
(Thompson et al., 2000) with permission.

In some sense this approach has the defects of its virtues: of the metal ions which bind
to the active site only Zn strongly promotes the binding of sulfonamides, which makes the
assay quite specific. Cobalt(II) is much less effective in this regard (Sven Lindskog &
Thorslund, 1968) and generally quenches most fluorophores, making it inaccessible for this
type of assay. Other metals which bind to the protein do not promote sulfonamide binding,
and cannot be measured by this approach; consequently, it is very specific for zinc.. Another
issue is that several of the fluorescent aryl sulfonamides we and others (Chen & Kernohan,
1967) have developed for carbonic anhydrase-based Zn sensing exhibit large increases in
their fluorescence lifetimes upon binding to the protein, which largely offset the increases in
rotational correlation time, and therefore lead to only modest increases in anisotropy. Thus
Dapoxyl sulfonamide exhibits a nearly twenty-fold increase in quantum yield and lifetime,
which more than offsets the increase in rotational correlation time, leading to a negligible
change in anisotropy. For best results the fluorescent sulfonamide should have only a small
change in lifetime accompanying binding.

Some fluorescent aryl sulfonamides exhibit a substantial shift in their excitation or
emission wavelengths, which allows the dynamic range of the assay to be expanded by choice
of wavelength excited or observed (Thompson, Maliwal, & Fierke, 1998). Thus the
sulfonamide ABD-M exhibits a sixty nanometer blue shift in its emission upon binding to the
protein, enabling one to selectively observe the emission of a small proportion of the bound
form in the presence of a large excess of the free form by looking on the blue side of the

emission, and *vice versa*. Therefore the range of analyte concentration which can be determined with a given level of accuracy may be adjusted by choice of excitation or emission wavelength. In the example illustrated **Figure 1.8**, free zinc concentrations ranging from 0.003 to 0.3 nM may be quantitated by observing emission at 450 nm, whereas concentrations from 0.2 to 3.0 nMmay be accurately quantitated at emission wavelength 590 nm. This is an advantage anisotropy-based sensing shares with lifetime-based sensing (Szmacinski & Lakowicz, 1993; Thompson & Patchan, 1995) which does not obtain with wavelength-ratiometric determinations.

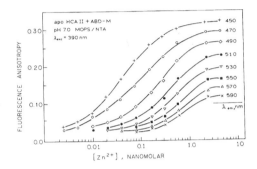

Figure 1.8.   Zinc-dependent fluorescence anisotropies of apo-carbonic anhydrase plus ABD-N observed at different emission wavelengths. Reproduced from (Thompson, Maliwal, & Fierke, 1998)with permission.

One of the drawbacks of this particular approach is the difficulty of configuring these assays as sensors: e.g., capable of being inserted into a solution and *continuously* measuring the analyte. The issue arises in that the aryl sulfonamide and zinc ion must both be free to diffuse to and from their respective binding sites on the protein, but the zinc ion must come from the aqueous matrix in which the transducer is inserted. The problem is that whereas the analyte may diffuse into and out of the transducer vicinity, the sulfonamide may do the same thing, with irretrievable loss (Thompson & Jones, 1993). While ionophores capable of selectively transporting the zinc ion through membranes are known (Ammann, 1986), the ion flux through these membranes in the absence of an applied potential is modest. Attaching the sulfonamide to a macromolecule to co-entrap it within a dialysis membrane perforce reduces its rotational diffusion and increases its anisotropy, reducing or eliminating the anisotropy change upon binding to the protein.

A related approach which does not use a separate, diffusible molecule takes advantage of the fact that analyte-induced quenching of a fluorophore may reduce its lifetime, and thereby increase its anisotropy. If the fluorophore is attached to a macromolecule (which may also selectively bind to the analyte, for instance), the rotational correlation time of the macromolecule may more closely match the (nanosecond) lifetime of the fluorophore, resulting in a sizable change in anisotropy upon binding of the analyte (and quenching). **Figure 1.9**.

ANISOTROPY - BASED SENSING

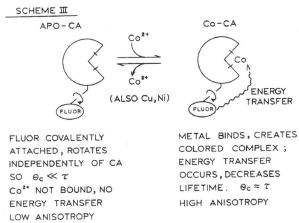

FLUOR COVALENTLY
ATTACHED, ROTATES
INDEPENDENTLY OF CA
SO $\theta_c \ll \tau$
Co^{2+} NOT BOUND, NO
ENERGY TRANSFER
LOW ANISOTROPY

METAL BINDS, CREATES
COLORED COMPLEX ;
ENERGY TRANSFER
OCCURS, DECREASES
LIFETIME. $\theta_c \approx \tau$
HIGH ANISOTROPY

Figure 1.9. Principle of anisotropy-based sensing of metal ions employing quenching by the metal ion.

It is important that the quenching of the label emission not be quantitative; e.g., there must be some emission from the bound form or no change in the anisotropy will be observed. It is convenient that the quenching in these cases is by Förster energy transfer (Forster, 1948), which is a through-space interaction. For instance, when Co, Cu, or Ni bind to apocarbonic anhydrase, they exhibit weak d-d absorption bands in the visible and near-IR which can serve as energy transfer acceptors for fluorophores whose emission overlaps the absorption (S. Lindskog et al., 1971). Using the theory of Forster transfer, we can predict the change in lifetime of the fluorophore upon binding of the metal ion based on the known absorbance of the metal ion in the bound state and the position of the fluorescent label with respect to the metal center, and thus the anisotropy response of a sensor transducer given the (known) rotational correlation time (Thompson, Maliwal, Feliccia, Fierke, & McCall, 1998) . The results of such a prediction using carbonic anhydrase are shown in **Figure 1.10**. Evidently, very usable changes in anisotropy can be expected for a properly positioned fluorescent label (which may be accomplished by site-selective mutagenesis of the protein and selective labeling (Thompson, Ge, Patchan, & Fierke, 1996) with the right lifetime (in comparison to the rotational correlation time of the protein). We note that there is a tradeoff in response of the system due to efficient quenching: greater quenching efficiency upon analyte binding reduces the lifetime more and increases the change in anisotropy, but also the relative contribution of the bound form to the total emission decreases. A happy medium is about 75 % quenching, which in this case results in a 55 % increase in anisotropy, and which is controlled by the positioning of the label with respect to the metal center.

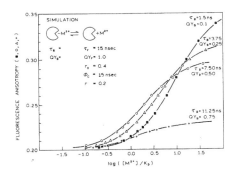

Figure 1.10. Simulated metal-dependent anisotropy of a 30 kD protein when binding of the metal results in 25% (+), 50 % (circles), 75% (triangles), or 90% (squares) decline in the lifetime of a 15 nsec lifetime label. Reproduced from (Thompson, Maliwal, Feliccia et al., 1998)with permission.

This approach works very well in practice, as illustrated in **Figure 1.11,** which illustrates the anisotropy response of a fluorescent-labeled apocarbonic anhydrase to a series of metal ions. The substantial changes in anisotropy indicate that the assay can be quite accurate. In this case the quenching of the fluorophore does not take place by energy transfer alone, as Cd and Zn do not exhibit d-d absorbance bands upon binding to the protein, but the quenching need only be partial and reduce the lifetime, it need not be by any particular mechanism. It is worthwhile noting that the selectivity and sensitivity of this sensor remains to be matched by any other (non-carbonic anhydrase-based) sensor.

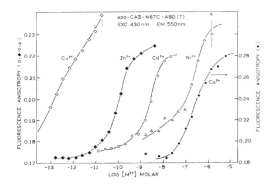

Figure 1.11. Cu- (open diamonds), Zn- (filled diamonds), Cd- (open circles), Ni- (triangles), and Co-dependent (filled circles) fluorescence anisotropies of apo-N67C-ABD-T. Reproduced from (Thompson, Maliwal, Feliccia et al., 1998) with permission.

It is also worth noting that the sensitivity and dynamic range of this (and other anisotropy-based approaches) is in part limited by the precision and accuracy with which anisotropy can itself be measured. For research-grade fluorometers and some microwell plate readers, accuracy and precision are usually within ± 0.002 . However, others have built instruments with much better precision and accuracy, and there seems no *ab initio* reason why the precision and accuracy of other instruments cannot be much better, particularly t-format instruments (Weber. 1956) or those using semiconductor sources and detectors.

Finally, we note that these approaches need not be limited to metal ions as analytes, although this has been the focus of our own interest. Thus the sulfonamide approach outlined above **in Figure 1.6** does not require the presence of a sulfonamide and is not necessarily limited to zinc or another metal, it only requires that the analyte's binding promote (or reduce) the binding of a fluorescent ligand to the macromolecule (which need not be a protein). Many examples of this are known or can be constructed (Kolb & Weber, 1975). Similarly, the second approach outlined in **Figure 1.9** does not require that the analyte be a metal or (evidently) quench by energy transfer. Energy transfer has the virtues of having a predictable, controllable response which is straightforward to engineer. Rather, the analyte (or a competitor in a competitive assay) need only partially quench the label on the macromolecule by binding to it. As interest grows in real time determination of a variety of analytes, particularly in imaging their concentrations within cells, the usefulness of these approaches will become more and more apparent.

1.6. ANISOTROPY-BASED SENSING OF OTHER ANALYTES USING PROTEINS AS TRANSDUCERS

The Lakowicz group has introduced enzymes and other proteins as anisotropy-based transducers for analytes other than metal ions, and developed a very clever approach that transduces the analyte level as a change in the observed fluorescence polarization of a device that occurs without a change in rotational diffusion or lifetime(J. R. Lakowicz et al., 1999). This approach effectively transduces an intensity change caused by the analyte to a fluorescent-labeled protein as a change in polarization by admixture of the emission from the protein with emission from a physically separate reference fluorophore having much higher anisotropy. The reference fluorophore is typified by an oblong fluorophore like Hoechst 33342 being entrapped in a polymer film which is stretched to orient the fluorophores, and which exhibits a polarization approaching 1.0. As the relative contribution of the fluorescence due to the analyte increases, the apparent net polarization declines. This is an effective approach, demonstrated using an iodo-anilinonaphthalene sulfonate-labeled variant of a glucose/galactose-binding protein (ANS-Q28C GGBP) isolated from *E. coli* to determine glucose: saturation with 6 uM glucose resulted in a 25 % decline in intensity in the ANS fluorescence, which induced a 0.15 increase in polarization. The value of this approach is that many examples of analytes or ligands inducing changes in fluorescent intensity of (for instance) labeled macromolecules are known, which could be converted to polarization based indicators by this method. In this sense it is akin to other external reference methods using separate fluorophores emitting (or being excited at) different wavelengths. The drawback in these approaches is that when the reference fluorophore is not the same molecule as the transducing fluorophore (as it is in classical ratiometric indicators or the quenching-based anisotropy indicators described above), assorted influences such as bleaching, precipitation, or pH may affect the emission of the one and not the other, defeating the purpose. Most of these influences can be curtailed in macroscopic (cuvette-sized) embodiments, but this would appear harder to arrange in microscopic specimens.

This group has used a similar approach but with a different binding protein, lactate dehydrogenase noncovalently labeled with ANS, to measure lactate in the 0 to 240 uM range (normal clinical values are 1.0 to 1.8 mM). In this case saturation with lactate results in a 40% decrease in intensity (D'Auria, Gryczynski, Gryczynski, Rossi, & Lakowicz, 2000) . Noteworthy in this paper was the very simple apparatus they used to measure the apparent polarization of the sample essentially using a rotating polarizer to balance (null) the apparent intensity of the sample and a reference chamber containing the same labeled protein unexposed to lactate.   This same nulling approach can even be performed by eye, with surprising accuracy. This improvement avoids the issue of the reference fluorophore behaving differentially to physical influences. More recently, the Lakowicz group used apoglucose dehydrogenase from a thermophilic organism with noncovalently bound ANS in a similar approach to measure glucose concentrations in the clinically relevant range (D'Auria, Cesare et al., 2000).

1.7. CONCLUSIONS

Based on the few results presented above, the field of anisotropy-based biosensors incorporating proteins as recognition molecules is obviously in its infancy. Yet the approach would appear promising, particularly for studies in cell biology where, in an approach akin to that which has proved so successful for calcium, one would wish to know the spatial and temporal distribution of an analyte within the cell.   Powerful approaches have been developed for delivering to or expressing protein biosensors in various compartments of the cell (Schwarze, Ho, Vocero-Akbani, & Dowdy, 1999; Zelphati et al., 2001), so this no longer represents a barrier (R. Bozym and R. Thompson, unpublished results). A key issue for this class of sensors remains, however. While Hellinga has recently shown that binding proteins for small molecule analytes can be constructed almost at will(Looger, Dwyer, Smith, & Hellinga, 2003), the ability to design or modify the recognition molecule such that some nearby fluorescent label undergoes a usable change in its emission upon binding some arbitrary analyte remains beyond the state of the art. For spectroscopically active analytes such as certain metal ions, this is straightforward (see above); for the majority of small molecule analytes of biomedical interest which lack spectroscopic "handles", it is far from easy. The most universal approach has been the use of covalent and non-covalent derivatives of "solvent-sensitive" fluorophores such as the ANS family (Godwin & Berg, 1996; J. R. Lakowicz et al., 1999; Li & Cass, 1991; Marvin & Hellinga, 1998; Thompson & Jones,

1993). Unfortunately, even placing the label close to the binding site by covalent attachment does not assure a usable change. Moreover, ANS and its cousins are far from ideal as fluorophores for use in complex matrices, particularly biological ones containing serum albumin and the like. It would appear that any general solution to this problem would have far-reaching impact in many fields.

1.8. ACKNOWLEDGMENTS

The author is grateful to the Office of Naval Research, the National Institutes of Health, and the National Science Foundation for support, and Prof. Carol Fierke for many fruitful discussions.

1.9. REFERENCES

Ammann, D. (1986). *Ion-Selective Microelectrodes: Principles, Design, and Application*. Berlin: Springer-Verlag.

Axelrod, D. (1989). Fluorescence polarization microscopy. In D. L. Taylor & Y.-L. Wang (Eds.), *Methods in Cell Biology: Fluorescence Microscopy of Living Cells in Culture, Part B: Quantitative Fluorescence Microscopy - Imaging and Spectroscopy* (Vol. 30, pp. 333 - 352). New York: Academic Press.

Bigelow, C. E., Conover, D. L., & Foster, T. H. (2003). Confocal fluorescence spectroscopy and anisotropy imaging system. *Optics Letters, 28*(9), 695 - 697.

Chen, R. F., & Kernohan, J. (1967). Combination of bovine carbonic anhydrase with a fluorescent sulfonamide. *Journal of Biological Chemistry, 242*, 5813-5823.

Dandliker, W. B., Kelly, R. J., Dandliker, J., Farquhar, J., & Levin, J. (1973). Fluorescence polarization immunoassay. Theory and experimental method. *Immunochemistry, 10*, 219-227.

Darwish, I. A., & Blake, D. A. (2002). Development and validation of a one-step immunoassay for determination of cadmium in human serum. *Analytical Chemistry, 74*(1), 52 - 58.

D'Auria, S., Cesare, N. D., Gryczynski, Z., Gryczynski, I., Rossi, M., & Lakowicz, J. R. (2000). A thermophilic apoglucose dehydrogenase as nonconsuming glucose sensor. *Biochemical and Biophysical Research Communications, 274*, 727 - 731.

D'Auria, S., Gryczynski, Z., Gryczynski, I., Rossi, M., & Lakowicz, J. R. (2000). A protein biosensor for lactate. *Analytical Biochemistry, 283*, 83 - 88.

Dinely, K. E., Malaiyandi, L. M., & Reynolds, I. J. (2002). A reevaluation of neuronal zinc measurements: artifacts associated with high intracellular dye concentration. *Molecular Pharmacology, 62*(3), 618 - 627.

Dix, J. A., & Verkman, A. S. (1990). Mapping of fluorescence anisotropy in living cells by ratio imaging: Application to cytoplasmic viscosity. *Biophysical Journal, 57*(2), 231-240.

Fernandez-Gutierrez, A., & Munoz de la Pena, A. (1985). Determinations of inorganic substances by luminescence methods. In S. G. Schulman (Ed.), *Molecular Luminescence Spectroscopy, Part I: Methods and Applications* (Vol. 77, pp. 371-546). New York: Wiley-Interscience.

Fierke, C. A., & Thompson, R. B. (2001). Fluorescence-based biosensing of zinc using carbonic anhydrase. *BioMetals, 14*, 205 - 222.

Forster, T. (1948). Intermolecular energy migration and fluorescence (Ger.). *Annalen der Physik, 2*, 55 - 75.

Godwin, H. A., & Berg, J. M. (1996). A fluorescent zinc probe based on metal induced peptide folding. *Journal of the American Chemical Society, 118*, 6514-6515.

Grynkiewicz, G., Poenie, M., & Tsien, R. Y. (1985). A new generation of calcium indicators with greatly improved fluorescence properties. *Journal of Biological Chemistry, 260*(6), 3440-3450.

Guo, X.-Q., Castellano, F. N., Li, L., & Lakowicz, J. R. (1998). Use of a long-lifetime Re(I) complex in fluorescence polarization immunoassays of high-molecular-weight analytes. *Analytical Chemistry, 70*(3), 632 - 637.

Huang, C.-c., Lesburg, C. A., Kiefer, L. L., Fierke, C. A., & Christianson, D. W. (1996). Reversal of the hydrogen bond to zinc ligand histidine-119 dramatically diminishes catalysis and enhances metal equilibration kinetics in carbonic anhydrase II. *Biochemistry, 35*(11), 3439-3446.

Jolley, M. E., Stroupe, S. D., Schwenzer, K. S., Wang, C. J., Lu-Steffes, M., Hill, H. D., et al. (1981). Fluorescence polarization immunoassay III: an automated system for drug determination. *Clinical Chemistry, 27*, 1575-1579.

Kimura, E., & Aoki, S. (2001). Chemistry of zinc(II) fluorophore sensors. *BioMetals, 14*(3 - 4), 191 - 204.

Kolb, D. A., & Weber, G. (1975). Cooperativity of binding of anilinonaphthalenesulfonate to serum albumin induced by a second ligand. *Biochemistry, 14*(20), 4476 - 4481.

Lakowicz, J. R. (1999). *Principles of Fluorescence Spectroscopy* (Second ed.). New York: Kluwer Academic / Plenum Publishers.

Lakowicz, J. R., Gryczynski, I., Gryczynski, Z., Tolosa, L., Randers-Eichhorn, L., & Rao, G. (1999). Polarization-based sensing of glucose using an oriented reference film. *Journal of Biomedical Optics, 4*(4), 443 - 449.

Li, Q. Z., & Cass, A. E. G. (1991). Periplasmic binding-protein based biosensors. 1. Preliminary study of maltose binding-protein as sensing element for maltose. *Biosensors and Bioelectronics, 6*(5), 445-450.

Lindskog, S., Henderson, L. E., Kannan, K. K., Liljas, A., Nyman, P. O., & Strandberg, B. (1971). Carbonic anhydrase. In P. D. Boyer (Ed.), *The Enzymes* (Vol. 5, pp. 587-665). New York: Academic Press.

Lindskog, S., & Thorslund, A. (1968). On the interaction of bovine cobalt carbonic anhydrase with sulfonamides. *European Journal of Biochemistry, 3*, 453-460.

Looger, L. L., Dwyer, M. A., Smith, J. J., & Hellinga, H. W. (2003). Computational design of receptor and sensor proteins with novel functions. *Nature, 423*, 185 - 190.

Maren, T. H. (1977). Use of inhibitors in physiological studies of carbonic anhydrase. *American Journal of Physiology, 232*(4), F291-F297.

Marvin, J. S., & Hellinga, H. W. (1998). Engineering biosensors by introducing fluorescent allosteric signal transducers: Construction of a novel glucose sensor. *Journal of the American Chemical Society, 120*(1), 7-11.

Nuccitelli, R. (Ed.). (1994). *A Practical Guide to the Study of Calcium in Living Cells* (Vol. 40). New York: Academic Press.

Schwarze, S. R., Ho, A., Vocero-Akbani, A., & Dowdy, S. F. (1999). In vivo protein transduction: Delivery of a biologically active protein into the mouse. *Science, 285*, 1569 - 1572.

Shinitzky, M., & Barenholz, Y. (1978). Fluidity parameters of lipid regions determined by fluorescence polarization. *Biochimica et Biophysica Acta, 515*, 367 - 394.

Steiner, R. F. (1991). Fluorescence anisotropy: theory and applications. In J. R. Lakowicz (Ed.), *Topics in Fluorescence Spectroscopy Volume 2: Principles* (Vol. 2, pp. 1-52). New York: Plenum Press.

Szmacinski, H., & Lakowicz, J. R. (1993). Optical measurements of pH using fluorescence lifetimes and phase-modulation fluorometry. *Analytical Chemistry, 65*, 1668-1674.

Szmacinski, H., & Lakowicz, J. R. (1994). Lifetime-based sensing. In J. R. Lakowicz (Ed.), *Topics in Fluorescence Spectroscopy Vol. 4: Probe Design and Chemical Sensing* (Vol. 4, pp. 295 - 334). New York: Plenum.

Terpetschnig, E., Szmacinski, H., & Lakowicz, J. R. (1995). Fluorescence polarization immunoassay of a high molecular weight antigen based on a long lifetime Ru-ligand complex. *Analytical Biochemistry, 227*, 140-147.

Thompson, R. B., Ge, Z., Patchan, M. W., & Fierke, C. A. (1996). Performance enhancement of fluorescence energy transfer-based biosensors by site-directed mutagenesis of the transducer. *Journal of Biomedical Optics, 1*(1), 131-137.

Thompson, R. B., & Jones, E. R. (1993). Enzyme-based fiber optic zinc biosensor. *Analytical Chemistry, 65*, 730-734.

Thompson, R. B., Maliwal, B. P., Feliccia, V. L., Fierke, C. A., & McCall, K. (1998). Determination of picomolar concentrations of metal ions using fluorescence anisotropy: biosensing with a "reagentless" enzyme transducer. *Analytical Chemistry, 70*(22), 4717-4723.

Thompson, R. B., Maliwal, B. P., & Fierke, C. A. (1998). Expanded dynamic range of free zinc ion determination by fluorescence anisotropy. *Analytical Chemistry, 70*(9), 1749-1754.

Thompson, R. B., Maliwal, B. P., & Zeng, H. H. (2000). Zinc biosensing with multiphoton excitation using carbonic anhydrase and improved fluorophores. *Journal of Biomedical Optics, 5*(1), 17-22.

Thompson, R. B., & Patchan, M. W. (1995). Fluorescence lifetime-based biosensing of zinc: origin of the broad dynamic range. *Journal of Fluorescence, 5*, 123-130.

Thompson, R. B., Peterson, D., Mahoney, W., Cramer, M., Maliwal, B. P., Suh, S. W., et al. (2002). Fluorescent zinc indicators for neurobiology. *Journal of Neuroscience Methods, 118*, 63 - 75.

Walkup, G. K., Burdette, S. C., Lippard, S. J., & Tsien, R. Y. (2000). A new cell-permeable fluorescent probe for Zn2+. *Journal of the American Chemical Society, 122*, 5644 - 5645.

Weber, G. (1956). Photoelectric method for the measurement of the polarization of the fluorescence of solutions. *Journal of the Optical Society of America, 46*(11), 962 - 970.

White, C. E., & Argauer, R. J. (1970). *Fluorescence Analysis: A Practical Approach*. New York: Marcel Dekker, Inc.

Zelphati, O., Wang, Y., Kitada, S., Reed, J. C., Felgner, P. L., & Corbeil, J. (2001). Intracellular delivery of proteins with a new lipid-mediated delivery system. *Journal of Biological Chemistry, 276*(37), 35103 - 35110.

GFP SENSORS

Peter M. Haggie[*†] and A.S. Verkman[*]

2.1. INTRODUCTION

The green fluorescent protein (GFP) and related genetically-encoded fluorescent proteins have had a major impact in cell biology. GFP has diverse applications in studies of protein localization, dynamics, interactions and regulation. GFP is targetable to specific cellular sites in cell culture models and *in vivo* in a wide variety of organisms. A unique application of fluorescent proteins is their engineering as real-time cellular sensors of pH, ion concentrations, second messengers, enzyme activities and other parameters. Compared to classical chemical probes, genetically-encoded fluorescent proteins permit stable non-invasive staining of specific subcellular sites in cell culture models and *in vivo* with little or no cellular toxicity. This chapter reviews the paradigms that have been developed for engineering of GFP-based sensors, applications of available sensors, and directions for further sensor development.

2.2. GENERAL PRINCIPLES OF ENGINEERING FLUORESCENT PROTEIN SENSORS

Figure 2.1 depicts several strategies that have been applied to the design of genetically-encoded fluorescent sensors. The simplest sensors (Class I) are fluorescent proteins that without modification (other than mutagenesis of primary sequence) sense a physiological parameter. Examples of intrinsic fluorescent sensors include GFP sensors of pH and chloride concentration. The steady-state fluorescence of the original (wildtype) GFP and most common GFP mutants is pH-dependent because of protonation-deprotonation of the GFP chromophore (1-4). Many yellow fluorescent proteins (YFPs) are intrinsically sensitive to halides by a mechanism involving halide-dependent shifts in pK_a (5, 6).

---

[*] Departments of Medicine and Physiology, Cardiovascular Research Institute, University of California, San Francisco, CA, 94143-0521.

[†] Correspondence to: Alan S. Verkman, M.D., Ph.D. 1246 Health Sciences East Tower, Cardiovascular Research Institute, University of California, San Francisco, CA 94143-0521, U.S.A., Phone: (415)-476-8530; Fax: (415)-665-3847, E-mail: verkman@itsa.ucsf.edu; Internet: http://www.ucsf.edu/verklab

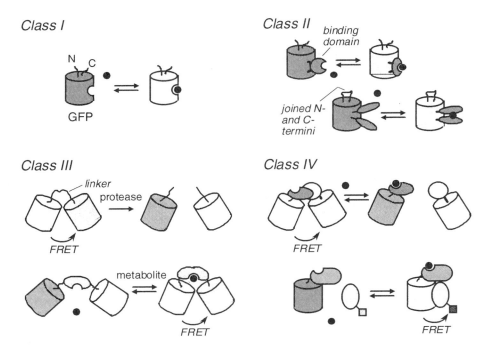

Figure 2.1. Classes of fluorescent protein-based sensors. Class I sensors consist of GFP (or other fluorescent proteins) that intrinsically respond to a parameter of interest (shown as ●). Halide and pH probes belong to this class. Class II sensors contain an additional domain or domains grafted onto GFP. Two examples of class II calcium sensors are shown: Camgaroo (*top*) contains a CaM domain (gray) inserted at residue Y145 of YFP; Pericam and G-CaMP sensors (*bottom*) contain CaM and M13 fragments joined to the new termini of circularly permuted GFP (see text for explanations). Class III sensors use intramolecular fluorescence resonance energy transfer (FRET). Donor/acceptor pairs are joined by an engineered linker that undergoes a conformational change to alter the separation/orientation of the GFP pair. Examples of class III sensors include protease sensors (*top*) that irreversibly lose FRET upon protease-mediate cleavage , and metabolite sensors (*bottom*) that that undergo conformational changes upon substrate binding. Class IV sensors are FRET-based using intermolecular interactions. A cAMP sensor has been described that is composed of regulatory and catalytic subunits of PKA separately joined to donor/acceptor pairs that dissociate upon cAMP binding (*top*). A PKC sensor contained a PKC-GFP fusion protein that upon activation and threonine phosphorylation binds to a specific Cy3-tagged antibody.

Class II sensors consist of fluorescent proteins linked to an additional protein domain or domains. Here the grafted domain(s) sense changes in analyte concentration or some other parameter, producing altered protein fluorescence by conformational changes or other interactions. The newer calcium sensors such as camgaroo, G-CaMP and pericam (7-9) and voltage sensors (10, 11) belong to this class. The grafted domains can reside within or outside of the GFP primary sequence. The development of sensors containing embedded peptide sequences has exploited circular permutation as a strategy to identify regions in GFP that are tractable to protein domain insertion. Proteins containing nearby amino- and carboxy-termini are amenable to circular permutation, where overall primary sequence connectivity of a protein is maintained whilst altering the protein termini (12). cDNA coding sequences are circularized, treated with low concentrations of DNAase to create randomly-permutated linear DNA fragments, and subcloned for expression and screening. Circular permutation of GFPs and the development of new fluorescent variants was reported in 1999 by two groups (7, 13). Using this approach with GFP, Baird *et al.* (7) determined that any of ten residues could serve as new starting amino acids, and that exogenous peptides could be inserted at specific sites without loss of fluorescence. In some cases permutation of fluorescent proteins has increased the magnitude of response of a sensor substantially, probably because the permuted fluorescent protein with new termini is better able to transduce conformational changes in grafted domains to the chromophore.

Class III sensors form the largest and most diverse family of probes and exploit changes in intramolecular fluorescence resonance energy transfer (FRET) efficiency between spectrally distinct fluorescent proteins. Intramolecular FRET sensors use an engineered linker that undergoes a conformational change to seperate and/or reorient the GFP moieties, having the advantage that stoichiometries of fluorophores are strictly maintained. A simple example of a class III FRET-based sensor is a donor/acceptor GFP pair linked by a protease-cleavable sequence (14-16). Protease cleavage of the sensor irreversibly abolishes FRET as the donor and acceptor GFPs separate. Class III sensors have been described for calcium (17, 18), metabolites (19, 20) and kinase activities (21, 22), each utilizing domains for analyte binding or phosphorylation inserted between GFP donor/acceptor pairs. Upon analyte binding or phosphorylation, separation and/or orientation of the GFP donor/acceptor pair is altered with consequent changes in FRET efficiency.

Class IV sensors rely on intermolecular FRET in which different proteins are fluorescently labeled, and physically associate or dissociate to generate or abolish a FRET signal. Class IV sensors have been developed to investigate the dynamics of small molecules such as the second messenger 3',5'-cyclic adenosine monophosphate (cAMP, ref. 23). GFP has also been used in the intermolecular FRET mode with other non-GFP fluorophores, such as in measurement of PKCα (24) and Rac 1 (25) activities.

2.3. NEW GREEN AND RELATED FLUORESCENT PROTEINS

2.3.1. GFP Mutants

The available repertoir of genetically-encoded autofluorescent proteins for sensor design has expanded rapidly. Early random and rational mutagenesis of GFP yielded spectrally distinct GFP variants, referred to as blue, cyan and yellow fluorescent proteins (BFP, CFP and YFP, reviewed in ref. 3). Recent GFP mutagenesis efforts have been directed at generating mutants with longer and different wavelengths, pH-insensitivity, improved brightness and photostability, and accelerated cellular maturation. For example, a GFP variant was generated with an emission wavelength between CFP and GFP called CGFP (cyan-green fluorescent protein, ref. 26). CGFP is essentially pH-insensitive in the physiological range, making it suitable for targeting to acidic organelles and an excellent partner in FRET-based sensors. New YFPs have been reported with improved characteristics for cellular applications (27-29). The YFP mutant Citrine (YFP-V68L/Q69M, ref. 27) is chloride-insensitive, more photostable, and less pH-sensitive than the original YFP. The YFP mutant Venus (P46L/F64L/M153T/V163A/S175G, ref. 28) has the favorable characteristics of Citrine as well as improved brightness and folding efficiency.

2.3.2. Novel Fluorescent Proteins

Naturally-occurring fluorescent proteins from non-bioluminescent Anthozoa species have been cloned which have high homology to GFP (30). Of the six native proteins cloned by Matz *et al.*, two had fluorescence emission maxima at longer wavelengths (up to 583 nm) than available GFP mutants; a red fluorescent protein (drFP583, λ_{em} 583 nm) was commercialized by Clontech as DsRed. Although extending the spectral range of genetically-encoded fluorescent proteins, DsRed undergoes tertramerization (31, 32) and its red fluorescence in cells develops slowly (up to 30 hours) as the green fluorescence from immature DsRed disappears (31). A DsRed mutant with ~10–fold more rapid maturation has been generated (33), as well as mutants with less propensity to aggregate (34, 35). Strategies to reduce DsRed aggregation have also been reported that involve the co-expression of the untagged or GFP-tagged protein of interest in addition to the DsRed-tagged protein, however, these methods may not be generally applicable to all systems (36-38).

In addition to the fluorescent proteins described by Matz *et al.*, (30) several fluorescent and non-fluorescent colored proteins have been cloned from other Anthozoa (39-41). A fluorescent protein from *Entacmaea quadricolour* (eqFP11) has several favorable properties, including red-shifted fluorescence emission (611 nm) with a large Stokes' shift (52 nm), low green fluorescence of its immature form, relatively rapid maturation (~12 hours), and pH-insensitivity (41); however, like DsRed, eqFP11 undergoes oligomerization. A non-fluorescent purple protein was cloned from the sea anemone *Anemonia sulcata*, which when mutated yielded a fluorescent protein with an emission wavelength of 595 nm (42). This strategy of mutating non-fluorescent proteins has been applied to other non-fluorescent GFP homologs to produce fluorescent proteins with emission wavelengths as long as 645 nm (43), one of which is available commercially from Clonetech as HcRed. Many new fluorescent proteins are becoming available for cell biological applications and development of new sensors (for review, see refs. 44, 45). The discovery and characterization of these proteins has also motivated novel applications such as fluorescent 'stopwatches' and photoactivatable proteins that emit at different wavelengths upon activation (46-48). A more complete biochemical and biophysical characterization of these proteins is probably required before they are ready for incorporation into fluorescent protein-based sensors.

2.4. GFP-BASED SENSORS

2.4.1. pH Sensors

The chromophore of many green fluorescent proteins is intrinsically pH sensitive. The GFP chromophore consists of three amino acids that are post-translationally modified by cyclization and oxidation (3, 49). The wildtype GFP chromophore, *p-*

hydroxybenzylideneimidazolinone, is created by modification of the genetically-encoded residues serine-tyrosine-glycine at positions 65–67. The resulting phenolic group within the chromophore can be titrated but is not directly accessible to solvent (50). Many GFP mutants are pH-sensitive, with different pK_a's depending on chromophore composition and the nature of surrounding amino acids. The intrinsic pH sensitivity of GFPs has been exploited to measure cytoplasmic and organellar pH non-invasively. Four GFP mutants containing the S65T mutation (which favors formation of the phenolate chromophore that absorbs at longer wavelengths than the neutral form) were characterized with pK_a values of 4.8–6.1 (1). A pH titration for purified recombinant GFP-S65T is shown in Figure 2.2A. Biophysical characterization of these proteins indicated that the pH response was rapid (<ms), reversible, and involved a ground-state effect. Calibration experiments in whole cells using ionophores indicated that the fluorescent proteins faithfully report pH *in vivo* (Figure 2.2B) and non-invasive measurements of cytoplasmic, mitochondrial and Golgi pH were carried out. Additional GFP mutants were also characterized and used to measure cellular pH changes. One of the GFP mutants used by Llopis *et al.* (2) for measurement of mitochondrial pH was YFP, which had an apparent pK_a value of 7.1; however, subsequent studies showed that YFP is also halide-sensitive and thus not suitable for cellular pH measurements (5, 6). A direct comparison of GFP with the established pH-sensitive dye BCECF provided direct validation for the use of GFP as a cytoplasmic pH indicator (51).

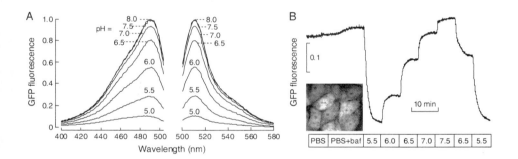

Figure 2.2. Sensitivity of intrinsic GFP fluorescence to pH. A. Fluorescence excitation and emission spectra of purified recombinant GFP-S65T as a function of solution pH. **B.** Demonstration that GFP-S65T fluorescence is pH sensitive in cells. GFP-S65T was expressed in cell cytoplasm and nucleoplasm of fibroblasts. Cell fluorescence was monitored continuously from a group of cells (inset) during perfusion with saline ('PBS') and saline containing ionophores ('PBS + baf') to equalize cytoplasmic and solution pH at indicated values. Adapted from ref. 1.

Ratioable probes are desirable for fluorescence microscopy since fluorescence ratios do not depend on expression levels, cell depth, illumination intensity, and non-uniform probe distribution. Miesenböck *et al.*, (52) reported ratioable pH sensors (called pHluorins) that were identified by random mutagenesis of residues that interact with the GFP chromophore. Fluorescence of the pHluorins is pH-sensitive with a reciprocal fluorescence response for the excitation wavelengths 395 and 475 nm in the pH range 5.5–7.5. Targeting of pHluorins to synaptic vesicles in neurons permitted the dynamic study of individual fusion events and synaptic vesicle cycling during neurotransmitter secretion at nerve terminals (52, 53). A family of ratioable, single excitation/dual-emission GFP pH sensors (called deGFPs) have also been described with pK_a values of 6.8–8.0 (54, 55). These sensors have cysteine residue substitutions at position 148 and/or 203 as well as threonine substitution at position 65. The fluorescence emission from deGFPs increases at ~460 nm and decreases at ~515 nm as pH decreases (54). Also, ratiometric pH sensors based on the fusion of two differentially pH-sensitive GFPs have been reported (56). These sensors (called GFpH and YFpH) had pK_a values of 6.1 and 6.8, respectively, and could be used either in single excitation or dual excitation FRET modes. These probes were used to measure cytoplasmic pH and local changes in pH at endosomal surfaces (when fused to the α_{1B}-adrenegic receptor) during norepinephrine-stimulated endocytosis.

2.4.2. Chloride/Halide Sensors

It was recognized that the fluorescence of YFP, but not other forms of GFP, was intrinsically sensitive to halides including chloride (5). The halide-sensing mechanism of one mutant, YFP-H148Q, was investigated and applications to living cells were developed (6). Halide binding produces an increase in the pK_a of YFP-H148Q (from ~7 to 8 for 0–400 mM chloride), such that its fluorescence at constant pH decreases with increasing halide concentration. When expressed in cell cytoplasm, YFP-H148Q fluorescence provided a useful read-out of cytoplasmic halide concentration suitable for measurement of halide transport processes. However, a limitation of YFP-H148Q was its relative weak chloride sensitivity, with 50% decrease in fluorescence at 100 mM Cl⁻ at pH 7.5. A subsequent structural study identified the putative YFP halide binding site (57).

To overcome this limitation, YFPs with greatly improved halide sensitivity (for Cl⁻ and I⁻), random mutagenesis was carried out of residue pairs in the vicinity of the halide binding site (58). A YFP mutant (H148Q/I152L) was identified with very high iodide sensitivity ($K_I \sim 3$ mM at pH 7.5, Figure 2.3A) and rapid fluorescence response to changes in halide concentration (figure inset). Like YFP-H148Q, the halide-sensing mechanism of mutants involved pK_a shifts (Figure 2.3B). Figure 2.3C shows large, reversible changes in cell fluorescence in response to Cl⁻/I⁻ exchange.

A useful application of YFPs has been in drug discovery for the identification of compounds that activate or inhibit the cystic fibrosis transmembrane conductance regulator (CFTR) chloride/halide channel. Figure 2.4A shows the strategy for identification of CFTR activators by high-throughput screening (59). Cells co-expressing human CFTR and a YFP indicator are subjected to an iodide gradient in a fluorescence plate reader. Pre-addition of an activator results in increased iodide entry and YFP fluorescence quenching. Figure 2.4B shows representative data from a 96-well plate. Fluorescence changed little in the saline (no activator) control. Addition of a known CFTR activator (apigenin) resulted in decreased fluorescence, as did some (~50 out of 60,000) of the test compounds (Figure 2.4B). More than a dozen novel CFTR activators were identified with novel chemical structures and sub-micromolar activating potencies, two of which are shown in Figure 2.4C. A similar screening strategy was applied recently to identify potent CFTR inhibitors, which function as antidiarrheals to prevent cholera toxin-induced intestinal fluid secretion (60).

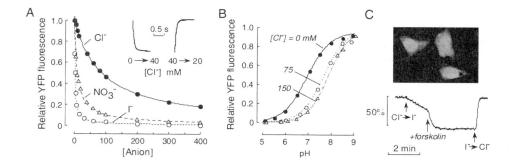

Figure 2.3. YFPs with improved halide sensitivity. A. Titration of purified YFP-H148Q/I152L with indicated anions at pH 7.4. Inset: Stopped-flow fluorescence kinetics showing rapid response to changes in [Cl⁻]. **B.** pH titration of YFP-H148Q/I152L at indicated [Cl⁻] showing pK_a shift. **C.** (top) Fluorescence micrograph of CFTR-expressing cells showing cytoplasmic YFP indicator. (bottom) Time course of cell fluorescence showing reversible decrease in fluorescence in response to I⁻ influx. Forskolin elevates cAMP and activates CFTR halide transport. Adapted from ref. 58.

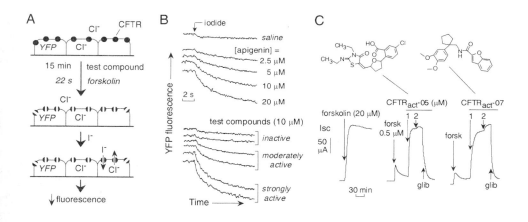

Figure 2.4. Identification of CFTR activators by high-throughput screening using YFP indicator. A. Screening strategy. FRT cells expressing YFP-H148Q and human CFTR were subjected to an inward I⁻ gradient after adding test compound (out of 60,000 small molecules) and a low concentration of forskolin. Activated CFTR permits I⁻ influx resulting in decreased fluorescence. **B.** Representative data from single wells of 96-well plates showing cell fluorescence in response to I⁻ addition without activators ('saline'), with a known activator ('apigenin') and with various test compounds having different levels of activity. **C.** Short-circuit current electrophysiology showing induction of strong CFTR currents by sub-micromolar concentrations of two CFTR activators discovered by the screening. Adapted from ref. 59.

A FRET-based ratiometric halide sensor has been generated by the fusion of CFP and YFP (61). The YFP moiety was chloride-sensitive as described above. The emission ratio of this sensor changed from ~2.0 at 0 mM chloride to ~0.5 at 500 mM chloride with

50% change at 160 mM chloride. This probe was used in neurons to measure temporal changes in chloride concentration during development and in response to GABA stimulation.

2.4.3. Sensors of Protease Activity

FRET-based GFP sensors for protease activity have been generated by genetically inserting protease consensus sequences between GFP donor/acceptor partners; protease-mediated cleavage of the consensus sequence irreversibly abolishes FRET. In a first report in 1996 a probe for Factor X_a protease activity was generated using this paradigm and validated for measurement of protease activity *in vitro* (14). Such sensors have been proven to provide robust readouts of protease activity using cell sorting (62), plate reader (63) and image (64) detection methods. Although FRET efficiency is generally deduced from steady-state donor/acceptor fluorescence intensities, cross-correlation analysis (15) and lifetime imaging (to obviate the requirement for fluorescent proteins to be spectrally resolvable, ref. 65) have also been applied to measure protease activity. Recently, sensors using CFP and the Venus variant of YFP (28) have been used to investigate the progression of caspase activity during apoptosis (16). Although a similar study had been reported (66), the newly designed probes did not suffer from pH- and halide- dependent changes in fluorescence. Caspase-3 activity in the cytosol was shown to precede activity in the nucleus, with maximal nuclear activity of caspase-3 at ~10 minutes. The kinetics of caspase-9 activation were slower and predominated in the cytosol; however, the progression of morphological changes after activation of both caspases were similar, suggesting concurrent activation during apoptosis.

2.4.4. Calcium Sensors

The use of fluorescent proteins to design FRET-based intramolecular sensors was stimulated by Tsien and colleagues with their design of the 'cameleon' sensors to investigate the dynamics of the second messenger calcium (17). The initial cameleons - so called because the sensor 'readily changes color and retracts and extends a long tongue (M13) into and out of the mouth of the calmodulin' - consisted of two spectrally distinct fluorescent proteins joined by a specialized calcium-sensing linker composed of calmodulin (CaM), a CaM-binding peptide (M13) and a short, flexible linker. Calcium binding to the CaM moiety and subsequent CaM binding to M13 results in a reorientation of the two fluorescent proteins, typically cyan and yellow, altering FRET efficiency. The calcium binding affinity of these sensors can be altered by mutagenesis of the CaM moiety. Cameleon-2 was used to detect calcium transients in HeLa cells in response to histimine and the purinergic agonist ATP, with ~25 % change in emission ratio (maximal cytoplasmic $[Ca^{2+}]$ of ~3 µM), nearly half the full dynamic range of this sensor (~60 %). Lower affinity cameleons targeted to the endoplasmic reticulum (ER) reported resting calcium concentrations ($[Ca^{2+}] \sim$ 60–400 µM) and calcium mobilization in response to histamine stimulation with similar changes in emission ratios.

To overcome the pH-sensitivity of the YFP moiety second generation cameleons were developed in which the mutations V68L and Q69K were introduced into YFP to reduce its pK_a from ~7 to 6 (18). More recently, cameleons were rationally re-engineered with a new linker region based on structural information from NMR studies of CaM bound to CaM-dependent kinase fragment (67). This new sensor had a two-fold increase in the FRET dynamic range for $[Ca^{2+}]$ in the range 50–1000 nM.

The general applicability of cameleons to detect changes in $[Ca^{2+}]$ has been demonstrated in diverse organisms. Calcium-mediated exocytosis was investigated by genetic fusion of cameleons to phogrin, a protein localized to the secretory granule membranes, which permitted the measurement of cytoplasmic $[Ca^{2+}]$ at the vesicle surface in pancreatic β cells (68). It was reported that vesicles near the plasma membrane (within ~1 μm) experienced greater apparent changes in $[Ca^{2+}]$ than vesicles far from the cell surface. Targeting of cameleons to the nucleus, ER and mitochondria have permitted investigation of calcium transients in these organelles in response to cell signaling, cell death and circadian rhythms (69-73). Allen *et al.* (74) investigated calcium signaling in Arabidopsis using cameleons, demonstrating stimulus-specific calcium oscillations accompanying stomatal closure. Two-photon fluorescence excitation has been used with cameleons in cell cultures (75), and in a study of calcium transients during muscle contraction in transgenic nematode worms expressing cameleons in the pharyngeal muscle (76). Further developments of cameleons will likely include the use of different FRET partners, such as red fluorescent proteins (77, 78), and use of environmentally-insensitive YFPs such as Citrine and Venus (27-29).

Calcium sensors using modified single GFPs (class II sensors) have also been reported. These sensors were developed using circularly permutated fluorescent proteins or information derived from circular permutation studies about the intrinsic structural malleability of fluorescent proteins (for example see ref. 7). An indicator containing a CaM domain at position Y145 of YFP, termed 'camgaroo 1' because it carries a smaller companion (CaM) inserted in its pouch, has an apparent K_d for calcium of 7 μM and a Hill coefficient of 1.6. Spectroscopic studies indicated that deprotonation of the chromophore with increasing $[Ca^{2+}]$ was responsible for the calcium-dependent changes in fluorescence intensity. In cell cultures camgaroo-1 reported changes in $[Ca^{2+}]$ in response to various maneuvers including histamine; however, despite a large dynamic range the response of camgaroo 1 to physiological stimuli was modest (~5 %, ref. 7). The general applicability of this insertional paradigm was further demonstrated by the insertion of a zinc finger motif into the same YFP location, giving a zinc sensor (7). First and second generation camgaroos (27) have been used to measure calcium transients in *Drosophila* mushroom bodies in response to depolarization and cholinergic stimulation (79) and to investigate the role of voltage-dependent anion channels in mitochondrial Ca^{2+} homeostasis (80).

Additional class II calcium sensors have been reported by two groups. Nakai *et al.* (8) generated a sensor (termed G-CaMP) using a circularly permuted GFP (with a new amino-terminus at residue 148) containing CaM at its carboxy-terminus and M13 at its amino-terminus. G-CaMP has a low K_d for calcium of 235 nM (~30 times lower that of camgaroo), large changes in fluorescence intensity in response to physiological changes in $[Ca^{2+}]$ (~5–fold change for 0–1 mM calcium) and rapid response kinetics. In an elegant study, G-CaMP was used with two-photon fluorescence microscopy to map odor perception in *Drosophila* brain (81). Miyawaki and colleagues (9) reported a similar calcium sensor called pericam (*per*muted YFP and a *CaM*) based upon YFP. Here, circularly permuted YFP (new amino-terminus at residue 144 with V68L/Q69K mutations to confer pH-insensitivity) contained amino-terminus M13 and carboxy-terminus CaM domains. Refinement of the pericam sensors by mutagenesis gave 'flash-pericam' and 'inverse-pericam' which become more and less bright, respectively, upon calcium binding, as well as a 'ratiometric-pericam' (mutations Y203F and H148D) suitable for dual excitation, single emission ratio imaging. These probes have apparent K_d for calcium of 0.2–1.7 μM and Hill coefficients of ~1. Calcium transients in the cytosol, nucleus and mitochondria were visualized with ratiometric-pericam in response to physiologically relevant stimuli (9). Pericams have also been used to investigate the relationship between cytosolic and mitochondrial calcium in beating ventricular myocytes from neonatal rats (82), and to measure calcium changes in pancreatic islet β cells in response to glucose and kinase activity (83).

2.4.5. Sensors of Calcium-Calmodulin

The group of Persechini developed a FRET-based sensors of free $(Ca^{2+})_4$-CaM concentration exploiting essentially the same components of calcium signaling as used in the cameleons: CaM and a CaM binding fragment (17 amino acids from smooth muscle myosin light chain kinase CaM binding domain, ref. 84). $(Ca^{2+})_4$-CaM binding to the CaM binding fragment linking donor/acceptor GFPs results in reduced FRET efficiency. Indicator protein and CaM were microinjected in an initial study which demonstrated that the activity of a typical CaM target (affinity ~1 nM) was responsive to physiological changes calcium concentration (84). The relationship between free calcium and $(Ca^{2+})_4$-CaM was investigated using stably transfected cell lines expressing similar indicator moieties (85). No $(Ca^{2+})_4$-CaM was detected below 0.2 mM free calcium and the maximum concentration of $(Ca^{2+})_4$-CaM was determined to be ~45 nM (85). Local intracellular differences in free $(Ca^{2+})_4$-CaM concentrations were found and it was speculated that these differences may provide a mechanism for selective activation of CaM dependent signaling process (86).

2.4.6. Sensors of Other Second Messengers

The main target of cAMP is the holotetrameric enzyme protein kinase A (PKA), which mediates process such as transcription, metabolism and secretion/absorption via protein phosphorylation. Binding to cAMP the regulatory subunit of PKA causes dissociation from the catalytic subunit, providing a strategy for engineering of a GFP-based cAMP FRET sensor (23). Two fusion proteins were generated, the first between the PKA catalytic subunit and GFP, and the second between the PKA regulatory subunit and BFP; cAMP binding results in subunit dissociation and reduced FRET efficiency. A similar strategy was described originally for a microinjectable cAMP sensor in which PKA subunits were chemically labeled with small organic molecules (87). The GFP-based cAMP sensor could detect increases in cAMP in cell culture models in response to cAMP agonists, giving ~30 % increase in the donor-to-acceptor emission ratio. Improved cAMP sensors were subsequently developed using CFP and YFP to study cAMP compartmentalization and organization in PKA-mediated signaling (88). Discrete regions of increased cAMP were found in cardiac myocytes (at transverse tubule/junctional sarcoplasmic reticulum) in response to β-adrenergic stimulation, resulting from restricted cAMP diffusion, PKA anchoring, and phosphodiesterase activity.

GFP-based sensors have been developed for the second messenger 3',5'-cyclic guanosine monophosphate (cGMP) (89, 90). Nitric oxide and various hormones and toxins stimulate guanylyl cyclases to convert guanosine triphosphate (GTP) to cGMP, which activates cGMP-dependent kinases, cyclic nucleotide-gated channels and phosphodiesterases. The first reported cGMP sensor consisted of the cGMP-binding domain from cGMP-dependent protein kinase Iα (with residues 1–47 removed) interposed between cyan and yellow fluorescent proteins (89). A similar probe was reported in which the first 77 amino acids of cGMP-dependent protein kinase Iα were removed (90). Exposure of cells expressing the sensor of Sato *et al.* (89) to 1 mM 8-Br-cGMP (a cell permeable and phosphodiesterase resistant cGMP analog) or nitric oxide (NO) release produced a 20–30 % decrease in the cyan-to-yellow emission ratio, indicating increased FRET efficiency. The probe of Honda *et al.* (90) gave a similar signal size in response to NO release, but in the opposite direction.

A GFP-based sensor for nitric oxide (NO), termed FRET-MT, was generated by inserting metallothionein (MT) between CFP and YFP (91). MT is a cysteine-rich metal binding protein implicated in intracellular redox and NO signaling. NO decreased the emission ratio (yellow-to-cyan) of FRET-MT due to conformational changes in MT upon formation of nitrosothiol groups with cysteines and displacement of MT-bound metal ions. In support of this mechanism, metal ion chelation from MT resulted in a similar reduction in FRET-MT emission ratio. Application of the NO donor S-nitrosylglutathione in endothelial cells reduced FRET-MT emission ratio. This system was used to investigate changes in NO after muscarinic agonists (carbochol and bradykinin) and calcium ionophores.

A class II sensor for the second messenger inositol 1,4,5-triphosphate (IP$_3$) has been reported (92). IP$_3$ mediates intracellular calcium mobilization and is generated by hormone and neurotransmitter stimulated phospholipase C metabolism of phosphatidyinositol-4,5-bisphosphate (PIP$_2$). In contrast to other GFP-based sensors, the fluorescence 'readout' of this probe is a spatial distribution profile within the cell. The IP$_3$ sensor is composed of GFP fused to the pleckstrin homology (PH) domain (termed GFP-PHD). PH domains bind both membrane-associated PIP$_2$ and cytoplasmic IP$_3$ but the affinity for IP$_3$ is ~20-fold higher. As such, generation of IP$_3$ is reported by translocation of GFP-PHD from the plasma membrane to the cytoplasm. Microinjection of IP$_3$ increased cytoplasmic GFP-PHD, whereas overexpression of IP$_3$ 5-phosphatase abolished agonist-induced cytoplasmic translocaion of GFP-PHD, verifying that GFP-PHD was reporting increased IP$_3$ and not PIP$_2$ disappearance. This sensor was used to correlate IP$_3$ dynamics and cytoplasmic calcium (using fura-2), identifying calcium and IP$_3$ waves in single cells and between cells.

2.4.7. Sensors of Protein Kinase Activity

Several groups have described class III GFP-based sensors to investigate the kinetics and spatial organization of protein kinase activity during signal transduction. The general strategy has been to insert a phosphorylation consensus sequence between GFP donor/acceptor pairs such that conformation changes upon protein phosphorylation alter FRET efficiency. A PKA activity sensor was engineered by fusing blue and green fluorescent proteins with the PKA target domain from the cAMP respone element binding protein (21). PKA-mediated phosphorylation yielded ~25 % changes in fluorescence emission ratio. The kinetics, spatial organization and effect of inhibitors on PKA-mediated signaling were investigated. Improved kinase sensors were designed by incorporating phophoamino-binding domains into the linker region along with the kinase substrate consensus sequences (22, 93, 94). Upon phosphorylation of the kinase consensus sequence, the phophoamino-binding domain induced conformational changes in the linker domain. For example, a PKA sensor was engineered using CFP and citrine joined by the 14–3-3τ phosphoserine/threonine binding domain and a 21 amino acid kinase substrate domain (22). The maximum change in emission ratio for this probe was ~30 % for physiological stimuli *in vivo*. Sensors for the specific tyrosine kinase activities of Src, Abl, epidermal growth factor (EGF) receptor, and insulin signaling have been generated using linkers composed of Src homology (SH) 2 domains (to bind phosphotyrosine) and specific kinase consensus substrates (93, 94).

An alternative strategy to generate a sensor of tyrosine kinase activity has been described involving insertion of the complete CrkII protein between CFP and YFP (95). CrkII functions as an 'adaptor protein' in signal transduction cascades initiated by EGF, nerve growth factor (NGF) and insulin-like growth factor-I. The CrkII adaptor contains phosphotyrosine-binding SH2 and enzyme-binding SH3 domains. In response to tyrosine kinase activation, CrkII delivers specific SH3-bound enzymes to activated phosphotyrosine-containing proteins at the plasma membrane via SH2-mediated interactions. CrkII is also phosphorylated at tyrosine 221 during signal transduction. An intramolecular interaction between a SH2 domain and phosphotyrosine-221 produces a conformational change in CrkII, giving an ~60 % change in emission ratio. Although not specific for a single kinase, this sensor could be used to investigate the subset of CrkII mediated signal transduction processes and was applied to investigate the spatial and temporal organization of EGF stimulation *in vivo*.

Fluorescence lifetime imaging (FLIM) has been used to study the spatial and temporal activation of protein kinases using GFP-tagged kinases. To investigate the activation of protein kinase Cα (PKCα) a two component assay system was developed consisting of GFP-tagged PKCα and a (microinjected) Cy3-labeled antibody specific for a phosphothreonine present in activated PKCα (24). Activation of PKCα and threonine phosphorylation resulted in antibody-PKCα interaction and reduction in GFP lifetime by GFP-Cy3 energy transfer. Treatment of fibroblasts with phorbol esters was shown to activate PKCα in punctate regions near the plasma membrane. Using this same FLIM technique Bastiaens and colleagues (96) investigated activation of the dimeric receptor tyrosine kinase ErbB1. Cells expressing GFP-tagged ErbB1 were microinjected with a Cy3-antibody specific for a phosphotyrosine present in the activated ErbB1 receptor. After cell stimulation with bead-immobilized EGF, rapid widespread ErbB1activation was observed as reported by decreased GFP lifetime. It was suggested that ErbB1 dimers were transient and that ligand-independent propagation of receptor activation was an important additional amplification step in ErbB1 signaling.

2.4.8. Sensors of G proteins

G proteins, such as those of the Ras family, are involved in many signal transduction pathways. G proteins cycle between active GTP bound and inactive GDP bound forms, with exchange controlled by activating guanine nucleotide exchange factor (GEF) and deactivating GTPase activating protein (GAP). GFP-based sensors have been engineered to investigate the spatial organization of G protein activation. Class III-type probes were designed in which CFP and YFP were linked using a Ras moiety (Ras or Rap 1) and the Ras-binding domain of Raf (Raf RBD). Ras activation by GEF increases FRET efficiency as Ras and Raf RBD interact, whereas Ras inactivation by GAP eliminates the

Ras-Raf RBD interaction and reduces FRET efficiency (97). The activation patterns of Ras and Rap 1 were found to differ in EGF-stimulated fibroblasts and NGF-stimulated PC12 cells: Rap 1 was activated intracellularly in perinuclear region, whereas Ras was activated at the cell periphery.

An alternative FRET strategy was developed to investigate another Ras-type GTPase called Rac 1 (25). A GFP-Rac 1 fusion was used in conjunction with fluorescently-tagged p21-binding domain (PBD, derived from p21-activated kinase 1) that binds selectively to the GTP-bound form of Rac 1. Cells expressing the GFP-Rac1 fusion were microinjected with the fluorescently tagged-PBD; Rac 1 activation and PBD-Rac 1 interaction resulted in GFP FRET to the acceptor dye (Alexa 546). This strategy was used to show that platelet derived growth factor stimulation of Rac 1 in fibroblasts was greatest in membrane ruffles (25). Rac 1 activity was also found to be important in mechano-sensitive signal transduction and chemotaxis (98, 99).

GFP-based sensors have also been engineered to investigate the activity of Ran, a member of the Ras family of GTPases with possible roles in mitosis and nuclear transport of importins (100). It was predicted that Ran-GTP is found only in interphase nuclei or close to mitotic chromosomes because of discrete localization of Ran-GEF and Ran-GAP. Two FRET-based sensors were developed to investigate Ran activities. The first sensor (YRC), designed to monitor the nucleotide-binding status of Ran, was composed of YFP and CFP joined by the Ran-binding domain (RBD) from the yeast Ran-GAP accessory factor Yrb1; RBD selectively binds Ran-GTP which reduces the FRET efficiency. The second sensor contained an importin binding domain in the linker such that importin sequestration by Ran-GTP produced increase FRET. These sensors were used to investigate Ran-GTP gradients in mitotic *Xenopus* extracts.

2.4.9. Metabolite Sensors

FRET-based intramolecular GFP sensors have been developed recently to measure metabolite concentrations and to investigate the subcellular distribution of metabolites (19, 20). Components of the gram-negative bacterial metabolite sensing apparatus, the periplasmic binding proteins (PBPs), were used to link CFP and YFP to generate sensors called FLIPs (fluorescent indicator proteins). FLIPs selective for maltose (FLIPmals) were engineered using mutated maltose binding protein inserts to yield sensors that are able to measure maltose concentrations from 0.3 to 2000 µM (19). The FLIPmal sensors were used to measure maltose concentrations in beer and uptake of maltose by yeast cells (19). A similar strategy was applied to measure glucose in fibroblasts using a sensor containing a glucose/galactose binding insert (20). Cytoplasmic glucose concentration was found to vary over ~2 orders of magnitude in response to changes in external glucose in the physiological range (0.5–10 mM), and was decreased in response 2-deoxyglucose and cytochalasin B.

2.4.10. Sensors of Reduction-Oxidation (Redox) Potential

A sensor of redox potential was generated by Østergaard *et al.* (101) by addition of redox-sensitive cysteines at positions 149 and 202 of YFP. Under oxidative conditions the fluorescence of this redox sensor was reduced by approximately 50 % as a consequence of decreased absorption by the chromophore. High-resolution crystallographic studies of the sensor indicated that formation of a disulfide bond between the reactive cysteine residues resulted in rearrangement of amino acids in the vicinity of the chromophore and deformation of the b -barrel structure, presumably causing the observed changes in fluorescence. The redox potential of the engineered cysteine residues, -261 mV, is within the physiological range for other redox-active cysteines. When expressed in *E. coli*, this sensor reported the oxidative state of the cytoplasm and distinguished between wildtype and an isogenic strain lacking thioredoxin reductase which resulted in a more oxidative cytoplasm.

2.4.11. Nitration Sensors

GFP has been used as an intrinsic sensor of protein nitration, a post-translational modification observed in several neurological diseases including Parkinson's Disease, as well as in inflammation (102-104). Nitration decreases GFP fluorescence as a result of the formation of a 3-nitrotyrosyl adduct in the chromophore (105); this process is irreversible, sensitive ($IC_{50} \sim 20 \ \mu M$) and dependent upon the concentration of nitration agent (as opposed to cumulative exposure effects). This property of GFP was used *in vivo* to investigate proposed mechanisms of protein nitration. It was concluded that oxidation and nitration are not necessarily coupled *in vivo*.

2.4.12. Voltage Sensors

GFP-based sensors that report changes in membrane potential have been developed. The first sensor described was constructed by ligating GFP in the Shaker potassium channel between the last membrane-spanning domain and the carboxy-terminal region (Figure 2.5A, ref. 10). The Shaker channel was mutated to prevent ion conductance whilst maintaining voltage-sensitive structural changes. When expressed in oocytes, the fractional change in fluorescence evoked by changes in membrane potential for this sensor was small (Figure 2.5B). It was proposed that structural rearrangements in Shaker were responsible for changes in fluorescence, as fluorescence changes correlated with channel gating. Different GFP variants were subsequently incorporated at the same position in Shaker and mutation of Shaker was performed to generate new sensors with altered spectra, kinetics and voltage dependence (106). For instance, the ecliptic GFP variant had much faster response kinetics (10 ms *vs*. 100 ms, Figure 2.5C) and the L336A mutant of Shaker responded better at negative potentials than the original sensor.

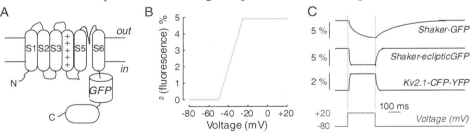

Figure 2.5. GFP-based sensors of cell membrane potential. A. Schematic of a Shaker-based membrane potential sensor. GFP was inserted after the last membrane-spanning region and before the carboxy-terminal region of Shaker. **B.** Relationship between percentage fluorescence change and membrane potential for Skaker-

based voltage sensor. **C.** Fluorescence responses to an applied depolarization pulse for Shaker-based sensors (*top two curves*) and FRET-based sensors containing a portion of Kv1.2. Adapted from refs. 10, 106 and 107.

A FRET based membrane potential sensor was generated by the tandem fusion of CFP and YFP to the carboxy-terminus of the Kv2.1 potassium channel (107). Membrane potential changes were proposed to result in structural rearrangements of the channel protein altering CFP/YFP relative orientation. This sensor has sub-millisecond response kinetics for depolarization (-80 to 20 mV), however, the magnitude response of this sensor to voltage changes was quite low (~2 % / 100 mV, Figure 2.5C, ref. 107). Recently a GFP-based membrane potential sensor was engineered using the voltage-gated rat μI skeletal sodium channel having rapid response but small voltage-dependent changes in fluorescence (11).

2.5. PERSPECTIVE AND FUTURE DIRECTIONS

Application of GFP and related genetically-encoded fluorescent proteins to the engineering of fluorescent sensors has significantly advanced our understanding of many cell biological process. As discussed, GFP-based sensors are non-invasive, stable and can be genetically targeted to specific cell types or organelles. These features offer distinct advantages over existing small molecule indicators. Recent advances have improved the specificity of GFP-sensors after appreciating the sensitivity of the fluorescence of some GFPs to pH, chloride and other factors.

To date, GFP-based sensors have been extensively applied to investigate signal transduction. Sensors of kinase activity, second messengers and G proteins have been used to investigate many aspects of cell signaling. In addition, sensors have been developed to measure enzyme activities, membrane potential, redox-state, protease activity, and metabolite concentrations. However, GFP-based sensors for other molecules of intermediary metabolism, such as ATP, phosphocreatine, and phosphate, which are not yet available, would be extremely useful in understanding metabolic organization in cells. Although ATP can be detected using luminescence assays (108-110), many of these molecules are at present only visible to nuclear magnetic resonance (NMR) which has poor sensitivity and spatial resolution. GFP-sensors have also been described for the metal ions calcium and zinc, as well as halides, but not for the many other small ions and molecules in cells.

Finally, technological advances should further exploit the unique advantages of GFP-sensors. Transgenic expression of GFP-based sensors in whole animals can provide a wealth of information about cell function *in vivo*. Two-photon fluorescence excitation has improved optical penetration depths for non-invasive *in vivo* measurements. New long-wavelength fluorescent proteins should further expand the possible sensor application in whole animals and plants.

2.6. ACKNOWLEDGMENTS

Supported by grants EB00415, HL73856, HL59198, DK35124 and EY13574 from the National Institutes of Health, and Research Development Program grant R613 and a CF Drug Discovery grant from the Cystic Fibrosis Foundation.

2.7. REFERENCES

1. M. Kneen, J. Farinas, Y. Li, and A.S. Verkman, Green fluorescent protein as a noninvasive intracellular pH indicator, *Biophys. J.* **74**, 1591-1599 (1997).
2. J. Llopis, J.M. McCaffery, A. Miyawaki, M.G. Farquhar, and R.Y. Tsien, Measurement of cytosolic, mitochondrial, and Golgi pH in single living cells with green fluorescent proteins, *Proc. Natl. Acad. Sci. USA* **95**, 6803-6808 (1998).
3. R.Y. Tsien, The green fluorescent protein, *Annu. Rev. Biochem.* **67**, 509-544 (1998).
4. M.-A. Elsliger, R.M. Wachter, G.T. Hanson, K. Kallio, and S.J. Remington, Structural and spectral repsonse of green fluorescent protein variants to changes in pH, *Biochemistry* **38**, 5296-5301 (1999).
5. R.M. Wachter and S.J. Remington, Sensitivity of the yellow variant of green fluorescent protein to halides and nitrate, *Curr. Biol.* **9**, R628-R629 (1999).
6. S. Jayaraman, P. Haggie, R.M. Wachter, S.J. Remington, and A.S. Verkman, Mechanism and cellular application of a green fluorescent protein-based halide sensor, *J. Biol. Chem.* **275**, 6047-6050 (2000).
7. G.S. Baird, D.A. Zacharias, and R.Y. Tsien, Circular permutation and receptor insertions within green fluorescent proteins, *Proc. Natl. Acad. Sci. USA* **96**, 11241-11246 (1999).
8. J. Nakai, M. Ohkura, and K. Imoto, A high signal-to-noise Ca^{2+} probe composed of a single green fluorescent protein, *Nat. Biotech.* **19**, 137-141 (2001).
9. T. Nagai, A. Sawano, E.S. Park, and A. Miyawaki, Circularly permuted green fluorescent proteins engineered to sense Ca^{2+}, *Proc. Natl. Acad. Sci. USA* **98**, 3197-3202 (2001).
10. M.S. Siegel and E.Y. Isacoff, A genetically encoded optical probe of membrane voltage, *Neuron* **19**, 735-741 (1997).
11. K. Ataka and V.A. Pieribone, A genetically targetable fluorescent probe of channel gating with rapid kinetics, *Biophys. J.* **82**, 509-516 (2002).
12. U. Heinemann and M. Hahn, Circular permutation of polypeptide chains: implications for protein folding and stability, *Prog. Biophys. Mol. Biol.* **64**, 121-143 (1995).
13. S. Topell, J. Henncke, and R. Glockshuber, Circularly permuted variants of the green fluorescent protein, *FEBS Lett.* **457**, 283-289 (1999).
14. R.D. Mitra, C.M. Silva, and D.C. Youvan, Fluorescence resonance energy transfer between blue-emitting and red-shifted excitation derivatives of the green fluorescent protein, *Gene* **173**, 13-17 (1996).
15. T. Kohl, K.G. Heinze, R. Kuhlemann, A. Koltermann, and P. Schwille, A protease assay for two-photon crosscorrelation and FRET analysis based on fluorescent proteins, *Proc. Natl. Acad. Sci. USA* **99**, 12161-12166 (2002).
16. K. Takemoto, T. Nagai, A. Miyawaki, and M. Miura, Spatio-temporal activation of caspase revealed by indicator that is insensitive to environmental effects, *J. Cell Biol.* **160**, 235-243 (2003).
17. A. Miyawaki, J. Llopis, R. Heim, J.M. McCaffery, J.A. Adams, M. Ikura, and R.Y. Tsien, Fluorescent indicators for Ca^{2+} based on green fluorescent proteins and calmodulin, *Nature* **388**, 882-887 (1997).
18. A. Miyawaki, O. Giesbeck, R. Heim, and R.Y. Tsien, Dynamic and quantitative Ca^{2+} measurements using improved cameleons, *Proc. Natl. Acad. Sci. USA* **96**, 2135-2140 (1999).
19. M. Fehr, W.B. Frommer, and S. Lalonde, Visulaization of maltose uptake in living yeast cells by fluorescent nanosensors, *Proc. Natl. Acad. Sci. USA* **99**, 9846-9851 (2002).
20. M. Fehr, S. Lalonde, L. I., M.W. Wolff, and W.B. Frommer, In vivo imaging of the dynamics of glucose uptake in the cytosol of COS-7 cells by fluorescent nanosensors, *J. Biol. Chem. In Press* (2003).
21. Y. Nagai, M. Miyazaki, R. Aoki, T. Zama, S. Inouye, K. Hirose, M. Iino, and M. Hagiwara, A fluorescent indicator for visualizing cAMP-induced phosphorylation in vivo, *Nat. Biotech.* **18**, 313-316 (2000).
22. J. Zhang, Y. Ma, S.S. Taylor, and R.Y. Tsien, Genetically encoded reporters of protein kinase A activity reveal impact of substrate tethering, *Proc. Natl. Acad. Sci. USA* **98**, 14997-15002 (2001).
23. M. Zaccolo, F.D. Georgi, C.Y. Cho, T.K. L. Feng, P.A. Negulscu, S.S. Taylor, R.Y. Tsien, and T. Pozzan, A genetically encoded, fluorescent indicator for cyclic AMP in living cells, *Nat. Cell Biol.* **2**, 25-29 (2000).
24. T. Ng, A. Squire, G. Hansra, F. Bornancin, C. Prevostel, A. Hanby, W. Harris, D. Barnes, S. Schmidt, H. Mellor, P.I. Bastiaens, and P.J. Parker, Imaging protein kinase Ca activation in cells, *Science* **283**, 2085-2089 (1999).
25. V.S. Kraynov, C. Chamberlain, G.M. Bekoch, M.A. Schwartz, S. Slabaugh, and K.M. Hahn, Localized Rac activation in living cells, *Science* **290**, 333-337 (2000).
26. A. Sawano and A. Miyawaki, Directed evolution of green fluorescent proteins by a new versatile PCR strategy for site-directed and semi-random mutagenesis, *Nuc. Acids Res.* **28**, E78 (2000).

27. O. Greisbeck, G.S. Baird, R.E. Campbell, D.A. Zacharias, and R.Y. Tsien, Reducing the environmental sensitivity of yellow fluorescent protein, *J. Biol. Chem.* **276**, 29188-29194 (2001).
28. T. Nagai, K. Ibata, E.S. Park, M. Kubota, K. Mikoshiba, and A. Miyawaki, A variant of yellow fluorescent protein with fast and efficient maturation for cell-biological applications, *Nat. Biotech.* **20**, 87-90 (2002).
29. A. Rekas, J.-R. Alattia, T. Nagai, A. Miyawaki, and M. Ikura, Crystal structure of Venus, a yellow fluorescent protein with improved maturation and reduced environmental sensitivity, *J. Biol. Chem.* **277**, 50573-50578 (2002).
30. M.V. Matz, A.F. Fradkov, Y.A. Labas, A.P. Savitsky, A.G. Zaraisky, M.L. Markelov, and S.A. Lukyanov, Fluorescent proteins from nonbioluminescent Anthozoa species, *Nat. Biotech.* **17**, 969-973 (1999).
31. G.S. Baird, D.A. Zacharias, and R.Y. Tsien, Biochemistry, mutagenesis, and oligomerization of DsRed, a red fluorescent protein from coral, *Proc. Natl. Acad. Sci. USA* **97**, 11984-11989 (2000).
32. D. Yarbough, R.M. Wachter, K. Kallio, M.V. Matz, and S.J. Remington, Refined crystal structure of DsRed, a red fluorescent protein from coral, at 2.0-Å resolution, *Proc. Natl. Acad. Sci. USA* **98**, 462-467 (2001).
33. B.J. Bevis and B.S. Glick, Rapidly maturing variants of the Discosoma red fluorescent protein (DsRed), *Nat. Biotech.* **20**, 83-87 (2002).
34. R.E. Campbell, O. Tour, A.E. Palmer, P.A. Steinbach, G.S. Baird, D.A. Zacharias, and R.Y. Tsien, A monomeric red fluorescent protein, *Proc. Natl. Acad. Sci. USA* **99**, 7877-82 (2002).
35. Y.G. Yanushevich, D.B. Staroverov, A.P. Savitsky, A.F. Fradkov, N.G. Gurskaya, M.E. Bulina, K.A. Lukyanov, and S.A. Lukyanov, A strategy for the generation of non-aggregating mutants of Anthozoa fluorescent proteins, *FEBS Lett.* **511**, 11-14 (2002).
36. P. Gavin, R.J. Devenish, and M. Prescott, An approach for reducing unwanted oligomerization of DsRed fusion proteins, *Bioch. Biophys. Res. Comm.* **298**, 707-713 (2002).
37. U. Lauf, P. Lopez, and M.M. Falk, Expression of fluorescently tagged connexins: a novel approach to rescue function of oligomeric DsRed-tagged proteins, *FEBS Lett.* **498**, 1-5 (2001).
38. A. Soling, A. Simm, and N. Rainov, Intracellular localization of Herpes simple virus type 1 thymidine kinase fused to different fluorescent proteins depends on choice of fluorescent tag, *FEBS Lett.* **527**, 153-158 (2002).
39. Y.A. Labas, N.G. Gurskaya, Y.G. Yanushevich, A.F. Fradkov, K.A. Lukyanov, S.A. Lukyanov, and M.V. Matz, Diversity and evolution of the green fluorescent protein family, *Proc. Natl. Acad. Sci. USA* **99**, 4256-4261 (2002).
40. J. Wiedenmann, C. Elke, K.-D. Spindler, and W. Funke, Cracks in the b-can: Fluorescent proteins from Anemonia sulcata (Anthozoa, Actinaria), *Proc. Natl. Acad. Sci. USA* **97**, 14091-14096 (2000).
41. J. Wiedenmann, A. Schenk, C. Röcker, A. Girod, K.-D. Spindler, and G.U. Nienhaus, A far-red fluorescent protein with fast maturation and reduced oligomerization tendancy from Entacmea quadricolor (Anthozoa, Actinaria), *Proc. Natl. Acad. Sci. USA* **99**, 11646-11651 (2002).
42. K.A. Lukyanov, A.F. Fradkov, N.G. Gurskaya, M.V. Matz, Y.A. Labas, A.P. Savitsky, M.L. Markelov, A.G. Zaraisky, X. Zhao, Y. Fang, W. Tan, and S.A. Lukyanov, Natural animal coloration can be determined by a nonfluorescent green fluorescent protein homolog, *J. Biol. Chem.* **275**, 25879-25882 (2000).
43. G. Gurskaya, A.Fradkov, A. Terskikh, M.V. Matz, Y.A. Labas, V.I. Martynov, Y.G. Yanushevich, K.A. Lukyanov, and S.A. Lukyanov, GFP-like chromoproteins as a source of far-red fluorescent proteins, *FEBS Lett.* **507**, 16-20 (2001).
44. M.V. Matz, K.A. Lukyanov, and S.A. Lukyanov, Family of green fluorescent protein: journey to the end of the rainbow, *BioEssays* **24**, 953-959 (2002).
45. A. Miyawaki, Green fluorescent protein-like proteins in reef Anthozoa animals, *Cell Struc. Func.* **27**, 343-347 (2002).
46. A. Terskikh, A.F. Fradkov, G. Ermakova, A.G. Zaraisky, P. Tan, A.V. Kajava, X. Zhao, S.A. Lukyanov, M.V. Matz, S. Kim, I. Weissman, and P. Siebert, "Fluorescent timer": Protein that changes color with time, *Science* **290**, 1585-1588 (2000).
47. V.V. Verkhusha, H. Otsuna, T. Awasaki, H. Oda, S. Tsukita, and K. Ito, An enhanced mutant of red fluorescent protein DsRed for double labeling and developmental timer of neural fiber bundle formation, *J. Biol. Chem.* **276**, 29621-29624 (2001).
48. D.M. Chudakov, V.V. Belousov, A.G. Zaraisky, V.V. Novoselov, D.B. Staroverov, D.B. Zorov, S.A. Lukyanov, and K.A. Lukyanov, Kindling fluorescent proteins for precise in vivo photolabeling, *Nat. Biotech.* **21**, 191-194 (2003).
49. B.G. Reid and G.C. Flynn, Chromophore formation in green fluorescent protein, *Biochemistry* **36**, 6786-6791 (1997).
50. M. Örmo, A.B. Cubitt, K. Kallio, L.A. Gross, R.Y. Tsien, and S.J. Remington, Crystal structure of the *Aequorea victoria* green fluorescent protein, *Science* **273**, 1392-1395 (1996).

51. R.B. Robey, O. Ruiz, A.V.P. Santo, J. Ma, F. Kear, L.-J. Wang, C.-J. Li, A.A. Bernado, and J.A.L. Arruda, pH-Dependent fluorescence of a heterologously expressed Aequorea green fluorescent protein mutant: in situ spectral characteristics and applicability to intracellular pH estimation, *Biochemistry* **37**, 9894-9901 (1998).

52. G. Miesenböck, D.A. De Angelis, and J.E. Rothman, Visualizing secretion and synaptic transmission with pH-sensitive green fluorescent proteins, *Nature* **394**, 192-195 (1998).

53. S. Sankaranarayanan, D.A. De Angelis, G. Rothman, and T.A. Ryan, The use of pHluorins for optical measurements of presynaptic activity, *Biophys. J.* **79**, 2199-2208 (2000).

54. G.T. Hanson, T.B. McAnaney, E.S. Park, M.E.P. Rendall, D.K. Yarbough, S. Chu, L. Xi, S.G. Boxer, M.H. Montrose, and S.J. Remington, Green fluorescent protein variants as ratiometric dual emission pH sensors. 1. Structural characterization and preliminary application, *Biochemistry* **41**, 15477-15488 (2002).

55. T.B. McAnaney, E.S. Park, G.T. Hanson, S.J. Remington, and S.G. Boxer, Green fluorescent protein variants as ratiometric dual emission pH sensors. 2. Excited-state dynamics, *Biochemistry* **41**, 15489-15494 (2002).

56. T. Awaji, A. Hirasawa, H. Shirakawa, G. Tsujimoto, and S. Miyazaki, Novel green fluorescent protein-based ratiometric indiators for monitoring pH in defined intracellular microdomains, *Bioch. Biophys. Res. Comm.* **289**, 457-462 (2001).

57. R.M. Wachter, D. Yarbough, K. Kallio, and S.J. Remington, Crystallographic and energetic analysis of binding of selected anions to the yellow variants of green fluorescent protein, *J. Mol. Biol.* **301**, 157-171 (2001).

58. L.J. Galietta, P.M. Haggie, and A.S. Verkman, Green fluorescent protein-based halide indicators with improved chloride and iodide affinities, *FEBS Letts.* **499**, 220-224 (2001).

59. T. Ma, L. Vetrivel, H. Yang, N. Pedemonte, O. Zegarra-Moran, L.J. Galietta, and A.S. Verkman, High-affinity activator of cystic fibrosis transmembrane conductance regulator (CFTR) chloride conductance identified by high-throughput screening, *J. Biol. Chem.* **277**, 37235-37241 (2002).

60. T. Ma, J.R. Thiagarajah, H. Yang, N.D. Sonawane, C. Folli, L.J. Galietta, and A.S. Verkman, Thiazolidinone CFTR inhibitor identified by high-throughput screening blocks cholera toxin-induced intestinal fluid secretion, *J. Clin. Invest.* **110**, 1651-1658 (2002).

61. T. Kuner and G. Augustine, A genetically encoded ratiometric indicator for chloride: capturing chloride transients in cultured hippocampal neurons, *Neuron* **27**, 447-459 (2000).

62. X. Xu, A.L.V. Gerard, B.C.B. Huang, D.C. Anderson, D.G. Payan, and Y. Luo, Detection of programmed cell death using fluorescence energy transfer, *Nuc. Acids Res.* **26**, 2034-2035 (1998).

63. J. Jones, R. Heim, J. Stack, and B.A. Pollok, Development and application of a GFP-FRET intracellular caspase assay for drug screening, *J. Biomol. Screen.* **5**, 307-318 (2000).

64. P.W. Vanderklish, L.A. Krushel, B.H. Holst, J.A. Gally, K.L. Crossin, and G.M. Edelman, Marking synaptic activity in dendritic spines with a calpain substrate exhibiting fluorescence resonance energy transfer, *Proc. Natl. Acad. Sci. USA* **97**, 2253-2258 (1999).

65. A.G. Harpur, F.S. Wouters, and P.I. Bastiaens, Imaging FRET between spectrally similar GFP molecules in single cells, *Nat. Biotech.* **19**, 167-169 (2001).

66. N.P. Mahajan, D.C. Harrison-Shostak, J. Michaux, and B. Herman, Novel mutant green fluorescent protein protease substrates reveal the activation of specific caspases during apoptosis, *Chem. Biol.* **6**, 401-409 (1999).

67. K. Truong, A. Sawano, H. Mizuno, H. Hama, K.I. Tong, T.K. Mal, A. Miyawaki, and F. M. Ikura, FRET-bases in vivo Ca^{2+} imaging by a new calmodulin-GFP fusion molecule, *Nat. Struct. Biol.* **8**, 1069-1073 (2001).

68. E. Emmanouilidou, A.G. Teschemacher, A.E. Pouli, L.I. Ncholls, E.P. Seward, and G.A. Rutter, Imaging Ca^{2+} concentration changes at the secretory vesicle surface with a recombinant targeted cameleon, *Curr. Biol.* **9**, 915-918 (1999).

69. D.N. Bowser, S. Petrou, R.G. Panchal, M.L. Smart, and D.A. Williams, Release of mitochondrial Ca^{2+} via the permeability transition activates endoplasmic reticulum Ca^{2+} uptake, *FASEB J.* **16**, 1105-1107 (2002).

70. R. Foyouzi-Youssefi, S. Arnaudeau, C. Borner, W.L. Kelley, J. Tschopp, D.P. Lew, N. Demaurex, and K.-H. Krause, Bcl-2 decreases the free Ca^{2+} concentration within the endoplasmic reticulum, *Proc. Natl. Acad. Sci. USA* **97**, 5723-5728 (2000).

71. M. Ikeda, T. Sugiyama, C.S. Wallace, H.S. Gompf, T. Yoshioka, A. Miyawaki, and C.N. Allen, Circadian dynamic of cytosolic and nuclear Ca^{2+} in single suprachiasmatic nucleus neurons, *Neuron* **38**, 253-263 (2003).

72. M. Jaconi, C. Bony, S.M. Richards, A. Terzic, S. Arnaudeau, G. Vassort, and M. Pucéat, Inositol 1,4,5-riphosphate directs Ca^{2+} flow between mitochondria and the endoplasmic/sacroplasmic reticulum: A role in regulating cardiac autonomic Ca^{2+} spiking, *Mol. Biol. Cell* **11**, 1845-1858 (2000).

73. R. Yu and P.M. Hinkle, Rapid turnover of calcium in the endoplasmic reticulum during signaling, *J. Biol. Chem.* **275**, 23648-23653 (2000).

74. G.J. Allen, S.P. Chu, K. Schumaker, C.T. Shimazaki, D. Vefeados, A. Kemper, S.D. Hawke, G. Tallman, R.Y. Tsien, J.F. Harper, J. Chory, and J.I. Schroeder, Alteration of stimulus-specific guard cell calcium oscillations and stomatal closing in *Arabidopsis det3* mutant, *Science* **289**, 2338-2342 (2000).

75. G.Y. Fan, H. Fujisaki, A. Miyawaki, R.-K. Tsay, R.Y. Tsien, and M.H. Ellisman, Video-rate scanning two-photon excitation fluorescence microscopy and ratio imaging with cameleons, *Biophys. J.* **76**, 2412-2420 (1999).

76. R. Kerr, V. Lev-Ram, G. Baird, P. Vincent, R.Y. Tsien, and W.R. Schafer, Optical imaging of calcium transients in neurons and pharyngeal muscle of C. elegans, *Neuron* **26**, 583-594 (2000).

77. A. Miyawaki, H. Mizuno, T. Nagai, and A. Sawano, Development of genetically encoded fluorescent indicators for calcium, *Meth. Enzymol.* **360**, 202-225 (2003).

78. H. Mizuno, A. Sawano, P. Eli, H. Hama, and A. Miyawaki, Red fluorescent protein from Discosoma as a fusion tag and partner for fluorescence energy transfer, *Biochemistry* **40**, 2502-2510 (2001).

79. D. Yu, G. Baird, R.Y. Tsien, and R.L. Davis, Detection of Calcium transients in Drosophila mushroom body neurons with Camgaroo reporters, *J. Neuro.* **23**, 64-72 (2003).

80. E. Rapizzi, P. Pinton, G. Szabadkai, M.R. Wieckowski, G. Vandecasteele, G. Baird, R.A. Tuft, K.E. Fogarty, and R. Rizzuto, Recombinant expression of the voltage-dependent anion channel enhances the transfer of Ca^{2+} microdomains to mitochondria, *J. Cell Biol.* **159**, 613-624 (2002).

81. J.W. Wang, A.M. Wong, J. Flores, L.B. Vosshall, and R. Axel, Two-photon calcium imaging reveals an odor-evoked map of activity in the fly brain, *Cell* **112**, 271-282 (2003).

82. V. Robert, P. Gurlini, V. Tosello, T. Nagai, A. Miyawaki, F. Di Lisa, and T. Pozzan, Beat-to-beat oscillations of mitochondrial $[Ca^{2+}]$ in cardiac cells, *EMBO J.* **20**, 4998-5007 (2001).

83. P. Pinton, T. Tsuboi, E.K. Ainscow, T. Pozzan, R. Rizzuto, and G.A. Rutter, Dynamics of glucose-induced membrane recruitment of protein kinase c bII in living pancreatic islet b-cells, *J. Biol. Chem.* **277**, 37702-37710 (2002).

84. V.A. Romoser, P.M. Hinkle, and A. Persechini, Detection in living cells of Ca^{2+}-dependent changes in the fluorescence emission of an indicator composed of two green fluorescent protein variants linked by a calmodulin-binding sequence. A new class of fluorescent indicators, *J. Biol. Chem.* **272**, 13270-13274 (1997).

85. A. Persechini and B. Cronk, The relationship between the free concentration of Ca^{2+} and Ca^{2+}-calmodulin in intact cells, *J. Biol. Chem.* **27**, 6827-6830 (1999).

86. M.N. Teruel, W. Chen, A. Persechini, and T. Meyer, Differential codes for free Ca^{2+}-calmodulin signals in the nucleus and cytosol, *Curr. Biol.* **10**, 86-94 (2000).

87. S.R. Adams, A.T. Harootunian, Y.J. Beuchler, S.S. Taylor, and R.Y. Tsien, Fluorescence ratio imaging of cAMP in single cells, *Nature* **349**, 694-697 (1991).

88. M. Zaccolo and T. Pozzan, Discrete Microdomains with high concentration of cAMP in stimulated rat neonatal cardiac myocytes, *Science* **295**, 1711-1715 (2002).

89. M. Sato, N. Hida, T. Ozawa, and Y. Umenzawa, Fluorescent indicators for cyclic GMP based on cyclic GMP-dependent protein kinase Ia and green fluorescent protein, *Anal. Chem.* **72**, 5918-5924 (2000).

90. A. Honda, S.R. Adams, C.L. Sawyer, V. Lev-Ram, R.Y. Tsien, and W.R.G. Dostman, Spatiotemporal dynamics of guanosine 3',5'-cyclic monophosphate revealed by a genetically encoded fluorescent indicator, *Proc. Natl. Acad. Sci. USA* **98**, 12437-12442 (2001).

91. L.L. Pearce, R.E. Gandley, W. Han, K. Wasserlsso, M. Stitt, A.J. Kanai, M.K. McLaughlin, B.R. Pitt, and E.S. Levitan, Role of metallothionein in nitric oxide signaling as revealed by green fluorescent fusion protein, *Proc. Natl. Acad. Sci. USA* **97**, 477-482 (2000).

92. K. Hirose, S. Kadowaki, M. Tanabe, H. Takeshima, and M. Iino, Spatiotemporal dynamics of inositol 1,4,5-triphosphate that underlies complex Ca^{2+} mobilization patterns, *Science* **284**, 1527-1530 (1999).

93. M. Sato, T. Ozawa, K. Inukai, T. Asano, and Y. Umezawa, Fluorescent indicators for imaging protein phosphorylation in single living cells, *Nat. Biotech.* **20**, 287-294 (2002).

94. A.Y. Ting, K.H. Kain, R.L. Klemke, and R.Y. Tsien, Genetically encoded fluorescent reporters of protein tyrosine kinase activities in living cells, *Proc. Natl. Acad. Sci. USA* **98**, 15003-15008 (2001).

95. K. Kurokawa, N. Mozchizuki, Y. Ohba, H. Mizuno, A. Miyawaki, and M. Matsuda, A pair of fluorescent energy transfer-based probes for tyrosine phosphorylation of the CrkII adaptor protein *in vivo*, *J. Biol. Chem.* **276**, 31305-31310 (2001).

96. P.J. Verveer, F.S. Wouters, A.R. Reynolds, and P.I. Bastiaens, Quantitative imaging of lateral ErbB1 receptor signal propagation in the plasma membrane, *Science* **290**, 1567-1570 (2000).

97. M. Mochizuki, S. Yamashita, K. Kurokawa, Y. Ohba, T. Nagai, A. Miyawaki, and M. Matsuda, Spatio-temporal images of growth-factor-induced activation of Ras and Rap1, *Nature* **411**, 1065-1068 (2001).

98. A. Katsumi, J. Milanini, W.B. Kiosses, M.A. del Pozo, R. Kaunas, S. Chien, K.M. Hahn, and M.A. Schwartz, Effect of cell tension on the small GTPase Rac, *J. Cell Biol.* **158**, 153-164 (2002).

99. E.M. Gardiner, K.N. Pestonjamasp, B.P. Bohl, C. Chamberlain, K.M. Hahn, and G.M. Bokoch, Spatial and temporal analysis of Rac activation during live neutrophil chemotaxis, *Curr. Biol.* **12**, 2029-2034 (2002).

100. P. Kalab, K. Weis, and R. Heald, Visualization of a Ran-GTP gradient in interphase and mitotic Xenopus egg extracts, *Science* **295**, 2452-2456 (2002).

101. H. Østergaard, A. Henriksen, F.G. Hansen, and J.R. Winther, Shedding light on disulfide bond formation: engineering a redox switch in green fluorescent protein, *EMBO J.* **20**, 5853-5862 (2001).

102. J. Beckman, A.G. Estevez, J.P. Crow, and L. Barbeito, Superoxide dismutase and death of motoneurons in ALS, *Trends Neurosci.* **24**, S15-S20 (2001).

103. B.I. Giasson, J.E. Duda, I.V. Murray, Q. Chen, J.M. Souza, H.I. Hurtg, H. Ischiropoulos, J.Q. Trojanowski, and V.M. Lee, Oxidative damage linked to neurodegeneration by selective a-synuclein nitration in synucleinopathy lesions, *Science* **290**, 985-989 (2000).

104. H. Ischiropoulos and J.S. Beckman, Oxidative stress and nitration in neurodegeneration: Cause, effect, or association?, *J. Clin. Invest.* **111**, 163-169 (2003).

105. M.G. Espey, S. Xavier, D.D. Thomas, K.M. Miranda, and D.A. Wink, Direct real-time evaluation of nitration with green fluorescent protein in solution and within human cells reveals the impact of nitrogen dioxide vs. peroxynitrite mechanisms, *Proc. Natl. Acad. Sci. USA* **99**, 3481-3486 (2002).

106. G. Guerrero, M.S. Siegel, B. Roska, E. Loots, and E.Y. Isacoff, Tuning FlaSh: redesign of the dynamics, voltage range and color of the genetically encoded optical sensor of membrane potential, *Biophys. J.* **83**, 3607-3618 (2002).

107. R. Sakai, V. Repunte-Canonigo, C.D. Raj, and T. Knöpfel, Design and characterization of a DNA-encoded, voltage sensitive fluorescent protein, *Eur. J. Neurosci* **13**, 2314-2318 (2001).

108. H.J. Kennedy, A.E. Pouli, E.K. Ainscow, L.S. Jouaville, R. Rizzuto, and G.A. Rutter, Glucose generates sub-plasma membrane ATP microdomians in single islet b-cells. Potential role for strategically located mitochondria, *J. Biol. Chem.* **274**, 13281-13291 (1999).

109. P. Magalhães and R. Rizzuto, Mitochondria and calcium homeostasis: a tale of three luminescent proteins, *Luminescence* **16**, 67-71 (2001).

110. A.M. Porcelli, P. Pinton, E.K. Ainscow, A. Chiesa, M. Rugolo, G.A. Rutter, and R. Rizzuto, Targeting of reporter molecules to mitochondria to measure calcium, ATP, and pH, *Methods Cell Biol.* **65**, 353-380 (2001).

FLUORESCENT SACCHARIDE SENSORS

Tony D James[†] and Seiji Shinkai[‡]

3.1. INTRODUCTION

> *"In the field of observation, chance favours only the prepared mind"*
> Louis Pasteur 1822-1895

In order to set the context for this chapter, we will outline why saccharide sensors are important. Saccharides and related molecular species are involved in the metabolic pathways of living organisms, therefore, the detection of biologically important sugars (D-glucose, D-fructose, D-galactose, etc.), is vital in a variety of medicinal and industrial contexts. The recognition of D-glucose is of particular interest, since the breakdown of glucose transport in humans has been correlated with a number of diseases: renal glycosuria,[1,2] cystic fibrosis,[3] diabetes[4,5] and also human cancer.[6] Industrial applications range from the monitoring of fermenting processes to establishing the enantiomeric purity of synthetic drugs.

This chapter will discuss fluorescent saccharide sensors constructed using 'boronic acid' receptor units. Current enzymatic detection methods of sugars offer specificity for only a few saccharides; additionally, enzyme based sensors are unstable under harsh conditions. Stable boronic acid based saccharide receptors offer the possibility of creating saccharide sensors which through design can be selective and sensitive for any chosen saccharide.

A growing number of excellent reviews exist in the literature covering the use of boronic acids in the development of saccharide receptors.[7-15]

[†] Department of Chemistry, University of Bath, Bath BA2 7AY UK. (t.d.james@bath.ac.uk)
[‡] Department of Chemistry and Biochemistry, Graduate School of Engineering, Kyushu University, Fukuoka 812-8581 JAPAN. (seijitcm@mbox.nc.kyushu-u.ac.jp)

Our aim with this review is to provide a specific and more personal perspective on current and future directions in the development of fluorescent saccharide sensors. We have structured the chapter to follow the same conceptual journey we ourselves take when developing new sensor systems. The two components required for a sensor are a selective interface and a read-out mechanism. The two main sections of this review are for that reason interface and read out. The read out or fluorescence will be discussed first since selectivity is only important when an output is possible.

Recognition of saccharides by boronic acids has a unique place in supramolecular chemistry. The pair-wise interaction energy is large enough to allow single-point molecular recognition, and the primary interaction involves the reversible formation of a pair of covalent bonds (rather than non-covalent attractive forces). Despite a long history - the first structural and quantitative binding constant data were reported in the1950's[16-18] - the structure of the boronic acid - saccharide complexes in aqueous solution continues to be discussed.[19-21] There is general agreement that boronic acids covalently react with 1,2 or 1,3 diols to form five or six membered cyclic esters. The adjacent rigid cis-diols of saccharides form stronger cyclic esters than simple acyclic diols such as ethylene glycol. With saccharides the choice of diol used in the formation of a cyclic ester is complicated by the possibility of pyranose to furanose isomerization of the saccharide moiety. Lorand and Edwards first determined the selectivity of phenylboronic acid towards saccharides and this selectivity order seems to be retained by all monoboronic acids (D-fructose > D-galactose > D-glucose).[18]

The equilibria involved in the phenylboronate binding of a diol are conventionally summarized as a set of coupled equilibria (equation 1). In aqueous solution phenylboronic acid reacts with water to form the boronate anion plus a hydrated proton thereby defining an acidity constant K_a.[¶] The formation of a diol boronate complex, defined by K_{tet} formally liberates two equivalents of water, but this stoichiometric factor is usually ignored as a constant in dilute aqueous solution. In a formal sense, phenylboronic acid could also bind diols to form a trigonal complex (K_{trig}), and this species would itself act as an acid according to K_a'. The "acidification" of solutions containing phenylboronic acid and diols is always discussed in terms of the trigonal complex being a stronger acid than the parent phenylboronic acid, i.e. $K_a' > K_a$.[14] As a result, $K_{tet} > K_{trig}$.

We have recently examined associations of the boronic acids with buffer conjugate bases (phosphate, citrate and imidazole).[23] What we discovered was that binary boronate - X complexes are formed with Lewis bases (X), together with ternary species (boronate-X-saccharide). The most important discovery as far as sensor design was the discovery of ternary complexes. In some cases, these previously unrecognized species persist into acidic solution and under some stoichiometric conditions they can be the dominant components of the solution. These complexes suppress the boronate and boronic acid concentrations leading to a decrease in the measured apparent formation constants.

[¶] Equation 1 shows an explicit water molecule "coordinated" to the trigonal boronic acids. There is undoubtedly water in rapid exchange on the Lewis acidic boron in the same way that hydrated Lewis acidic metal ions exchange bound water. A good analogy is Zn^{2+} (aq), which ionizes in water to give a $pK_a = 8.8$, i.e. $Zn\text{-}OH_2 \rightarrow Zn\text{-}OH + H^+$.[22] Coordinated water is omitted in subsequent structures.

The more important implications for future work are clear; avoid buffers with significant concentrations of strongly Lewis basic components! Conditional or apparent constants may be suitable for the purposes of developing useful systems, so long as any competing equilibria with the buffer do not overwhelm the system. In any case, it is simple to inspect the influence of the buffer components to allow a judicious choice of buffer concentration and pH to generate an optimal signal from the system under study.

As a consequence of this work, the scope of the "simple" diol-boronate recognition system is greatly expanded over the simple picture of equation 1.

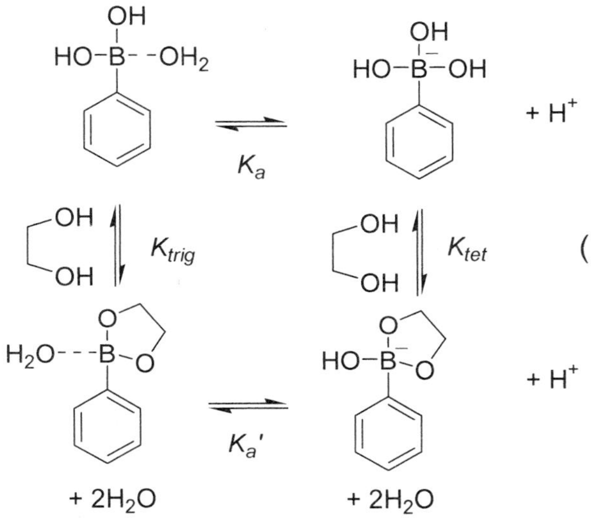

$$(1)$$

3.2. READ-OUT

Fluorescent sensors for saccharides are of particular interest in a practical sense. This is in part due to the inherent sensitivity of the fluorescence technique. Only small amounts of a sensor are required (typically 10^{-6} M) offsetting the synthetic costs of such sensors. Also, fluorescence spectrometers are widely available and inexpensive.

Fluorescence sensors have also found applications in continuous monitoring using an optical fiber and intracellular mapping using confocal microscopy.

As discussed in the introduction the selectivity of monoboronic acids towards saccharides follows the same trend as that discovered by Lorand and Edwards for phenylboronic acid (D-fructose > D-galactose > D-glucose).[18] Consequently, monoboronic acid systems should be used to evaluate potential read out units for use in saccharide selective sensors.

3.2.1. Internal Charge Transfer (ICT)

The first fluorescent sensors for saccharides were based on fluorophore boronic acids. Czarnik showed that 2- and 9- anthryboronic acid[24] **1** and **2** could be used to detect saccharides. The fluorophore boronic acids pK_a was lowered by saccharide present in the medium. Therefore, since the pK_a is lowered the amount of quenching at a fixed pH increases.

Aoyama has also shown that 5-indolylboronic acid **3** undergoes fluorescence quenching upon complexation with oligosaccharides.[25] The stability constants of monosaccharides were as expected.[18] However, higher oligomers of saccharides enjoy increased stabilization relative to lower oligomers of saccharides due to a secondary interaction with the indole N-H.

The 2-anthrylboronic acid used by Czarnik displays only a small fluorescence change. However, we have shown through the screening of eight aromatic boronic acids that **4** and **5** produce large fluorescence responses on saccharide binding and as such are more suitable candidates for saccharide detection.[26,27]

Fluorescence quenching from the boronate anion through PET was thought to be the source of the fluorescence quenching for the systems described above. However, a more reasonable explanation of the fluorescence properties comes from our investigation of stilbene boronic acid **6a**[28] and the detailed investigation of several stilbene and longer polyene boronic acids **6a-f** by Lakowicz.[29,30] Essentially, the neutral form of the boron group acts as an electron withdrawing group while the anionic form acts as an electron-donating group. Hence, the change in the electronic properties of the boron group are what cause the spectral changes of the fluorophore.

Accordingly, in this chapter the original fluorescence systems have been classified as internal charge transfer ICT fluorophores where the acceptor is the boronic acid (these systems have no defined donor).

However, the first systematic attempt to couple a donor and acceptor in the construction of an ICT system was through the coumarine boronic acid **7**.[31] Here both fluorescence intensity and wavelength are affected since the nitrogen is directly connected with the chromophore. Sadly, this system shows only a small shift in intensity and wavelength, in spite of its clever molecular design.

Lakowicz quickly recognized the importance of stilbene boronic acid **6d** and has since prepared a number of analogous ICT fluorophore systems, including oxazoline[32] **8**, chalcones[33] **9a,b** and boron-dipyrromethene (BODIPY)[34] **10**. The oxazoline and chalcone systems produce large fluorescence changes, however, the BODIPY system displays only a small change in fluorescence. Although the BODIPY system **10** did not work particularly well, this class of fluorophore does require further exploration. The BODIPY chromophore possess many advantages as a fluorescence probe: for example they possess high extinction coefficients, high fluorescence quantum yields, good photostability, a narrow emission band and their building block synthesis allows the development of many different analogues showing emission maxima from 500 to 700 nm. Long-wavelength fluorescent probes for glucose are highly desirable for transdermal glucose monitoring and/or for measurements in whole blood. In addition, narrow emission bands are desirable for high signal to noise ratio.

We have also been interested in developing new ICT fluorescence sensors that show both a large shift in wavelength and intensity. While investigating the development of colour sensors for saccharides,[35-37] we discovered that one of the components behaves like an ICT fluorescent saccharide sensor **11**. We are currently investigating the fluorescent properties of this and related systems in detail. Although this system may be of limited practical use (aniline is not the best fluorophore) we believe that understanding this simple unit will allow us to design and develop improved ICT fluorescent systems.[38]

1

2

3

4

5

6a(n=1, X=H) **6b** (n=1, X=CN) **6c** (n=1, X=OCH_3) **6d** (n=1, X=$N(CH_3)_2$) **6e** (n=2, X=$N(CH_3)_2$) **6f** (n=3, X=$N(CH_3)_2$)

7

8

9a (n=1)
9b (n=2)

10

11

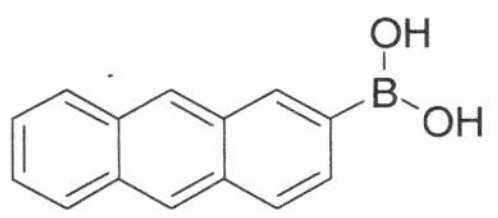

3.2.2. Photoinduced electron transfer (PET)

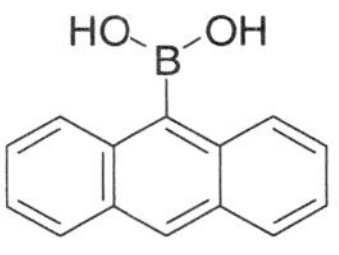

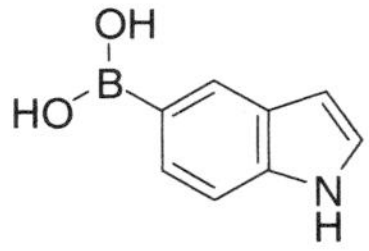

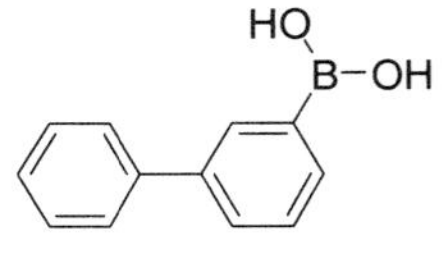

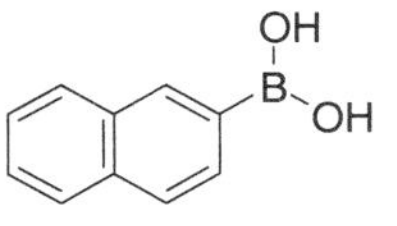

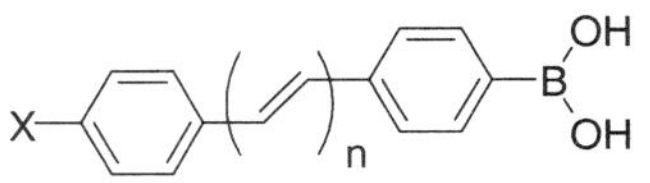

Photoinduced electron transfer (PET) has been widely used as the preferred tool in fluorescent sensor design for atomic and molecular species.[39-42] PET sensors generally consist of a fluorophore and a receptor linked by a short spacer. The changes in oxidation/reduction potential of the receptor upon guest binding can alter the PET process creating changes in fluorescence.

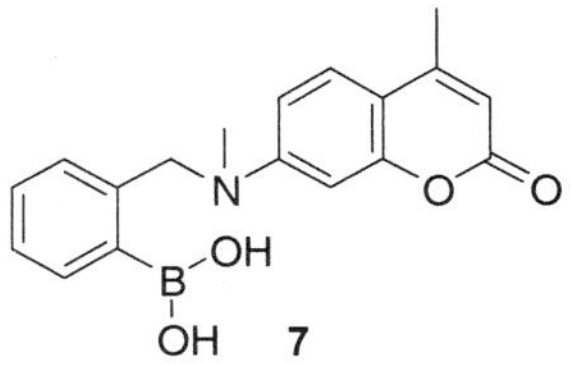

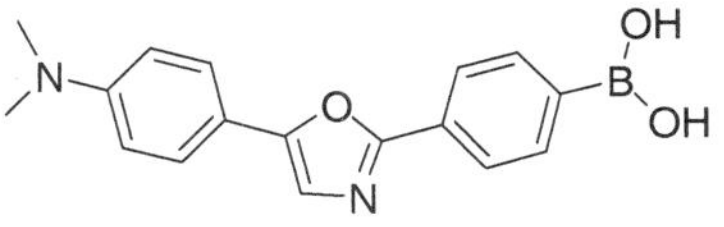

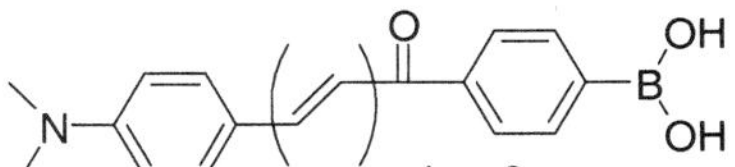

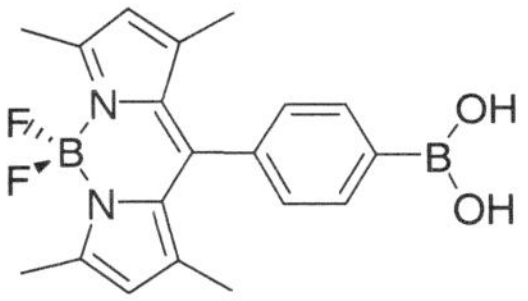

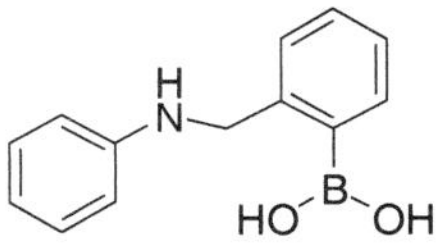

The first rationally designed fluorescent PET saccharide sensor compound **12** was prepared in 1994.[43,44] This fluorescence sensor contains a boronic acid group and an amine group. The boronic acid group is required to bind with and capture sugar molecules in water. The amine group plays two roles in the system. (1) Boronic acids with a neighbouring amine facilitate the binding of saccharides at neutral pH. (2) The fluorescence intensity is controlled by the amine.

With **12** the 'free' amine reduces the intensity of the fluorescence (quenching by PET). This is the 'off' state of the fluorescent sensor. When sugar is added, the amine becomes 'bound' to the boron center. The boron bound amine cannot quench the fluorescence and hence a strong fluorescence is observed. This is the 'on' state of the fluorescent sensor. The system described above illustrates the basic concept of an 'off-on' fluorescent sensor for sugars.

Two other systems are worthy of special mention since they differ from the system outlined above. The simple monoboronic acid **13** which can selectively signal the furanose form of saccharides[45] and the 'on-off' PET system **14** where on saccharide binding steric crowding breaks the B-N bond found in the 'free' receptor.[46]

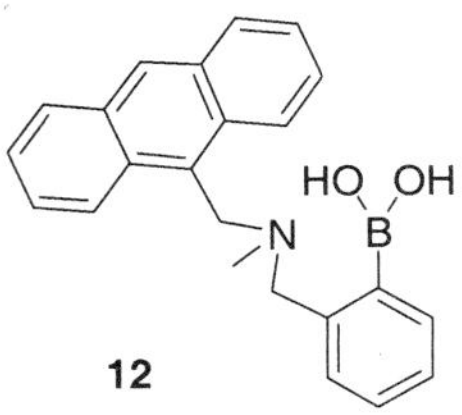

3.2.3. Others

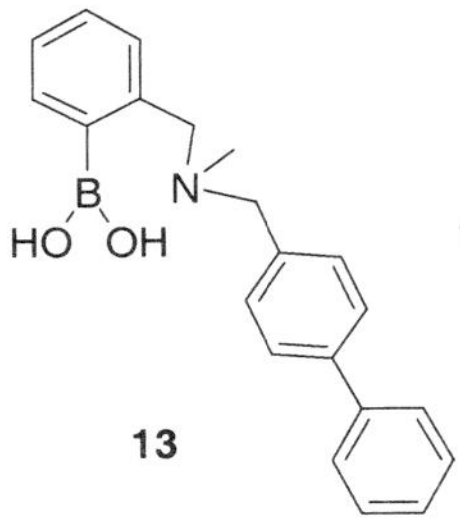

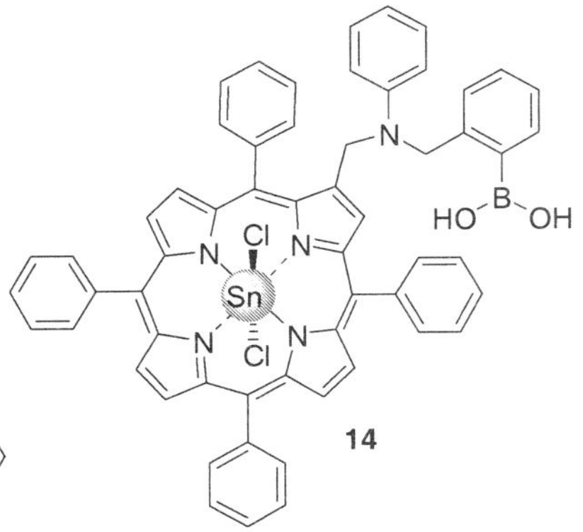

Wang has recently reported that 8-qinoline boronic acid **15** responds to the binding of saccharides with over 40-fold increases in fluorescence intensity.[47] The authors ascribe the fluorescence changes to environmental factors and not a change in the hybridization of the boron. We believe that further investigation into the fluorescence signaling mechanism is required since it seems more reasonable to associate large fluorescence changes to differential hybridization of the boron.

Heagy and Lakowicz have been investigating *N*-phenylboronic acid derivatives of 1,8-naphthalimide **16a,b** with these systems the fluorescence is substantially quenched (*ca.* 5-fold) on saccharide binding.[48,49] The fluorescence change has been ascribed to PET from the boronate to the naphthalimide fluorophore. The nitro derivative **16c** was particularly interesting because it displayed dual fluorescence and was particularly sensitive for D-glucose.[50]

Hayashita and Teramae have prepared an interesting fluorescent ensemble comprising of compound **17** and β–cyclodextrin.[51] The system displays fluorescence enhancement on saccharide binding and as expected for a monoboronic acid the highest binding was observed with D-fructose.

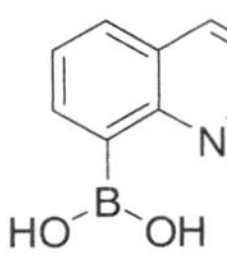

15

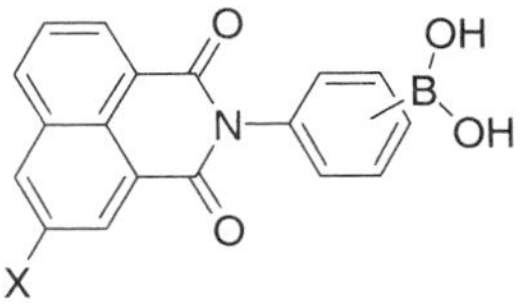

16a (X=H, meta)
16b (X=H, ortho)
16c (X=NO$_2$, meta)

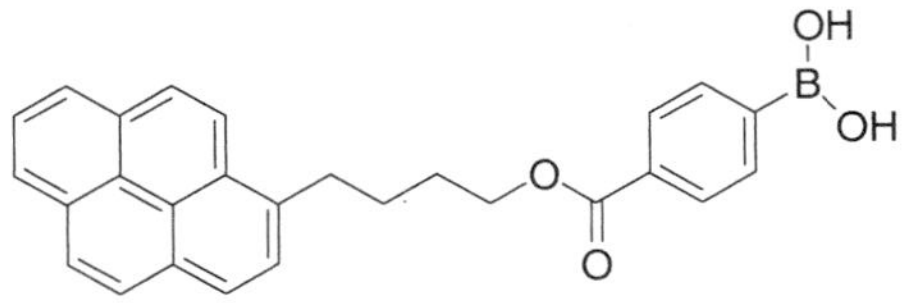

17

3.3. INTERFACE

So far we have only discussed monoboronic acid systems which can be used to evaluate the read out units in the construction of selective fluorescent saccharide sensors. The next step is to incorporate a better interface so that selectivity (other than the inherent selectivity) can be controlled. Factors that will affect saccharide selectivity are the number and type of receptor units and the orientation and position of these receptor units within the sensor. More boronic acid units should, if positioned correctly, improve selectivity for particular saccharides, for example a larger spacing will favour polysaccharides. Also, additional non boronic acid receptors can be incorporated to allow selectivity amongst derivatives such as amino, carboxy and phosphorylated saccharides.

3.3.1. Internal Charge Transfer (ICT)

Having successfully designed an ICT fluorescent signaling unit compound **11**, our subsequent goal was to prepare a D-glucose selective receptor containing the same signaling unit. As outlined above selectivity can be achieved through the correct spacing of two boronic acids. Combining the requirements of the signaling unit with those of a D-glucose selective receptor we designed the diboronic acid sensor **18**.[52] The relative stability constants of the diboronic acid **18** to the monoboronic acid **11** shows how effective our molecular design is at enhancing the D-glucose binding. Cooperative binding of the two boronic acid groups is clearly observed as illustrated by the stability constant differences between the mono- and diboronic acid compounds. The stability constant K of diboronic acid sensor **18** with D-glucose is 14 times greater than with monoboronic sensor **11**. Whereas, the stability constant K of diboronic acid sensor **18** with D-fructose is 0.6 times weaker than monoboronic acid sensor **11**. This result can be explained since it is well known that D-glucose readily forms 1:1 cyclic complexes with diboronic acids, whereas D-fructose tends to form 2:1 acyclic complexes with diboronic acids.

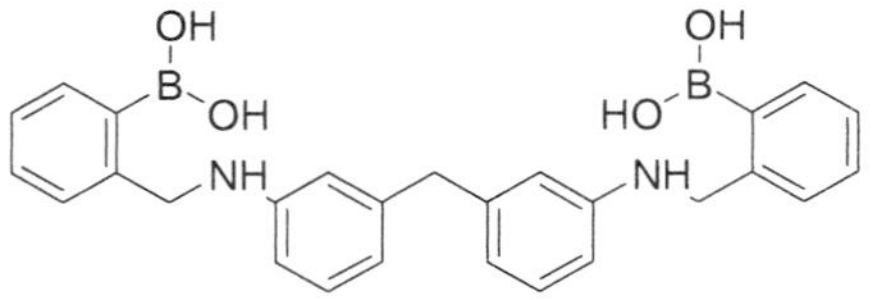

18

3.3.2. Photoinduced electron transfer (PET)

The simple 'off-on' PET system **12** was improved with the introduction of a second boronic acid group **19**.[44,53] For compound **19** two possible saccharide binding modes can inhibit the electron transfer process so giving higher fluorescence: the 2:1 complex and 1:1 complex. Due to fortuitous spacing of the boronic acid groups the diboronic acid was selective for D-glucose over other monosaccharides. Lakowicz has recently demonstrated that these two systems could be used as fluorescence lifetime sensors for saccharides.[54] These lifetime measurements were also able to unequivocally confirm that the boronic acid saccharide interaction is reversible.

Norrild has carried out a more detailed investigation of this system in order to confirm the structure of the bound glucose.[55] Norrild was interested in the system since the ^{1}H NMR reported[44,53] indicated that D-glucose bound to the receptor in its pyranose form. Norrild with Eggert had previously shown that simple boronic acids selectively bind with the furanose form of D-glucose.[19] From ^{1}H NMR observations it was concluded that the diboronic acid initially binds with the pyranose form of D-glucose and over time the bound glucose converts to the furanose form.

Chiral recognition of saccharides by **20** (R or S) utilizes both steric and electronic factors.[56] The asymmetric immobilization of the amine groups relative to the binaphthyl moiety upon 1:1 complexation of saccharides by D- or L-isomers creates a difference in PET. This difference is manifested in the maximum fluorescence intensity of the complex. Steric factors arising from the chiral binaphthyl building block are chiefly represented by the stability constant of the complex. However, the interdependency of electronic and steric factors upon each other is not excluded. This new molecular cleft, with a longer spacer unit compared to the anthracene based diboronic acid **19**, gave the best recognition for fructose. The chiral recognition of saccharides employing hydrogen bonding by polyhydroxyl molecules has been reported.[57] However, discriminative detection of isomers in aqueous media by fluorescence, as far as we are aware, had not been achieved before. In this system steric factors and electronic factors bimodally discriminate the chirality of the saccharide. Competitive studies with D- and L-monosaccharides show the possibility of selective detection of saccharide isomers. The availability of both R and S isomers of this particular molecular sensor is an important advantage, since concomitant detection by two probes is possible.

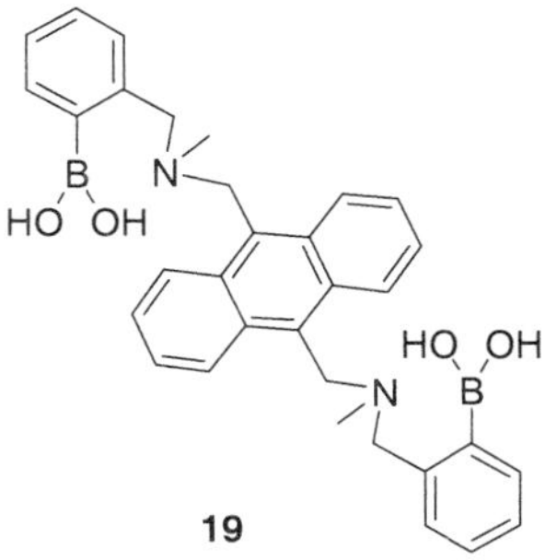

19

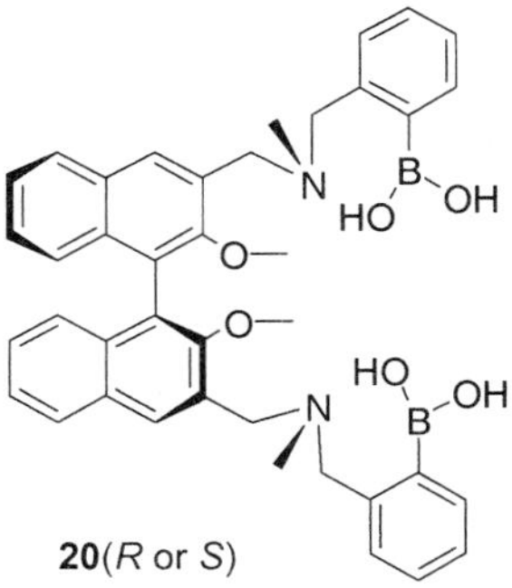

20(*R* or *S*)

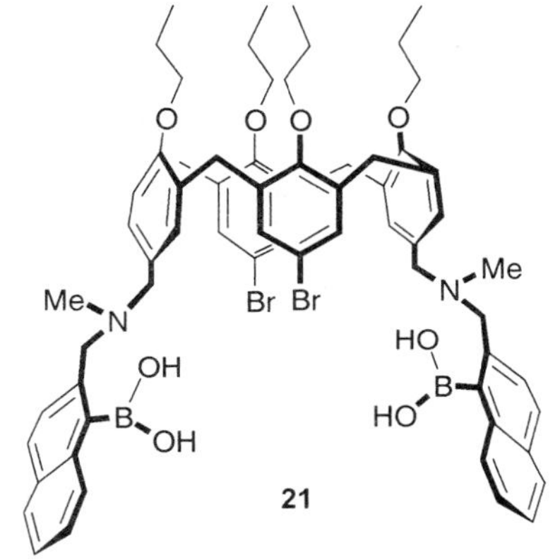

21

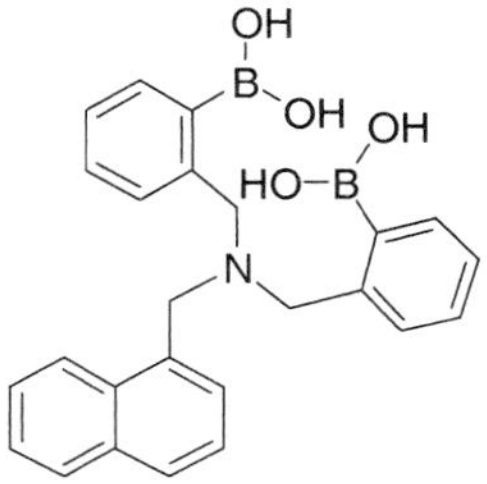

22

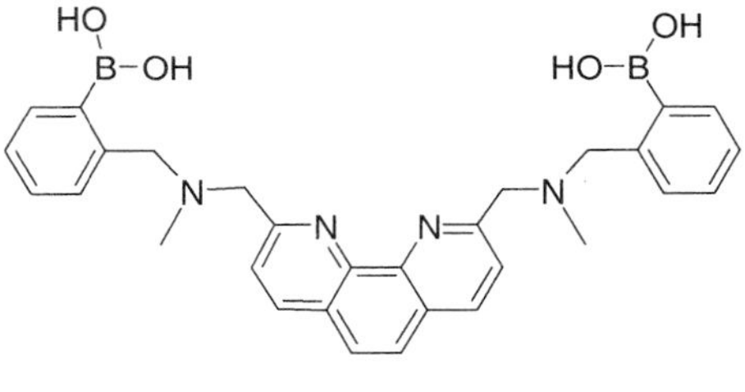

23

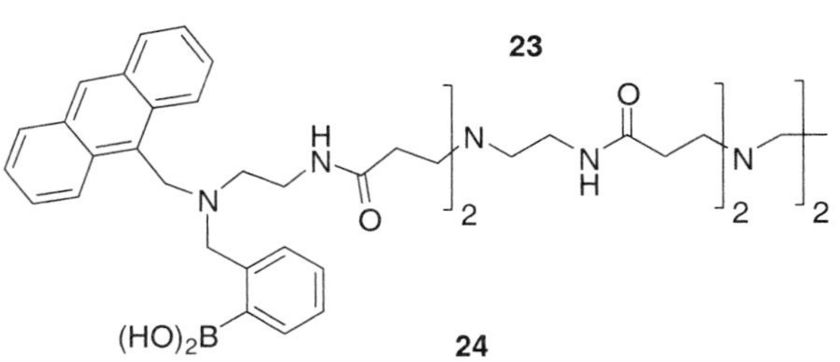

24

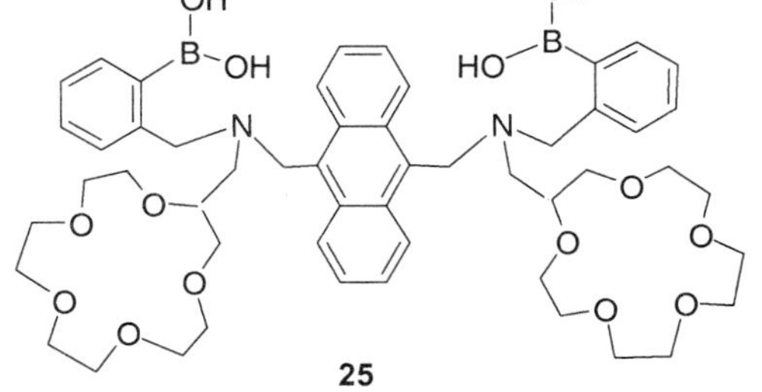

25

Attempts to use the calixarene framework **21** as a core on which to develop novel saccharide selective systems have been made.[58] Compound **21** shows the strongest binding with D-fructose. The calixarene unit has also been used to develop novel luminescent systems.[59] Although success has been limited with these calixarene systems, we are confident that the concept of using the versatile calixarene framework upon which to build saccharide selective systems is a good one.

Diboronic acid **22** with a small bite angle has been synthesized and has been shown to be selective for small saccharides such as D-sorbitol.[60] Conversely, diboronic acid **23** with a larger spacing of the boronic acid groups looses selectivity and sensitivity.[61] However the system may be useful for the detection of saccharides in concentrated solutions, such as those encountered in the brewing and confectionery industries. Dendritic boronic acid **24**, shows enhanced binding affinities but the selectivity amongst the monosaccharides is reduced.[62]

An allosteric diboronic acid **25** has been prepared where formation of a 1:2 metal/crown sandwich causes the release of bound saccharide due to a metal induced conformational change.[63]

With the two dimensional PET sensor **26**, the amount of eximer can be directly correlated with the amount of non-cyclic saccharide complex formed[64] The effect of the linker length on this system has been investigated by Appleton, who found that as the linker length was increased selectivity for glucose was lost.[65]

Each of the sensors described above have been designed and synthesised individually. This is a very time and resource intensive approach to sensor design. For that reason, we decided to look for more convergent strategies towards the construction of sensor systems. On inspection of the sensor molecules we noted that in all cases the receptor unit was constant for an *ortho* aminomethyl boronic acid. Taking the receptor unit as one module new sensors could be constructed using this and other modules. Thus the modular approach to new saccharide selective fluorescent sensors was born. The basic idea was to break a sensor into three components; receptor units, spacer units, and read-out units. The approach can be illustrated by describing the D-glucose selective fluorescence sensor **26** which contains two boronic acid units (receptors), hexamethylene unit (D-glucose selective spacer), and pyrene unit (fluorophore - read-out). Using compound **26** as a model any new saccharide selective sensor requires at least two boronic acid units, one spacer unit, and a "read-out" unit. Two or more boronic acid units as with sensor **26** are required because only through two points binding can saccharide selectivity be controlled. One "read-out" unit is required to report on saccharide binding. We realised that two fluorophores are not required and may in fact be detrimental to an operational sensor since the fluorescence spectra of sensor **26** are complicated by excimer emission due to stacking of the two-pyrene units. Also, a spacer is required and the choice of spacer is very important since it will determine the selectivity of the sensor. With compound **26** a hexamethylene spacer results in D-glucose selectivity. Based on the above criteria we designed the sensors **27a-f**.[66,67]

Compound **27d** with a hexamethylene linker and pyrene fluorophore displays D-glucose selectivity, whilst systems with longer linker units **27e-f** display enhanced selectivity for D-galactose.†

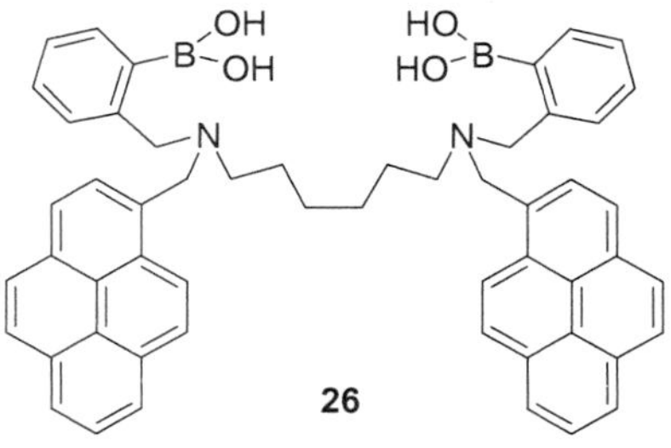
OH
B
OH
HO
HO
B
N
N

26

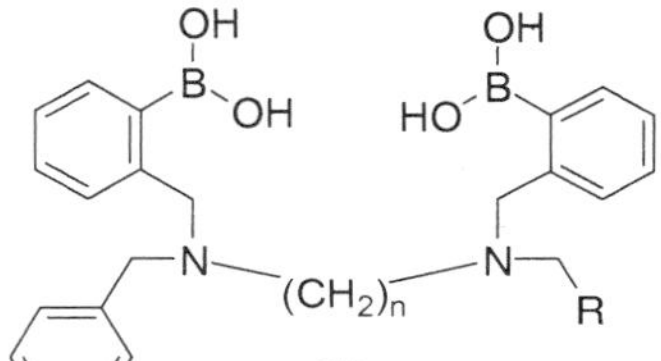
OH
B
OH
OH
B
HO
N
(CH2)n
N
R

27

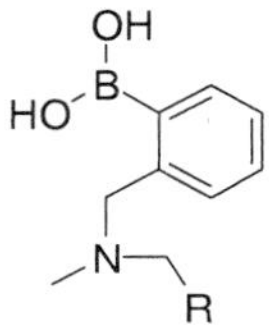
OH
B
HO
N
R

28

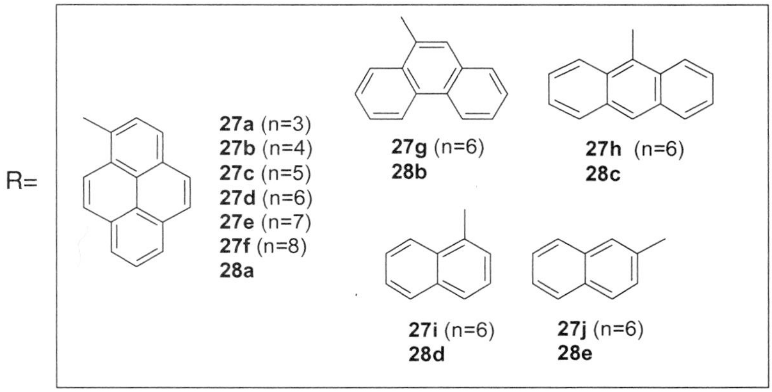
R=
27a (n=3)
27b (n=4)
27c (n=5)
27d (n=6)
27e (n=7)
27f (n=8)
28a
27g (n=6)
28b
27h (n=6)
28c
27i (n=6)
28d
27j (n=6)
28e

† To help visualize the trends in the observed stability constants, the stability constants of the diboronic acid sensors **27** were compared with the stability constants of the equivalent monoboronic acid analogues **28**. The relative stability illustrates that an increase in selectivity is obtained by cooperative binding through the formation of 1:1 cyclic systems. The large enhancement of the relative stability observed for the 1:1 cyclic systems (D-glucose, D-galactose) are clearly contrasted with the small two fold enhancement observed for the 2:1 acyclic systems (D-fructose, D-mannose).

Having determined the effect of the linker on saccharide selectivity we set out to probe further the factors affecting saccharide selectivity. The next logical component to vary is the fluorophore or 'read-out' unit sensors **27d, g-j**.[68] Although not directly involved in saccharide binding, the nature of this unit will directly influence both the solvation and steric crowding of the binding site.

Sensors **27d, g-h** show enhanced selectivity for D-glucose over D-galactose.[†] While with sensors **27i-j** the selectivity switches from D-glucose to D-galactose.[†] These results demonstrate that in a PET saccharide sensor with two phenylboronic acid groups, a hexamethylene linker and a fluorophore, the choice of the fluorophore is crucial. Selectivity is fluorophore dependent and careful choice of the fluorophore, such that it complements the polarity of the chosen guest species, is imperative.

Wang has also prepared diboronic acid systems with two anthracene fluorophores with variable spacers **29a-j**.[69,70] They have demonstrated that **29e** is selective for Sialyl Lewis X[69] and that **29f** is selective for D-glucose.[70] Compound **29f** is clearly D-glucose selective, however, the reported stability constant determined for D-fructose is lower than that evident by visual inspection of the titration curves. This discrepancy is important because the authors claim 43 fold selectivity for D-glucose over D-fructose. Whereas, our interpretation of the published titration curve for fructose gives an apparent K for D-fructose of about 100 M^{-1} which would give a D-glucose over D-fructose selectivity of only 15.

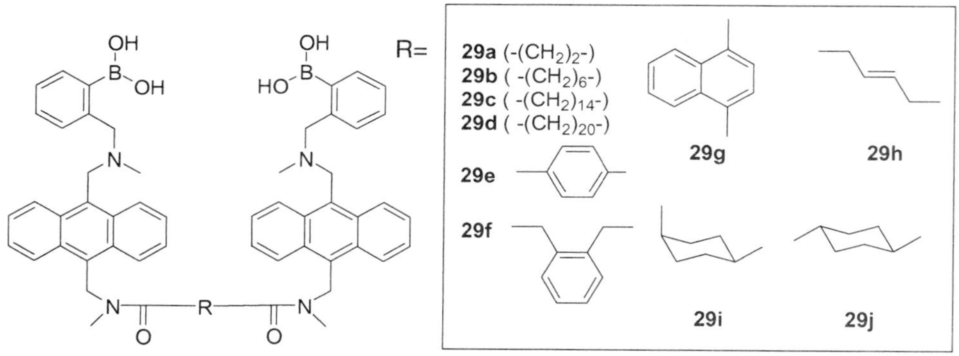

Our aim with compound **30** was to apply the modular design to prepare a saccharide sensing system using fluorescence energy transfer. Fluorescence energy transfer is the transfer of excited-state energy from a donor to an acceptor. Our idea with this system was to investigate the efficiency of energy transfer (ET) from phenanthrene to pyrene as a function of saccharide binding.[71]

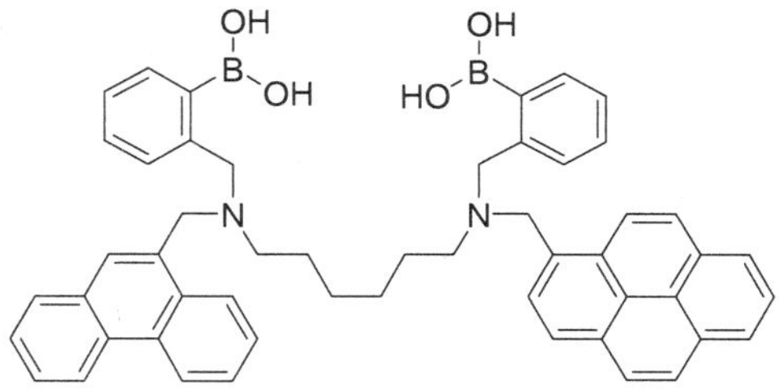

30

Sensor **30** is particularly interesting in that the differences between the observed fluorescence enhancements obtained when excited at phenanthrene (299 nm) and pyrene (342 nm) can be correlated with the molecular structure of the saccharide-sensor complex. The fluorescence enhancement of sensor **30** with D-glucose is 3.9 times greater when excited at 299 nm and 2.4 times greater when excited at 342 nm. Whereas, with D-fructose the enhancement was 1.9 times greater when excited at 299 nm and 3.2 times greater when excited at 342 nm. These results indicate that the energy transfer from phenanthrene (donor) to pyrene (acceptor) in a rigid 1:1 cyclic D-glucose complex is more efficient than in a flexible 2:1 acyclic D-fructose complex. The more efficient energy transfer leads to an enhanced fluoresence response to D-glucose. Our ongoing research is directed toward the development of other fluorescent sensors employing energy transfer as a method to enhance sensitivity and selectivity.

The boronic acid PET system has also been used in combination with other binding sites. The D-glucosamine selective fluorescent systems **31a** and **31b** based on a boronic acid and aza crown ether has been explored.[72,73] Sensors **31a** and **31b** consist of monoaza-18-crown-6 ether or monoaza-15-crown-5 as a binding site for the ammonium terminal of D-glucosamine hydrochloride, while a boronic acid serves as a binding site for the diol (carbohydrate) part of D-glucosamine hydrochloride. The nitrogen of the azacrown ether unit can participate in PET with the anthracene fluorophore, ammonium ion binding can then cause fluorescence recovery. This recovery is due to hydrogen bonding from the ammonium ion to the nitrogen of the azacrown ether. The strength of this hydrogen-bonding interaction modulates the PET from the amine to anthracene. As explained above, the boronic acid unit can also participate in PET with the anthracene fluorophore, and diol binding can also cause fluorescence recovery. The anthracene unit serves as a rigid spacer between the two-receptor units, with the appropriate spacing for the glucose guest. This system behaves like an **AND** logic gate,[42,74] in that fluorescence recovery is only observed when two chemical inputs are supplied, for this system the two chemical inputs are an ammonium cation and a diol group.

A D-glucuronic acid selective fluorescent system **32** based on a boronic acid and metal chelate has also been developed.[75,76] The behaviour of the system has been analysed with and without Zn(II). The fluorescence measurements indicate that binding with common monosaccharides (D-fructose, D-glucose and D-galactose) are not affected by incorporation of a Zn(II), whereas the binding with uronic (D-glucuronic and D-galacturonic) acids and sialic (*N*-acetylneuraminic) acid are greatly enhanced.

A D-glucarate system consisting of boronic acid and guanidinium receptor units has recently been reported by Wang.[77] The novelty of this work lies in the enhanced stability of **33** over our original saccharide PET system **12** with glucarate. However, as with the

D-glucose system discussed previously the reported stability constants do not seem to correlate with the given titration curves. The stability constant of **12** with D-glucarate is reported as 846 M^{-1}, however, visual inspection of the titration curve gives an estimated value of about 1700 M^{-1}. Based on our value obtained from the titration curve compound **12** could also be considered as a D-glucarate sensor.

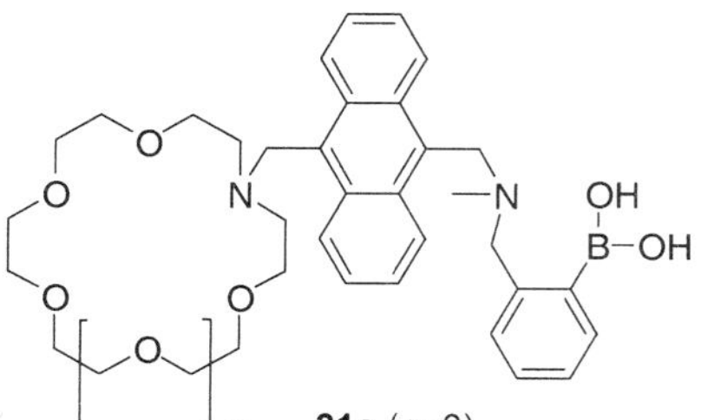

31a (n=0)
31b (n=1)

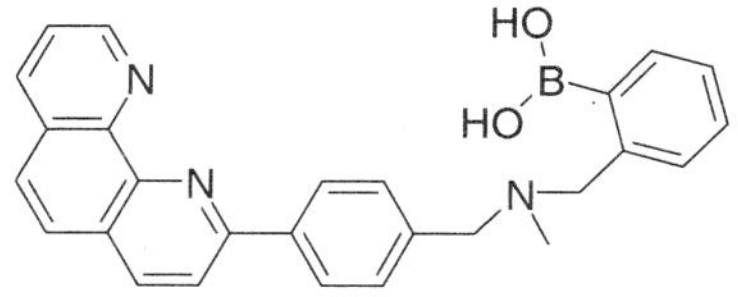

32

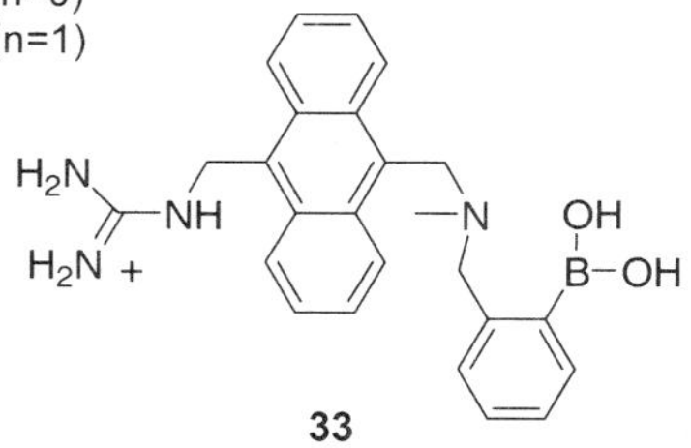

33

3.3.3. Others

As shown above, diboronic acids can bind monosaccharides selectively, where the 1:1 binding creates a rigid molecular complex.[78-84] This rigidification effect can also be utilized in designing fluorescent sensors for disaccharides. Diboronic acid species **34** selectively complexes with disaccharides in basic aqueous media to create cyclic complexes which alter the fluorescence properties of the molecule.[81] It is known that excited stilbene is quenched by radiationless decay via rotation of the ethylene double bond. Obstruction of this rotation leads to increased fluorescence.[85] The rigidification of **34** by disaccharide binding increases the stilbene fluorescence. In particular, the disaccharide melibiose shows higher selectivity for **34** than other common disaccharides.

Molecular rigidificaton has been used to generate a fluorescence increase with cyanine diboronic acid **35** on saccharide binding.[86] Rigidification has also been used with a diboronic acid appended binaphthyl **36** to develop a chiral discriminating system.[87]

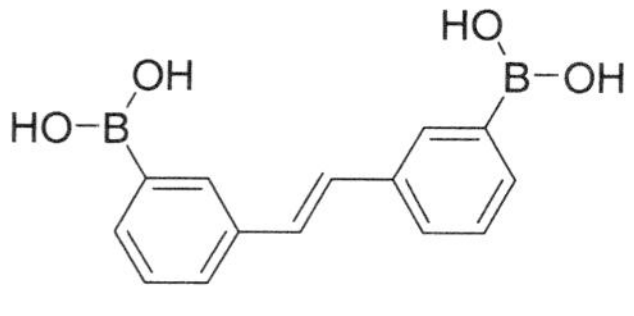

34

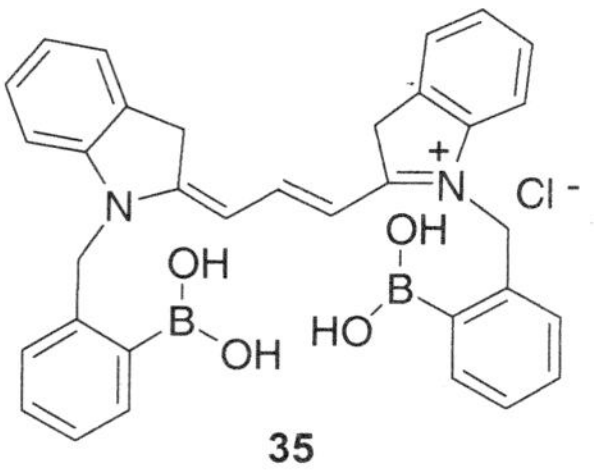

35

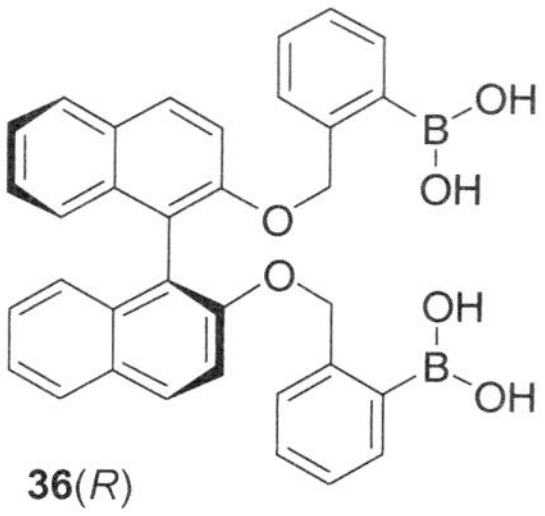

36(R)

A D-lactulose selective system **37** based on a diboronic acid porphyrin has been developed.[88] The spatial disposition of the two boronic acids in **37** produces the perfect 'cleft' for the disaccharide D-lactulose.

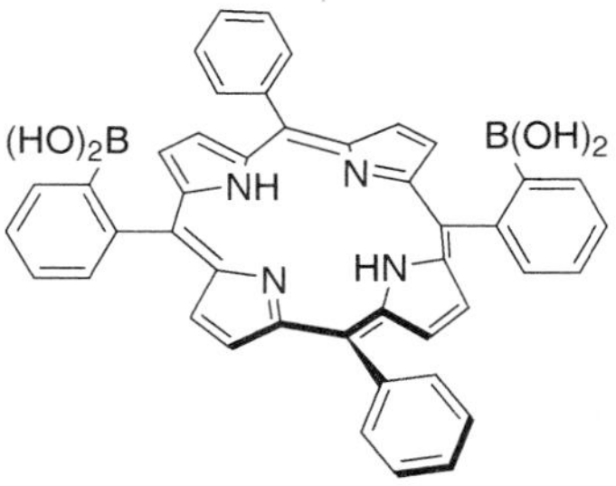

37

Norrild has developed an interesting diboronic acid system **38**.[89] The system works by reducing the quenching ability of the pyridine groups of **38** on saccharide binding. The system selectively binds D-glucose with a log K of 3.4. The structure of the complex was determined to be to a 1,2:3,5 bound α-D-glucofuranose. Evidence for the furanose structure was obtained from ^{1}H and ^{13}C NMR data with emphasis on the information from $^1J_{C-C}$ coupling constants.

All of the systems described above for the selective binding of D-glucose have been designed using the approximate positioning of two boronic acid units. Many of these systems bind D-glucose strongly and selectively. In systems where the structure of the D-glucose complex has been determined the furanose rather than pyranose form of D-glucose is favoured. Norrild has stated that "In our opinion binding of the more abundant α-pyranose form of glucose by boronic acids is not to be considered in the future design of boronic acid based sensors for aqueous systems."[89]

However, Drueckhammer has purposely set out to design a system selective for the pyranose form of D-glucose from first principles using computational methods to define the exact placing of two boronic acid groups. The approach resulted in the design and synthesis of compound **39** which shows very high binding towards D-glucose.[90] Compound **39** exhibited a 400 fold affinity for D-glucose over any of the other saccharides (D-galactose, D-mannose and D-fructose). ^{1}H NMR was used to confirm that the bound D-glucose was captured in the pyranose and not furanose form.

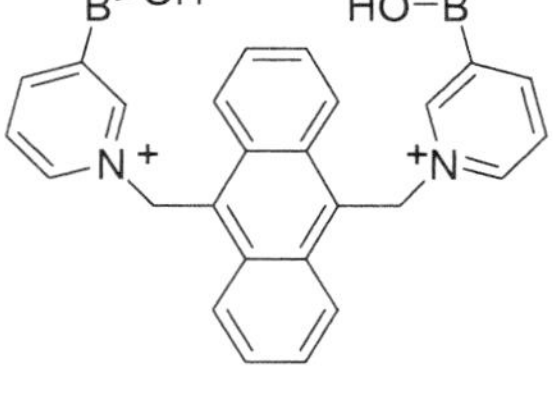

38

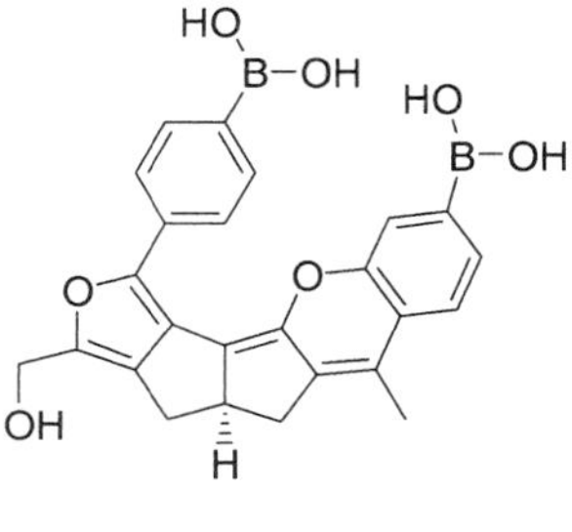

39

3.4. FLUORESCENT ASSAY

Over the last few years we have been interested in developing molecular sensors using boronic acids. The systems we are developing contain a receptor and reporter (fluorophore or chromophore) as part of a discrete molecular unit. However, another approach towards boronic acid based sensors is also possible where the receptor and a reporter unit are separate – a competitive assay. A competitive assay requires that the receptor and reporter (typically a commercial dye) associate under the measurement conditions. The receptor-reporter complex is then selectively dissociated by the addition of the appropriate guests. When the reporter dissociates from the receptor, a measurable response is produced.

The competitive assay approach to novel chemosensors has been pioneered by Anslyn.[91] We are also interested in such competitive systems because they reduce the synthetic complexity of the receptor. Towards that goal we have been exploring the use of coloured and fluorescent dyes to signal binding with boronic acids. Our goal was to discover effective dye molecules for competitive assays.[92] During our investigation we proposed that alizarin red S was a suitable candidate. Wang subsequently confirmed our belief by demonstrating that alizarin red S and phenyl boronic acid could be used in competitive assays for saccharides.[93,94] The system is D-fructose selective, which is the expected selectivity for a monoboronic acid system.[18]

Anslyn has recently reported two very interesting systems based on boronic acid receptors. Although the Anslyn systems involves a competitive colourimetric assay, there is no reason why the system cannot be extended to a fluorimetric assay through the choice of an appropriate dye molecules. The first system is a receptor for glucose-6-phosphate **40**.[95] The binding of glucose-6-phosphate is measured through the competitive displacement of 5-carboxyfluorescein. The second is a system where the binding of heparin and **41**, is monitored through displacement of pyrocatechol violet.[96]

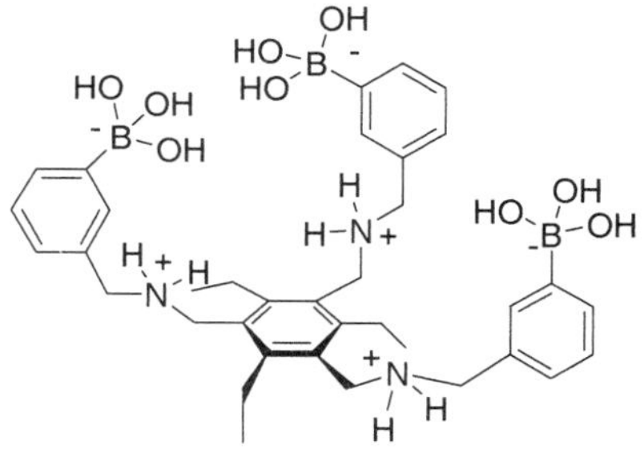

40

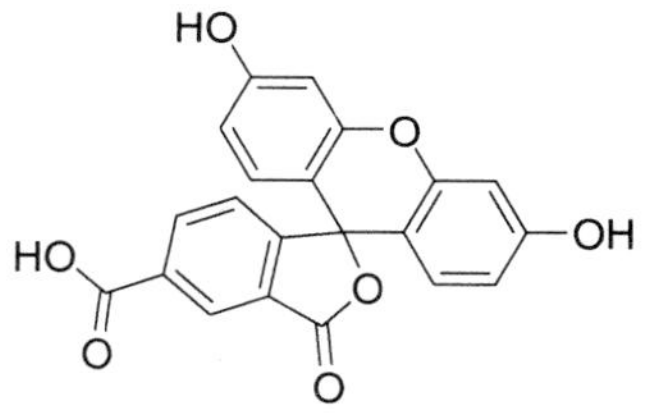

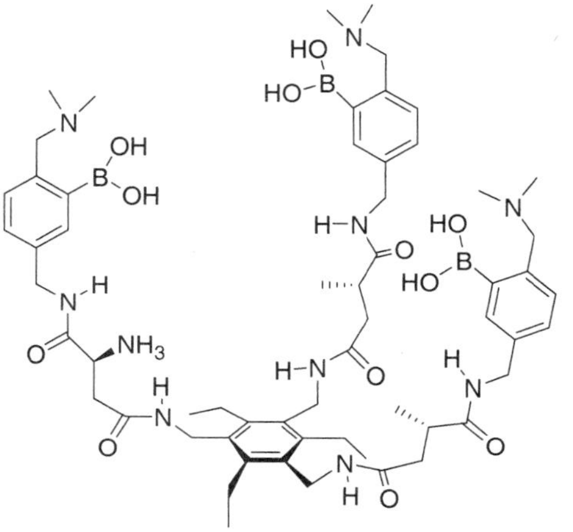

41

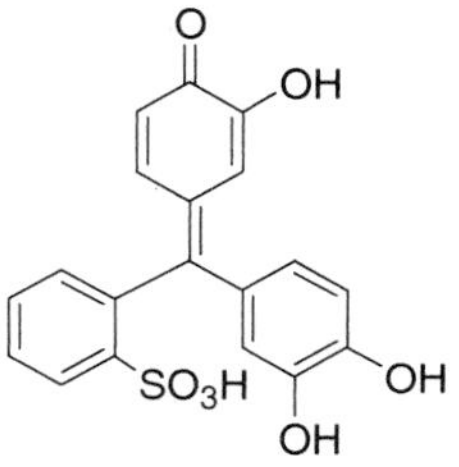

Compounds **42** and **43** have been used with 1,5- or 2,6 – anthraquinone disulfonates (ADS) as a competitive system for the fluorescence detection of D-fructose.[97] 1,5- or 2,6 - ADS binds with **42** or **43** and quenches the fluorescence, addition of D-fructose causes decomplexation and fluorescence recovery.

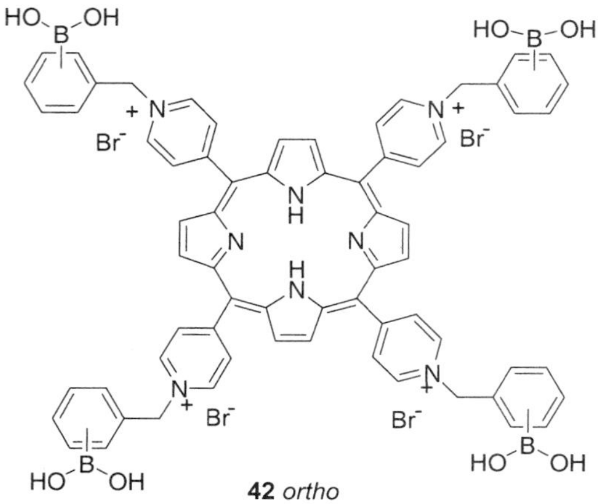

42 *ortho*
43 *meta*

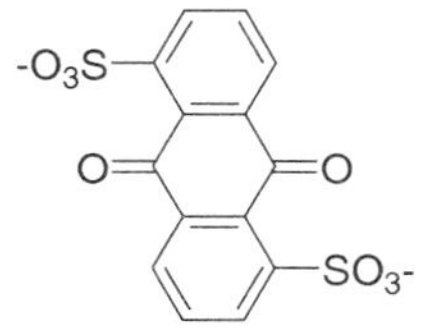

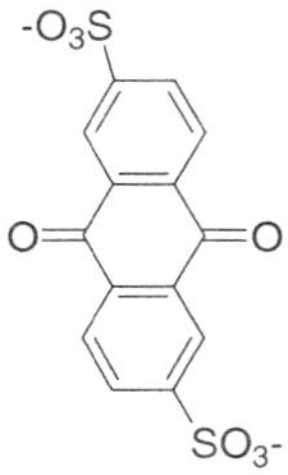

Lakowicz has also used competitive interactions between a ruthenium metal ligand complex, a boronic acid derivative and glucose.[98] The metal-ligand complex forms a reversible complex with 2-toluylboronic acid or 2-methoxyphenyl boronic acid. Complexation is accompanied by a several-fold increase in the luminescent intensity of the ruthenium complex. Addition of glucose results in decreased luminescent intensity, which appears to be the result of decreased binding between the metal-ligand complex and the boronic acid. Ruthenium metal-ligand complexes are convenient for optical sensing because their long luminescent decay times allow lifetime-based sensing with simple instrumentation.

An interesting multicomponent system has also be devised by Singaram where quenching of a pyranine dye by bisboronic acid viologen units **44** and **45** is modulated by added saccharide.[99,100] Compound **44** binds well with D-fructose ($K = 2600$ M^{-1}) and weakly with D-glucose ($K = 43$ M^{-1}), also, the system only produces a 4% fluorescence recovery.[99] Whereas, compound **45** binds well with both D-fructose ($K = 3300$ M^{-1}) and D-glucose ($K = 1800$ M^{-1}). Together with the enhanced selectivity for D-glucose this system also produces a 45% fluorescence recovery on addition of saccharides.[100]

Lakowicz has also examined the quenching and recovery of on a sulfonated poly(phenylene ethynylene) by a bisboronic acid viologen **46** on addition of saccharides.[101] The system is D-fructose selective and produces up to 70 fold fluorescence enhancement on addition of saccharides.

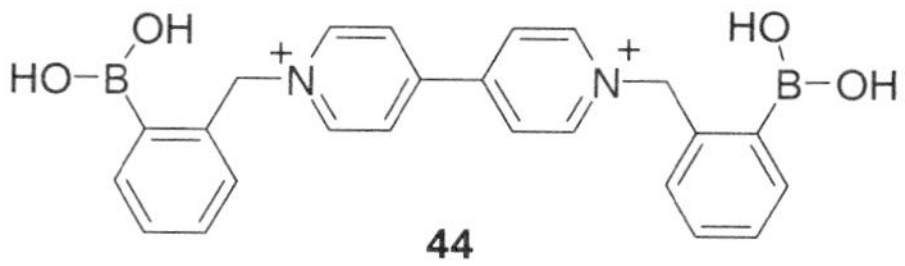

44

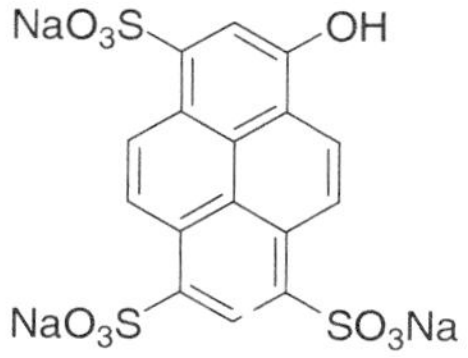

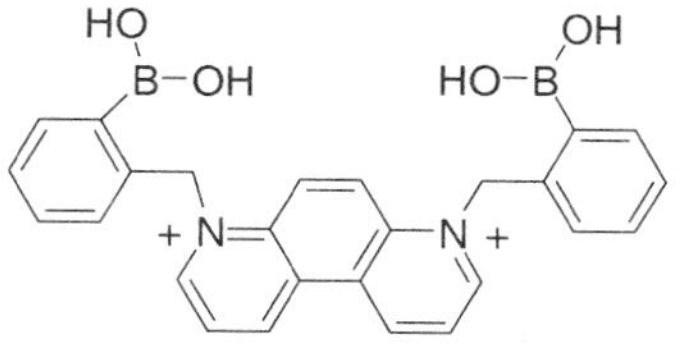

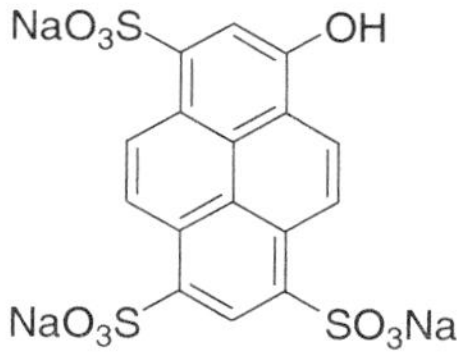

45

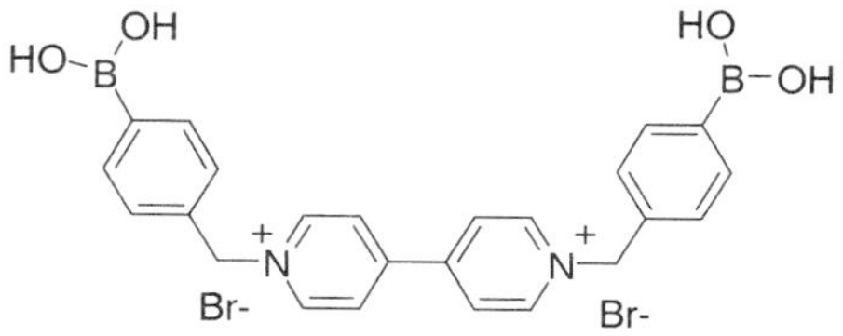

46

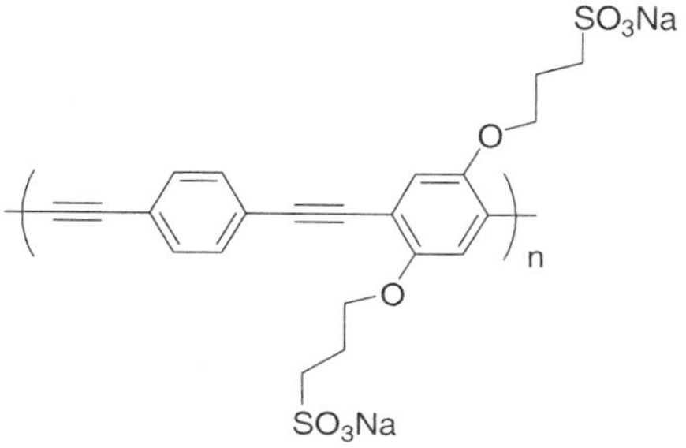

As mentioned earlier we are also interested in developing saccharide selective assays. Therefore, we decided to employ the modular concept to design a fluorescent assay for D-glucose. We based our new receptor compound **47** on the successful fluorescent PET sensor **27d**. The receptor components of **47** are identical to those of **27d**, but the difference is the lack of a fluorescent signaling unit (pyrene).

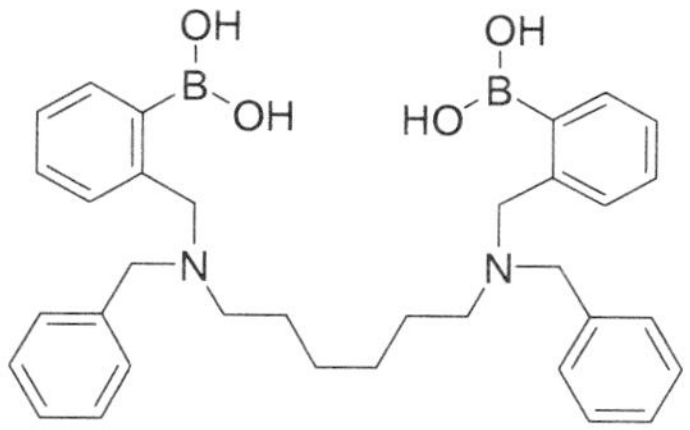

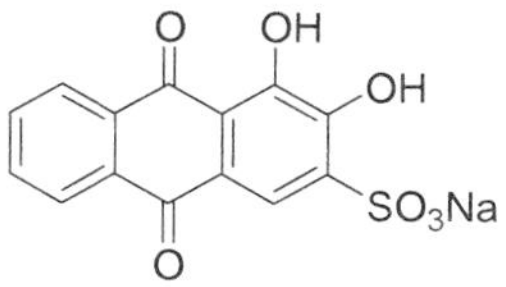

47

We were able to demonstrate that **47** and alizarin red S produces a very efficient D-glucose assay.[102] Sensor **47** and alizarin red S shows a four-fold enhancement over phenyl boronic acid (PBA) for D-glucose. Sensor **47** can also be used at a ten times much lower concentration than PBA.

From what we have described above the fluorescent assay method seems to represent one of the best ways forward in the design for saccharide selective sensors. However, since in a competitive assay, all competition must be controlled so that the signal can used to produce an analytical outcome. The presence of previously unrecognized interactions between boronic acids and buffer conjugate bases (phosphate, citrate and imidazole)[23] to create ternary complexes (boronate-X-saccharide) will need to be considered in future assay design.

3.5. POLYMER SUPPORTED SENSORS

If practically useful sensors are to be developed from the boronic acid sensors described above, then they will need to be integrated into a device. One way to help achieve this goal is to incorporate the saccharide selective interface into a polymer support.

Smith has prepared a grafted polymer a ribonucleoside 5'-triphosphate selective sensor. The polymers were prepared using poly(allylamine) (PAA) to which 10% of boronic acid monomer unit **48** was grafted.[103] Also, a library of potential sialic acid receptors were prepared.[104] In this case the polymers were prepared using poly(allylamine) (PAA) to which 2% of boronic acid monomer unit **48** was grafted. The final polymers also contained various amounts of 4-hydroxybenzoic acid, 4-imidazolacetic acid, octanoic acid and/or succinic anhydride.

Polymers of poly(lysine) with boronic acids appended to the amine residue have been used as saccharide receptors.[105-107] On saccharide complexation these polymers are converted from neutral sp^2 boron to anionic sp^3 hybridised boron. The anionic polymer thus formed interact with added cyanine dye. Saccharide binding can then be 'read-out' by changes in the absorption and ICD spectra of the cyanine dye molecule.

Wang has employed the template approach using monomer **49** to prepare a fluorescent polymer with enhanced selectivity towards D-fructose.[108,109] Appleton has used a similar approach using monomer **50** to prepare a D-glucose selective polymer.[65] The Appleton polymer clearly shows the value of the imprinting technique, selectivity for D-fructose over D-glucose by the monomer has been reversed in the polymer.

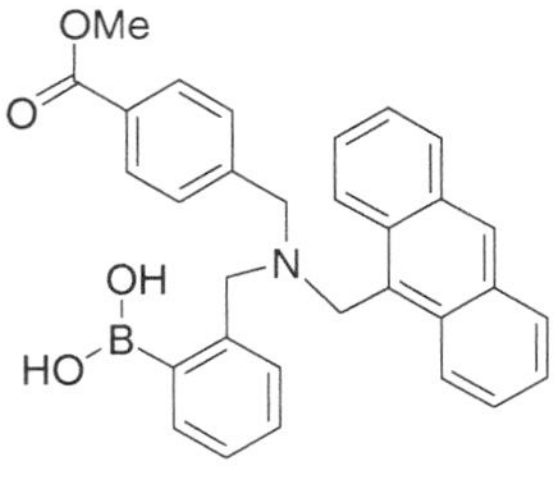

48 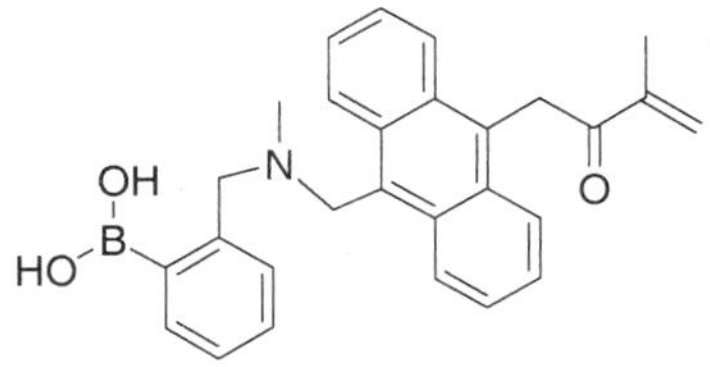   49 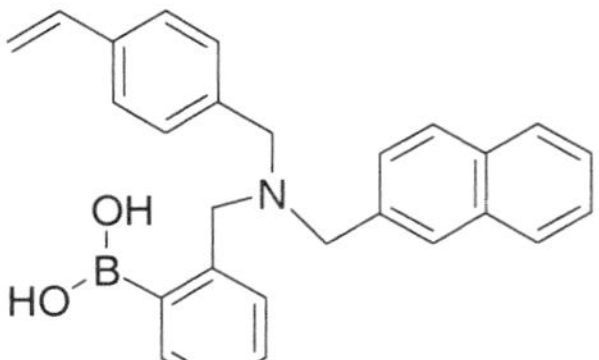   50

We also liked the idea of creating polymer bound fluorescent sensors, but wanted more control over the receptor. Our approach was to take a successful solution based D-glucose selective receptor **27d** and link the receptor to a polymer support.[66] The major difference between the polymer bound system **51** and solution based system **27d** is the D-glucose selectivity. The D-glucose selectivity drops for compound **51** whereas the selectivity with other saccharides is similar to those observed for compound **27d**.

However, it should be pointed out that the polymeric system still has a nine fold enhancement for D-glucose over the monoboronic acid model compound. We believe these differences between **27d** and **51** are due to the proximity of receptor to the polymer backbone. We are currently investigating this phenomenon by modifying the linker to the polymer support.

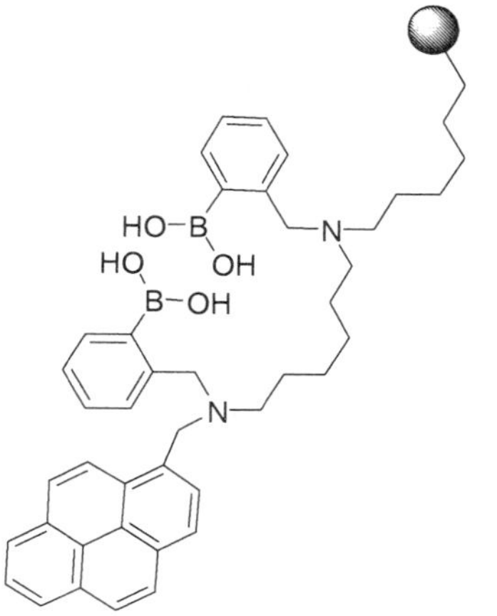

51

More recently, a membrane in which a PET-based glucose sensing system is immobilized has been developed **52**. The amide group was introduced not only as a linker to the membrane but also to shift the excitation and emission maximum to longer wavelengths. We had also attached the fluorescent boronic unit to the membrane through two linkers (via the two amino nitrogens). However, it was discovered that both changes in fluorescence intensity and affinity for D-glucose were reduced. This result suggests that single chain immobilization is superior to double chain immobilization. We have also confirmed that **52** is applicable for the sensing of glucose in blood.[110]

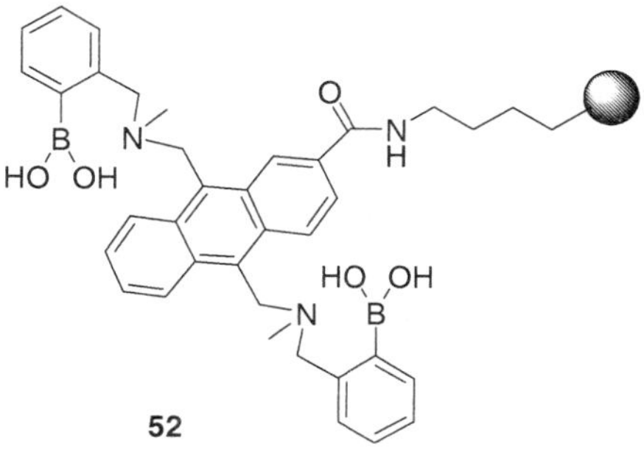

52

3.6. CONCLUSIONS

"If we knew what it was we were doing, it would not be called research, would it?"
Albert Einstein 1879-1955

The aim with this review was to provide a personal perspective on current and future directions in the development of fluorescent saccharide sensors. The idea was that the reader would follow the same conceptual journey we ourselves take when developing new sensor systems. Having exposed how we develop a new sensor system we hope that the reader will be inspired to develop their own systems.

In order to facilitate that development we have distilled what we believe to be the two most important considerations when designing new saccharide selective fluorescent sensors.

A selective interface requires at least two appropriately positioned receptors.

The geometry and choice of the receptors used will be determined by the target saccharide and synthetic constraints.

Signal read-out can be provided by an integrated fluorophore-receptor or fluorophore/receptor ensemble.

The choice of integrated or assay system depends on the application; integrated systems can more difficult to synthesize while with ensembles obtaining an analytical output is non trivial. For continuous monitoring an integrated sensor is probably the best choice, while using an assay system could be cheaper and hence more suitable for use in disposable sensors.

With this review we limited our coverage of the literature to concentrate on saccharide selective systems based on the boronic acid interface. However, the same maxims given above can also be applied in the development of other saccharide selective systems using different interfaces. For example saccharide selective interfaces can also be developed using hydrogen bonding receptors.[57] These hydrogen bonding receptors can be either synthetic[111-121] or based on the or the modification of selective saccharide binding proteins.[122]

3.7. REFERENCES

1. S. De Marchi, E. Cecchin, A. Basil, G. Proto, W. Donadon, A. Jengo, D. Schinella, A. Jus, D. Villalta, P. De Paoli, G. Santini, and F. Tesio, Close genetic linkage between HLA and renal glycosuria., *J. Nephrol.* **4**, 280-286 (1984).
2. L. J. Elsas, and L. E. Rosenberg, Familial renal glycosuria: a genetic reappraisal of hexose transport by kidney and intestine, *J. Clinic. Invest.* **48**, 1845-1854 (1969).
3. P. Baxter, J. Goldhill, P. T. Hardcastle, and C. J. Taylor, Enhanced intestinal glucose and alanine transport in cystic fibrosis., *Gut.* **31**, 817-820 (1990).
4. R. N. Fedorak, M. D. Gershon, and M. Field, Induction of intestinal glucose carriers in streptozocin-treated chronically diabetic rats., *Gasteroenterology* **96**, 37-44 (1989).
5. H. Yasuda, T. Kurokawa, Y. Fuji, A. Yamashita, and S. Ishibashi, Decreased D-glucose transport across renal brush-border membrane vesicles from streptozotocin-induced diabetic rats., *Biochim. Biophys. Acta.* **1021**, 114-118 (1990).
6. T. Yamamoto, Y. Seino, H. Fukumoto, G. Koh, H. Yano, N. Inagaki, Y. Yamada, K. Inoue, T. Manabe, and H. Imura, Over-expression of facilitative glucose transporter genes in human cancer., *Biochem. Biophys. Res. Commun.* **170**, 223-230 (1990).
7. T. D. James, H. Kawabata, R. Ludwig, K. Murata, and S. Shinkai, Cholesterol as a versatile platform for chiral recognition, *Tetrahedron* **51**, 555-566 (1995).

8. T. D. James, K. R. A. S. Sandanayake, and S. Shinkai, Recognition of sugars and related compounds by "reading-out" type interfaces, *Supramol Chem.* **6**, 141-157 (1995).
9. T. D. James, K. R. A. S. Sandanayake, and S. Shinkai, Saccharide sensing with molecular receptors based on boronic acid, *Angew. Chem., Int. Ed. Engl.* **35**, 1911-1922 (1996).
10. T. D. James, P. Linnane, and S. Shinkai, Fluorescent saccharide receptors: A sweet solution to the design, assembly and evaluation of boronic acid derived PET sensors, *Chem. Commun.* 281-288 (1996).
11. K. R. A. S. Sandanayake, T. D. James, and S. Shinkai, Molecular design of sugar recognition systems by sugar-diboronic acid macrocyclization, *Pure Appl. Chem.* **68**, 1207-1212 (1996).
12. W. Wang, X. Gao, and B. Wang, Boronic acid-based sensors, *Curr. Org. Chem* **6**, 1285-1317 (2002).
13. M. Granda-Valdes, R. Badia, G. Pina-Luis, and M. E. Diaz-Garcia, Photoinduced electron transfer systems and their analytical application in chemical sensing, *Quim. Anal.* **19**, 38-53 (2000).
14. T. D. James, and S. Shinkai, Artificial receptors as chemosensors for carbohydrates, *Top. Curr. Chem.* **218**, 159-200 (2002).
15. S. Striegler, Selective carbohydrate recognition by synthetic receptors in aqueous solution, *Curr. Org. Chem* **7**, 81-102 (2003).
16. H. G. Kuivila, A. H. Keough, and E. J. Soboczenski, Areneboronates from diols and polyols, *J. Org. Chem.* **19**, 780-783 (1954).
17. G. L. Roy, A. L. Laferriere, and J. O. Edwards, A comparative study of polyol complexes of arsenite, borate, and tellurate ions, *J. Inorg. Nucl. Chem.* **114**, 106-114 (1957).
18. J. P. Lorand, and J. O. Edwards, Polyol complexes and structure of the benzeneboronate Ion, *J. Org. Chem.* **24**, 769-774 (1959).
19. J. C. Norrild, and H. Eggert, Evidence for monodentate and bisdentate boronate complexes of glucose in the furanose form - application of (1)J(C-C)-coupling-constants as a structural probe, *J. Am. Chem. Soc.* **117**, 1479-1484 (1995).
20. J. C. Norrild, and H. Eggert, Boronic acids as fructose sensors. Structure determination of the complexes involved using (1)J(CC) coupling constants, *J. Chem. Soc., Perkin Trans. 2* 2583-2588 (1996).
21. J. Rohovec, T. Maschmeyer, S. Aime, and J. A. Peters, The structure of the sugar residue in glycated human serum albumin and its molecular recognition by phenylboronate, *Chem. Eur. J.* **9**, 2193-2199 (2003).
22. A. E. Martell, and R. M. Smith *Critical stability constants*; (Plenum Press, New York, 1976)
23. L. I. Bosch, T. M. Fyles, and T. D. James, Boronates, *submitted*
24. J. Yoon, and A. W. Czarnik, Fluorescent chemosensors of carbohydrates - a means of chemically communicating the binding of polyols in water based on chelation- enhanced quenching, *J. Am. Chem. Soc.* **114**, 5874-5875 (1992).
25. Y. Nagai, K. Kobayashi, H. Toi, and Y. Aoyama, Stabilization of sugar-boronic esters of indolylboronic acid in water via sugar indole interaction - a notable selectivity in oligosaccharides, *Bull. Chem. Soc. Jpn.* **66**, 2965-2971 (1993).
26. H. Suenaga, M. Mikami, K. R. A. S. Sandanayake, and S. Shinkai, Screening of fluorescent boronic acids for sugar sensing which show a large fluorescence change, *Tetrahedron Lett.* **36**, 4825-4828 (1995).
27. H. Suenaga, H. Yamamoto, and S. Shinkai, Screening of boronic acids for strong inhibition of the hydrolytic activity of alpha-chymotrypsin and for sugar sensing associated with a large fluorescence change, *Pure Appl. Chem.* **68**, 2179-2186 (1996).
28. H. Shinmori, M. Takeuchi, and S. Shinkai, Spectroscopic sugar sensing by a stilbene derivative with push (Me(2)N-)-pull ((HO)(2)B-)-type substituents, *Tetrahedron* **51**, 1893-1902 (1995).
29. N. DiCesare, and J. R. Lakowicz, Spectral properties of fluorophores combining the boronic acid group with electron donor or withdrawing groups. Implication in the development of fluorescence probes for saccharides, *J. Phys. Chem. A* **105**, 6834-6840 (2001).
30. N. Di Cesare, and J. R. Lakowicz, Wavelength-ratiometric probes for saccharides based on donor-acceptor diphenylpolyenes, *J. Photochem. Photobiol., A* **143**, 39-47 (2001).
31. K. R. A. S. Sandanayake, S. Imazu, T. D. James, M. Mikami, and S. Shinkai, Molecular fluorescence sensor for saccharides based on amino coumarin, *Chem. Lett.* 139-140 (1995).
32. N. DiCesare, and J. R. Lakowicz, A new highly fluorescent probe for monosaccharides based on a donor-acceptor diphenyloxazole, *Chem. Commun.* 2022-2023 (2001).
33. N. DiCesare, and J. R. Lakowicz, Chalcone-analogue fluorescent probes for saccharides signaling using the boronic acid group, *Tetrahedron Lett.* **43**, 2615-2618 (2002).
34. N. DiCesare, and J. R. Lakowicz, Fluorescent probe for monosaccharides based on a functionalized boron-dipyrromethene with a boronic acid group, *Tetrahedron Lett.* **42**, 9105-9108 (2001).
35. C. J. Ward, P. Patel, P. R. Ashton, and T. D. James, A molecular colour sensor for monosaccharides, *Chem. Commun.* 229-230 (2000).
36. C. J. Ward, P. Patel, and T. D. James, Molecular color sensors for monosaccharides, *Org. Lett.* **4**, 477-479 (2002).

37. C. J. Ward, P. Patel, and T. D. James, Boronic acid appended azo dyes-color sensors for saccharides, *J. Chem. Soc., Perkin Trans. 1* 462-470 (2002).

38. S. Arimori, L. I. Bosch, C. J. Ward, and T. D. James, Fluorescent internal charge transfer (ICT) saccharide sensor, *Tetrahedron Lett.* **42**, 4553-4555 (2001).

39. A. P. de Silva, H. Q. N. Gunaratne, T. Gunnlaugsson, A. J. M. Huxley, C. P. McCoy, J. T. Rademacher, and T. E. Rice, Signaling recognition events with fluorescent sensors and switches, *Chem. Rev.* **97**, 1515-1566 (1997).

40. A. P. de Silva, T. Gunnlaugsson, and C. P. McCoy, Photoionic supermolecules: Mobilizing the charge and light brigades, *J. Chem. Educ.* **74**, 53-58 (1997).

41. A. P. de Silva, D. B. Fox, T. S. Moody, and S. M. Weir, The development of molecular fluorescent switches, *Trends Biotechnol.* **19**, 29-34 (2001).

42. G. J. Brown, A. Prasanna de Silva, and S. Pagliari, Focus Article: Molecules that add up, *Chem. Commun.* 2461-2464 (2002).

43. T. D. James, K. Sandanayake, and S. Shinkai, Novel photoinduced electron-transfer sensor for saccharides based on the interaction of boronic acid and amine, *J. Chem. Soc., Chem. Commun.* 477-478 (1994).

44. T. D. James, K. R. A. S. Sandanayake, R. Iguchi, and S. Shinkai, Novel saccharide-photoinduced electron-transfer sensors based on the interaction of boronic acid and amine, *J. Am. Chem. Soc.* **117**, 8982-8987 (1995).

45. C. R. Cooper, and T. D. James, Selective fluorescence signalling of saccharides in their furanose form, *Chem. Lett.* 883-884 (1998).

46. H. Kijima, M. Takeuchi, A. Robertson, S. Shinkai, C. Cooper, and T. D. James, Exploitation of a novel 'on-off' photoinduced electron-transfer (PET) sensor against conventional 'off-on' PET sensors, *Chem. Commun.* 2011-2012 (1999).

47. W. Yang, J. Yan, G. Springsteen, S. Deeter, and B. Wang, A novel type of fluorescent boronic acid that shows large fluorescence intensity changes upon binding with a carbohydrate in aqueous solution at physiological pH, *Bioorg. Med. Chem. Lett.* **13**, 1019-1022 (2003).

48. D. P. Adhikiri, and M. D. Heagy, Fluorescent chemosensor for carbohydrates which shows large change in chelation-enhanced quenching, *Tetrahedron Lett.* **40**, 7893-7896 (1999).

49. N. DiCesare, D. P. Adhikari, J. J. Heynekamp, M. D. Heagy, and J. R. Lakowicz, Spectroscopic and photophysical characterization of fluorescent chemosensors for monosaccharides based on N-phenylboronic acid derivatives of 1,8-naphthalimide, *J. Fluorescence* **12**, 147-154 (2002).

50. H. Cao, D. I. Diaz, N. DiCesare, J. R. Lakowicz, and M. D. Heagy, Monoboronic acid sensor that displays anomalous fluorescence sensitivity to glucose, *Org. Lett.* **4**, 1503-1505 (2002).

51. A. J. Tong, A. Yamauchi, T. Hayashita, Z. Y. Zhang, B. D. Smith, and N. Teramae, Boronic acid fluorophore/beta-cyclodextrin complex sensors for selective sugar recognition in water, *Anal. Chem.* **73**, 1530-1536 (2001).

52. S. Arimori, L. I. Bosch, C. J. Ward, and T. D. James, A D-glucose selective fluorescent internal charge transfer (ICT) sensor, *Tetrahedron Lett.* **43**, 911-913 (2002).

53. T. D. James, K. Sandanayake, and S. Shinkai, A glucose-selective molecular fluorescence sensor, *Angew. Chem., Int. Ed. Engl.* **33**, 2207-2209 (1994).

54. N. DiCesare, and J. R. Lakowicz, Evaluation of two synthetic glucose probes for fluorescence-lifetime-based sensing, *Anal. Biochem.* **294**, 154-160 (2001).

55. M. Bielecki, H. Eggert, and J. C. Norrild, A fluorescent glucose sensor binding covalently to all five hydroxy groups of alpha-D-glucofuranose. A reinvestigation, *J. Chem. Soc., Perkin Trans. 2* 449-455 (1999).

56. T. D. James, K. R. A. S. Sandanayake, and S. Shinkai, Chiral discrimination of monosaccharides using a fluorescent molecular sensor, *Nature* **374**, 345-347 (1995).

57. A. P. Davis, and R. S. Wareham, Carbohydrate recognition through noncovalent interactions: A challenge for biomimetic and supramolecular chemistry, *Angew. Chem., Int. Ed. Engl.* **38**, 2979-2996 (1999).

58. P. Linnane, T. D. James, and S. Shinkai, The synthesis and properties of a calixarene-based sugar bowl, *J. Chem. Soc., Chem. Commun.* 1997-1998 (1995).

59. H. Matsumoto, A. Ori, F. Inokuchi, and S. Shinkai, Saccharide control of energy-transfer luminescence of lanthanide ions encapsulated in calix[4]arenes: A novel discrimination method for the energy-transfer route, *Chem. Lett.* 301-302 (1996).

60. T. D. James, H. Shinmori, and S. Shinkai, Novel fluorescence sensor for 'small' saccharides, *Chem. Commun.* 71-72 (1997).

61. P. Linnane, T. D. James, S. Imazu, and S. Shinaki, A sweet toothed saccharide (PET) sensor, *Tetrahedron Lett.* **36**, 8833-8834 (1995).

62. T. D. James, H. Shinmori, M. Takeuchi, and S. Shinkai, A saccharide 'sponge'. Synthesis and properties of a dendritic boronic acid, *Chem. Commun.* 705-706 (1996).

63. T. D. James, and S. Shinkai, Diboronic acid glucose cleft and a biscrown ether metal sandwich are allosterically coupled, *J. Chem. Soc., Chem. Commun.* 1483-1485 (1995).

64. K. Sandanayake, T. D. James, and S. Shinkai, 2-Dimensional photoinduced electron-transfer (PET) fluorescence sensor for saccharides, *Chem. Lett.* 503-504 (1995).

65. B. Appleton, and T. D. Gibson, Detection of total sugar concentration using photoinduced electron transfer materials: development of operationally stable, reusable optical sensors, *Sens. Actuators, B* **65**, 302-304 (2000).

66. S. Arimori, M. L. Bell, C. S. Oh, K. A. Frimat, and T. D. James, Modular fluorescence sensors for saccharides, *Chem. Commun.* 1836-1837 (2001).

67. S. Arimori, M. L. Bell, C. S. Oh, K. A. Frimat, and T. D. James, Modular fluorescence sensors for saccharides, *J. Chem. Soc., Perkin Trans. 1* 803-808 (2002).

68. S. Arimori, G. A. Consiglio, M. D. Phillips, and T. D. James, Tuning saccharide selectivity in modular fluorescent sensors, *Tetrahedron Lett.* **44**, 4789-4792 (2003).

69. W. Yang, S. Gao, X. Gao, V. V. R. Karnati, W. Ni, B. Wang, W. B. Hooks, J. Carson, and B. Weston, Diboronic acids as fluorescent probes for cells expressing sialyl lewis X, *Bioorg. Med. Chem. Lett.* **12**, 2175-2177 (2002).

70. V. V. Karnati, X. Gao, S. Gao, W. Yang, W. Ni, S. Sankar, and B. Wang, A glucose-selective fluorescence sensor based on boronic acid-diol recognition, *Bioorg. Med. Chem. Lett.* **12**, 3373-3377 (2002).

71. S. Arimori, M. L. Bell, C. S. Oh, and T. D. James, A Modular Fluorescence Intramolecular Energy Transfer Saccharide Sensor, *Org. Lett.* **4**, 4249-4251 (2002).

72. C. R. Cooper, and T. D. James, Selective D-glucosamine hydrochloride fluorescence signalling based on ammonium cation and diol recognition, *Chem. Commun.* 1419-1420 (1997).

73. C. R. Cooper, and T. D. James, Synthesis and evaluation of D-glucosamine-selective fluorescent sensors, *J. Chem. Soc., Perkin Trans. 1* 963-969 (2000).

74. A. P. de Silva, H. Q. N. Gunaratne, and C. P. McCoy, A molecular photoionic AND gate based on fluorescent signaling, *Nature* **364**, 42-44 (1993).

75. M. Takeuchi, M. Yamamoto, and S. Shinkai, Fluorescent sensing of uronic acids based on a cooperative action of boronic acid and metal chelate, *Chem. Commun.* 1731-1732 (1997).

76. M. Yamamoto, M. Takeuchi, and S. Shinkai, Molecular design of a PET-based chemosensor for uronic acids and sialic acids utilizing a cooperative action of boronic acid and metal chelate, *Tetrahedron* **54**, 3125-3140 (1998).

77. W. Yang, J. Yan, H. Fang, and B. Wang, The first fluorescent sensor for D-glucarate based on the cooperative action of boronic acid and guanidinium groups, *Chem. Commun.* 792-793 (2003).

78. G. Deng, T. D. James, and S. Shinkai, Allosteric interaction of metal-ions with saccharides in a crowned diboronic acid, *J. Am. Chem. Soc.* 4567-4572 (1994).

79. T. D. James, Y. Shiomi, K. Kondo, and S. Shinkai In *XVIII International Symposium on Macrocyclic Chemistry*: University of Twente, Enschede, The Netherlands, 1993.

80. K. Kondo, Y. Shiomi, M. Saisho, T. Harada, and S. Shinkai, Specific complexation of disaccharides with diphenyl-3,3'-diboronic acid that can be detected by circular-dichroism, *Tetrahedron* 8239-8252 (1992).

81. K. Sandanayake, K. Nakashima, and S. Shinkai, Specific recognition of disaccharides by trans-3,3'-stilbenediboronic acid - rigidification and fluorescence enhancement of the stilbene skeleton upon formation of a sugar-stilbene macrocycle, *J. Chem. Soc., Chem. Commun.* 1621-1622 (1994).

82. Y. Shiomi, K. Kondo, M. saisho, T. Harada, K. Tsukagoshi, and S. Shinkai, Specific complexation of saccharides with dimeric phenylboronic acid that can be detected by circular dichroism, *Supramol. Chem.* **2**, 11-17 (1993).

83. Y. Shiomi, M. Saisho, K. Tsukagoshi, and S. Shinkai, Specific complexation of glucose with a diphenylmethane-3,3'- diboronic acid-derivative - correlation between the absolute- configuration of monosaccharide and disaccharide and the circular dichroic activity of the complex, *J. Chem. Soc., Perkin Trans. 1* 2111-2117 (1993).

84. K. Tsukagoshi, and S. Shinkai, Specific complexation with monosaccharides and disaccharides that can be detected by circular-dichroism, *J. Org. Chem.* 4089-4091 (1991).

85. N. J. Turro *Modern molecular photochemistry*; (Benjamin, Menlo Park, CA, USA, 1978)

86. M. Takeuchi, T. Mizuno, H. Shinmori, M. Nakashima, and S. Shinkai, Fluorescence and CD spectroscopic sugar sensing by a cyanine-appended diboronic acid probe, *Tetrahedron* **52**, 1195-1204 (1996).

87. M. Takeuchi, S. Yoda, T. Imada, and S. Shinkai, Chiral sugar recognition by a diboronic-acid-appended binaphthyl derivative through rigidification effect, *Tetrahedron* **53**, 8335-8348 (1997).

88. H. Kijima, M. Takeuchi, and S. Shinkai, Selective detection of D-lactulose by a porphyrin-based diboronic acid, *Chem. Lett.* 781-782 (1998).

89. H. Eggert, J. Frederiksen, C. Morin, and J. C. Norrild, A new glucose-selective fluorescent bisboronic acid. First report of strong alpha-furanose complexation in aqueous solution at physiological pH, *J. Org. Chem.* **64**, 3846-3852 (1999).
90. W. Yang, H. He, and D. G. Drueckhammer, Computer-guided design in molecular recognition: Design and synthesis of a glucopyranose receptor, *Angew. Chem., Int. Ed. Engl.* **40**, 1714-1718 (2001).
91. S. L. Wiskur, H. Ait-Haddou, J. J. Lavigne, and E. V. Anslyn, Teaching old indicators new tricks, *Acc. Chem. Res.* **34**, 963-972 (2001).
92. S. Arimori, C. J. Ward, and T. D. James, The first fluorescent sensor for boronic and boric acids with sensitivity at sub-micromolar concentrations-a cautionary tale, *Chem. Commun.* 2018-2019 (2001).
93. G. Springsteen, and B. Wang, A detailed examination of boronic acid-diol complexation, *Tetrahedron* **58**, 5291-5300 (2002).
94. G. Springsteen, and B. Wang, Alizarin Red S. as a general optical reporter for studying the binding of boronic acids with carbohydrates, *Chem. Commun.* 1608-1609 (2001).
95. L. A. Cabell, M. K. Monahan, and E. V. Anslyn, A competition assay for determining glucose-6-phosphate concentration with a tris-boronic acid receptor, *Tetrahedron Lett.* **40**, 7753-7756 (1999).
96. Z. Zhong, and E. V. Anslyn, A colorimetric sensing ensemble for heparin, *J. Am. Chem. Soc.* **124**, 9014-9015 (2002).
97. S. Arimori, H. Murakami, M. Takeuchi, and S. Shinkai, Sugar-Controlled Association and Photoinduced Electron-Transfer in Boronic-Acid-Appended Porphyrins, *J. Chem. Soc., Chem. Commun.* 961-962 (1995).
98. Z. Murtaza, L. Tolosa, P. Harms, and J. R. Lakowicz, On the possibility of glucose sensing using boronic acid and a luminescent ruthenium metal-ligand complex, *J. Fluorescence* **12**, 187-192 (2002).
99. J. N. Camara, J. T. Suri, F. E. Cappuccio, R. A. Wessling, and B. Singaram, Boronic acid substituted viologen based optical sugar sensors: modulated quenching with viologen as a method for monosaccharide detection, *Tetrahedron Lett.* **43**, 1139-1141 (2002).
100. J. T. Suri, D. B. Cordes, F. E. Cappuccio, R. A. Wessling, and B. Singaram, Monosaccharide detection with 4,7-phenanthrolinium salts: Charge-induced fluorescence sensing, *Langmuir* **19**, 5145-5152 (2003).
101. N. DiCesare, M. R. Pinto, K. S. Schanze, and J. R. Lakowicz, Saccharide detection based on the amplified fluorescence quenching of a water-soluble poly(phenylene ethynylene) by a boronic acid functionalized benzyl viologen derivative, *Langmuir* **18**, 7785-7787 (2002).
102. S. Arimori, C. J. Ward, and T. D. James, A D-glucose selective fluorescent assay, *Tetrahedron Lett.* **43**, 303-305 (2002).
103. S. Patterson, B. D. Smith, and R. E. Taylor, Fluorescence sensing of a ribonucleoside 5'-triphosphate, *Tetrahedron Lett.* **38**, 6323-6326 (1997).
104. S. Patterson, B. D. Smith, and R. E. Taylor, Tuning the affinity of a synthetic sialic acid receptor using combinatorial chemistry, *Tetrahedron Lett.* **39**, 3111-3114 (1998).
105. T. Nagasaki, T. Kimura, S. Arimori, and S. Shinkai, Influence of added saccharides on the conformation of boronic-acid- appended poly(l-lysine) - attempts to control a helix-coil transition by sugars, *Chem. Lett.* 1495-1498 (1994).
106. T. Kimura, S. Arimori, M. Takeuchi, T. Nagasaki, and S. Shinkai, Sugar-induced conformational-changes in boronic acid-appended poly(L- lysine) and poly(D-lysine) and dugar-controlled orientation of a cyanine dye on the polymers, *J. Chem. Soc., Perkin Trans.* 2 1889-1894 (1995).
107. T. Kimura, M. Takeuchi, T. Nagasaki, and S. Shinkai, Sugar-induced color and orientation changes in a cyanine dye bound to boronic-acid-appended poly(L-lysine), *Tetrahedron Lett.* **36**, 559-562 (1995).
108. W. Wang, S. Gao, and B. Wang, Building fluorescent sensors by template polymerization: the preparation of a fluorescent sensor for D-fructose, *Org. Lett.* **1**, 1209-1212 (1999).
109. S. Gao, W. Wang, and B. Wang, Building fluorescent sensors for carbohydrates using template-directed polymerizations, *Bioorg. Chem.* **29**, 308-320 (2001).
110. T. Kawanishi, M. Holody, M. Romey, and S. Shinkai, *submitted*
111. M. Segura, B. Bricoli, A. Casnati, E. M. Munoz, F. Sansone, R. Ungaro, and C. Vincent, A prototype calix[4]arene-based receptor for carbohydrate recognition containing peptide and phosphate binding groups, *J. Org. Chem.* **68**, 6296-6303 (2003).
112. R. Welti, and F. Diederich, A new family of C3-symmetrical carbohydrate receptors, *Helv. Chem. Acta* **86**, 494-503 (2003).
113. G. Lecollinet, A. P. Dominey, T. Velasco, and A. P. Davis, Highly selective disaccharide recognition by a tricyclic octaamide cage, *Angew. Chem., Int. Ed. Engl.* **41**, 4093-4096 (2002).
114. M. Mazik, M. Radunz, and W. Sicking, High α/β-anomer selectivity in molecular recognition of carbohydrates by artificial receptors, *Org. Lett.* **4**, 4579-4582 (2002).

115. O. Rusin, K. Lang, and V. Kral, 1,1'-binaphthyl-substituted macrocycles as receptors for saccharide recognition, *Chem. Eur. J.* **8**, 655-663 (2002).
116. M. He, J. R. Johnson, J. O. Escobedo, P. A. Beck, K. K. Kim, N. N. St. Luce, C. J. Davis, P. T. Lewis, F. R. Fronczeck, B. J. Melancon, A. A. Mrse, W. D. Treleaven, and R. M. Strongin, Chromophore formation in resorcinarene solutions and the visual detection of mono- and oligosaccharides, *J. Am. Chem. Soc.* **124**, 5000-5009 (2002).
117. D. W. P. M. Lowik, and C. R. Lowe, Synthesis of macrocyclic, triazine-based receptor molecules, *Eur. J. Org. Chem.* 2825-2839 (2001).
118. S. Tamaru, M. Yamamoto, S. Shinkai, A. Khasanov, and T. W. Bell, A hydrogen-bonding receptor that binds cationic monosaccharides with high affinity in methanol, *Chem. Eur. J.* **7**, 5270-5276 (2001).
119. J. Bitta, and S. Kubik, Cyclic hexapeptides with free carboxylate groups as new receptors for monosaccharides., *Org. Lett.* **3**, 2637-2640 (2001).
120. H.-J. Kim, Y.-H. Kim, and J.-I. Hong, Sugar recognition by C3-symmetric oxazoline hosts, *Tetrahedron Lett.* **42**, 5049-5052 (2001).
121. J. M. Benito, M. Gomez-Garcia, J. L. J. Blanco, C. O. Mellet, and J. M. G. Fernandez, Carbohydrate-based receptors with multiple thiourea binding sites. Multi-point hydrogen bond recognition of dicarboxylates and monosaccharides, *J. Org. Chem.* **66**, 1366-1372 (2001).
122. T. Nagase, S. Shinkai, and I. Hamachi, Post-photoaffinity labeling modification using aldehyde chemistry to produce a fluorescent lectin toward saccharide-biosensors, *Chem. Commun.* 229-230 (2001).

FLUORESCENT PEBBLE NANO-SENSORS AND NANOEXPLORERS FOR REAL-TIME INTRACELLULAR AND BIOMEDICAL APPLICATIONS

Hao Xu[*], Sarah M. Buck[*], Raoul Kopelman[*†], Martin A. Philbert[‡],
Murphy Brasuel[‡], Eric Monson[*], Caleb Behrend[*], Brian Ross[†], Alnawaz
Rehemtulla[†], and Yong-Eun Lee Koo[*]

4.1. INTRODUCTION

PEBBLEs (Probes Encapsulated By Biologically Localized Embedding) are sub-micron sized optical sensors specifically designed for minimally invasive analyte monitoring in viable, single cells with applications for real time analysis of drug, toxin, and environmental effects on cell function. PEBBLE nanosensor is a general term that describes a family of matrices and nano-fabrication techniques used to miniaturize many existing optical sensing technologies. The main classes of PEBBLE nanosensors are based on matrices of cross-linked polyacrylamide, cross-linked poly(decyl methacrylate), and sol-gel silica. These matrices have been used to fabricate sensors for H^+, Ca^{2+}, K^+, Na^+, Mg^{2+}, Zn^{2+}, Cu^{2+}, Cl^-, O_2, NO, and glucose that range from 20 nm to 600 nm in diameter. A number of delivery techniques have been used successfully to deliver PEBBLE nanosensors into mouse oocytes, rat alveolar macrophages, rat C6-glioma, and human neuroblastoma cells. For majority of this chapter, we will focus on the fabrication, characterization and applications of all the different kinds of PEBBLE sensors developed up to date. In the remainder of the chapter, we will introduce a new family of PEBBLEs with several emerging directions in PEBBLE design and applications, from intracellular imaging to in-vivo actuating and targeting.

4.1.1. Background and History

In medical and biochemical research, as the domain of the sample is reduced to micrometer dimensions, e.g. living cells or their sub-compartments, the real-time measurement of chemical and physical parameters with high spatial resolution and

---

[*] Department of Chemistry, University of Michigan, Ann Arbor, Michigan, 48109-1055
[†] Molecular Therapeutics Inc., Ann Arbor, Michigan, 48109
[‡] Department of Environmental Health Sciences, University of Michigan, Ann Arbor, Michigan, 48109-1055

negligible perturbation of the sample becomes extremely challenging. In cellular research one strives to become a "silent observer", measuring and observing processes without interfering with normal cell behavior. However, achieving that goal becomes difficult and nearly impossible as probes must be introduced to the cell and consequently cause slight perturbation to the cell and/or are perturbed themselves by the cellular function (i.e. protein binding, sequestration). The minimization of interactions between the cell and probes is a common goal for all involved in developing novel cellular sensors. A traditional strength of chemical sensors (optical, electrochemical, etc.) is the minimization of chemical interference between sensor and sample, achieved with the use of inert, "biofriendly" matrices or interfaces. However, chemical sensors create significant physical interference, simply by the large insertion volume, in the cellular environment, resulting in serious biological damage and abnormal biochemical responses.

Significant progress in miniaturization of electrodes and fiber-optic optodes occurred over the last decade. [1-10] Common methods for single cell analysis include pulled optical fibers (micro and nano-optodes) and miniaturized electrodes. While the tips of these sensors may have nanometer dimensions, mechanical and physical perturbation of the cell is caused by punching holes in the cellular membrane during insertion. Additionally, the penetration volume of the sensor is a significant percentage of intracellular space; the perturbation has a high potential of interfering significantly with the study. The mouse oocyte is a good example of such perturbations (Figure 4.1); the tip of the inserted fiber optic sensor is as small as 200 nm, but because the penetration depth is 50 mm, the penetration volume is about 20,000 mm^3 (this cone-shaped volume grows as the third power of the penetration depth). The cell volume of a 100 mm diameter oocyte is about 524,000 mm^3; thus, the penetration volume is nearly 4% of the cell volume. Most mammalian cells are even smaller than the mouse oocyte, increasing the significance of the large fiber optode insertion volume. Maintaining cell viability, while monitoring more than one analyte in a single cell using pulled nano-optodes or pulled capillary microelectrodes, is an even more significant challenge as several optodes must be inserted.

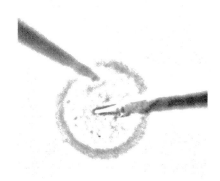

Figure 4.1. Two potassium selective fiber optic sensors inserted in a 100 μm mouse oocyte.

Injection of fluorescent indicator "naked" dyes into a cell, combined with ratiometric imaging, confocal microscopy, two-photon fluorescence, or fluorescence lifetime imaging, has provided insights into concentrations and spatial locations of ions throughout a single cell. In contrast to the pulled fiber optical nano-optodes, these individual molecular probes ("naked" sensing dyes) are physically insignificant but usually suffer from the chemical interference between probe and cellular components. Also, the free dye itself may be toxic to the cell or to specific cellular organelles. Thus, each dye used for intracellular measurements must be carefully chosen and assessed for its ability to enter a cell and provide accurate and reliable information from within the cell. Problems such as toxicity to the cell, intracellular sequestration to specific organelles, non-ratiometric properties, protein binding and intracellular buffering must be evaluated. [11-14] While some indicator dyes work well within a cell, many fluorescent probes suffer from the above problems, limiting the indicator dyes available for reliable intracellular measurements. For example, some calcium indicators, commonly used for other applications, buffer a cell when used in high concentrations intracellularly. [15] Also, some pH indicators, such as carboxyfluorescein and BCECF (a polar fluorescein derivative), are affected by self-quenching or by binding to proteins while inside cells. [16]

A recent development in sensor design combines the advantages of sensor tips and molecular probes, i.e. simultaneously avoids both physical and chemical interference between sensor (probe) and sample (cell or organelle). These spherical nanosensors or PEBBLEs (Probes Encapsulated By Biologically Localized Embedding) are optical sensors of nanometer dimensions with optical sensing components entrapped in an inert matrix. The small size of the PEBBLE sensors enables them to be inserted non-invasively into a living cell. The porous, transparent nature of the matrix allows the analyte to interact with the indicator dye, which reports this interaction via a change in emitted fluorescence.

Variation in the design and fabrication of the PEBBLEs has led to enhancements of the original purpose of chemical sensing and broadened applications to include nanoparticle devices used in drug delivery and medical imaging. These enhancements and variations will be discussed in more detail later in the chapter.

As mentioned earlier, though pulled fiber optic nano-optodes are useful in the monitoring of single analytes in live cells, insertion of several fibers would endanger the cell and compromise the integrity of the measurements. PEBBLEs were developed to use the same protected sensing technologies of fiber optics, while significantly reducing the mechanical perturbation of the cells. For example, poly(decyl methacrylate) (PDMA)-based PEBBLEs have a total volume of only 0.2 μm^3 ($\leq$600 nm in diameter). This is a mere 1 ppm of the oocyte cell volume. Up to 22,000 individual PEBBLEs could be inserted, while maintaining the same insertion volume as one pulled nano-optode. In reality, far fewer PEBBLEs are typically inserted. With even smaller PEBBLEs, e.g. 20-40 nm polyacrylamide PEBBLEs, one PEBBLE constitutes about 1 ppb of the volume of a cell. The small size of the PEBBLEs results in minimal physical perturbations to the cell, even when a large number of PEBBLEs are inserted into a single cell at one time. The small size of the PEBBLE also shortens response time of the sensor.

The prime advantages of PEBBLE nanosensors over "naked" fluorescent indicator dyes in intracellular sensing is that their matrix provides a protective coating for the indicator dyes, protecting the response from interferences such as protein binding and/or membrane/organelle sequestration in the biological sample. [11] Conversely, the nanosensor matrix also provides protection to the cellular contents, enabling the use of dyes that would usually be toxic. Thus, by entrapping the indicator dyes within a polymer matrix, PEBBLEs provide a method for minimizing many of the undesired interactions that occur between fluorescent probes and the cell. For example, the matrix allows ions or neutral analyte species to diffuse through and bind with the indicator, but precludes the diffusion of the indicator dyes into and throughout the cell, thus avoiding sequestration and self-quenching. In short, the matrix protects the cell from the dyes and the dyes from the cell components.

Another important trait of PEBBLEs that increases their utility for intracellular sensing is that the PEBBLE creates a separate sensing phase (the PEBBLE matrix) distinct from the cellular environment. Multiple dyes, ionophores, and other components (i.e. enzymes) can be combined within this sensing phase to create complex sensing schemes. These complex sensing schemes can include reference dyes that allow ratiometric sensing/imaging, or ionophore/chromoionophore combinations that allow for the use of highly selective, non-fluorescent ionophores. Furthermore, these schemes can include enzyme molecules together with the fluorescent dyes to allow enzyme-based intracellular sensing for biologically important small molecules, i.e. glucose. In this specific example, enzyme-based intracellular glucose sensing using PEBBLE sensors, [17] the sensing system enables a synergistic approach in which there is a steady state with local depletion of oxygen due to enzymatic oxidation of glucose by the enzyme glucose oxidase, which is necessary for the glucose sensing mechanism. However, this cannot be achieved by separately introducing free enzymes and fluorescent sensing dyes into a cell. [17]

In summary, PEBBLEs were developed specifically for intracellular applications and have the advantages, without the disadvantages, of both pulled optical fiber sensors and free molecular probes ("naked" indicator dye molecules). The small size and lack of deleterious response due to interferents suggest that these optical nanosensors are well suited for intracellular applications. The protection of both dye and cell and the powerful sensing flexibility come in a nano-package, which, in terms of minimal mechanical and physical perturbation, is closer to "free molecular dyes" than most other sensing platforms. However the nanosensor preserves the excellent chemical sensing and biocompatibility of macro-sensors and surpasses their performance in terms of response time and absolute detection limit.

4.2. PEBBLE MATRICES: DESIGN, PRODUCTION, AND QUALITY CONTROL

The science of nano-optode production relies on advances in nano-scale production, using emulsion and dispersion fabrication techniques. PEBBLEs, representing a new form of nano-optode technology, are also dependent on progress in fabrication. The nano-emulsion/dispersion processes for preparing PEBBLEs are subtle and there is no universal method for making hydrophilic, hydrophobic, and amphiphilic nanospheres that contain the right matrix and right chemical components in their proper proportions. Thus, switching from single dye containing hydrophilic polyacrylamide nanospheres to multi-component, hydrophobic, liquid polymer sensors, or to inert silica sol-gel sensors is not yet a routine procedure. However, the production methods, once optimized for a given matrix and its constituents, are based on relatively simple wet chemistry techniques, as opposed to many complicated physical and chemical nanotechnology schemes.

4.2.1. Polyacrylamide PEBBLEs

Polyacrylamide PEBBLEs are synthesized in a water-in-oil (W/O) microemulsion system. These W/O microemulsions are transparent, isotropic, thermodynamically stable liquid media, with a continuous oil domain and an aqueous domain thermodynamically compartmentalized by a surfactant as nanometer-sized liquid entities (also referred to as reverse micellar solutions). These surfactant-covered entities offer a unique microenvironment for the polymerization reactions of acrylamide monomers; that is, they act not only as nano-reactors for hosting the reaction but also as steric barriers for inhibiting the polymerization of reacting species among different droplets during the reaction period. [18] The PEBBLE components are hydrophilic and stay in the aqueous core of the micellar droplets. Both the polymerization reaction and the subsequent formation of nanoparticles occur within these aqueous cores. As a result, the size of the polyacrylamide PEBBLEs formed in these nano-reactors approaches that of the aqueous core of the micellar droplet, which is typically in the nanometer range. Previous studies have shown that the size of the polyacrylamide nanospheres can be varied by changing the amount of surfactant, solvent, monomer or the temperature of the reaction. [19] It was determined previously in our laboratory that the surfactant concentration plays the largest role in the control of sensor size achieved in the emulsion. [16] The use of two types of surfactant (Brij 30 and AOT), in high concentration, keeps the initial monomer micellar size very small. [19] This small initial size prevents the spheres from becoming too large during polymerization, and nanometer-sized sensors are produced. As mentioned earlier, this small size of the PEBBLEs would result in only minimal physical perturbations to the cell even when a large number of PEBBLEs are inserted into a single cell at one time.

The batch-to-batch reproducibility of polyacrylamide matrix PEBBLE sensors is very good. The standard deviation between batches matched the deviation found within a single batch. One can have confidence that once a procedure is optimized for a particular application, the protocol will be reliable for producing the required sensors. [16] However, each batch should be calibrated to ensure that reproducibility was maintained. To date consistent polyacrylamide PEBBLE sensors for pH, calcium, oxygen, glucose, zinc and magnesium have been produced.

In our laboratory, polyacrylamide PEBBLEs are prepared using a microemulsion polymerization technique modified from the work of Daubresse et. al. [20] Briefly, a polymer solution (9 ml) containing the main monomer acrylamide (2.7 g) and the cross-linking agent N, N'- methylene-bisacrylamide (0.8 g) dissolved in phosphate buffer (10 mM, pH 7.2) is prepared as the aqueous phase. The oil phase of the microemulsion consists of sodium AOT (1.6 g) and Brij 30 (3.08 g) dissolved in deoxygenated hexane (43 ml). 2 ml of the aqueous phase solution is added to the oil phase and stirred to produce a microemulsion system. Argon is passed over the microemulsion mixture to ensure that it is oxygen free. The polymerization reaction is initiated by addition of N, N, N', N'-tetramethylethylenediamine (TEMED) (15 ml) and ammonium persulfate (APS) (24 ml), and the reaction is allowed to proceed under stirring in an Ar atmosphere for two hours.·

Once the polymerization reaction is complete, the hexane is removed by rotary evaporation yielding an aqueous phase consisting of unreacted monomers, cross-linkers and surfactants (AOT and Brij 30). This mixture is dissolved in ethanol (200 proof) and transferred to an ultrafiltration cell (Millipore Corp., Bedford, MA). A 100 kDa membrane is used to separate the reacted polymer beads from the unreacted monomer, cross-linkers and surfactants, under a pressure of 10 psi. The polymer PEBBLEs are rinsed with at least 500 ml ethanol to ensure all unreacted components had been removed from the PEBBLEs. The PEBBLEs are then collected by a suction filtration system (Fisher, Pittsburgh, PA) and dried under vacuum.

TEM images and results of asymmetric field-flow fraction (AFFF)-static light scattering measurements of polyacrylamide PEBBLEs produced by the microemulsion process described above are shown in Figure 4.2.

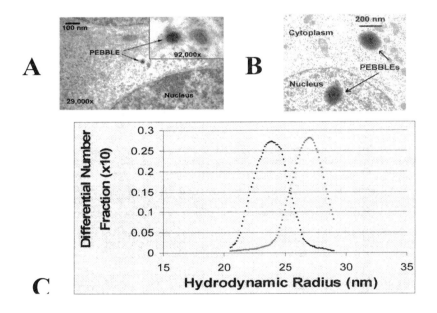

Figure 4.2. (A and B) Transmission electron micrographs of polyacrylamide PEBBLE sensors embedded in the cytoplasm of neuroblastoma cells (via gene gun): (A) One 20 nm PEBBLE next to a lysosome of similar size. (B) Two 200 nm PEBBLEs near and inside the cell nucleus. The orginal magnification/scale bars are indicated in the figures. (C) Static light scattering result showing the size distributions of polyacrylamide PEBBLEs (left curve) and 50nm reference polystyrene beads (right curve). The average size of the PEBBLEs is about 45-50 nm in diameter.

4.2.2. Poly(decyl methacrylate) PEBBLEs

The use of fluorescent indicator molecules in encapsulated form (polyacrylamide PEBBLEs), has proven valuable in the study of a number of intracellular analytes[16, 21-26] (H^+, Ca^{2+}, Mg^{2+}, Cu^{2+} Zn^{2+}, O_2), however, there are many ions for which no fluorescent indicator dye is sufficiently selective or even available. A new class of optical nano-sensors using an alternate sensing scheme is therefore required.

Polyacrylamide PEBBLEs do not take advantage of the rich history of electrochemical sensors which has identified a host of highly selective, hydrophobic, ionophores. In many cases the selectivity of these ionophores has yet to be matched by hydrophilic indicator dyes (hydrophilic chromoionophores). Highly selective intracellular (and extracellular) hydrophilic indicator dyes are limited to a small set of analytes, such as pH and calcium. While the use of PEBBLEs, instead of traditional free ("naked") dye molecule indicators, results in protection beneficial to both the cell and the dye, it does not solve the selectivity problems. For instance, hydrophilic potassium indicators will not work in the presence of significantly higher sodium concentrations, and conversely, sodium indicators will not work in the presence of high potassium concentrations. [27] Obviously, this has serious implications for both intracellular (i.e. high potassium/sodium ion ratios) and extracellular (i.e. high sodium/potassium) applications. Moreover, no satisfactory color indicators are available for many important analyte ions (e.g. nitrite). The above problem has been solved in optodes by using in tandem a highly selective, optically silent ionophore and an adjacent, optically visible agent that plays the role of a spectator or reporter dye. While the principles of such tandem sensing schemes were worked out by Bakker and coworkers, [1, 28, 29] Suzuki, [30, 31] and Wolfbeis, [32, 33] the first demonstration of such a sensing scheme on the nanoscale occurred with the pulled optodes developed by Shortreed et al. [7,8] The extension of these principles to PEBBLEs required the optimization of a new liquid polymer matrix, poly(decyl methacrylate) (PDMA). [34] The result was a 600 nm PEBBLE that is selective for potassium, based on the potassium ionophore 2-Dodecyl-2-methyl-1,3-propanediylbis[N-[5'nitro(benzo-15-crown-5)-4'-yl]carbamate] (BME-44) and chromoionophore III or 9-(Diethylamino)-5-[(2-octoyldecyl)imino]benzo[a]phenoxazine (ETH 5350). The first application of the liquid polymer class of PEBBLEs was the observation of potassium uptake in Rat C6-glioma cells. [34] Micron-sized liquid polymer beads based on either polyvinyl chloride (PVC) or dodecyl acrylate have also been developed. [35, 36]

PDMA-based sensors, with much higher selectivity for most ions, rely on having a chemical equilibrium (or steady state) among different sensor components. These sensors act as "bulk optodes" or ion selective optodes (ISO) where the matrix (hydrophobic liquid polymer) contains a selective lipophilic ionophore ("optically silent"), a fluoroionophore and an ionic additive. [34] The operation of the entire system is based on having a thermodynamic equilibrium that controls ion exchange (cation sensors) or ion co-extraction (anion sensors), i.e. an equilibrium-based correlation between different ion species. To achieve sensor miniaturization, fluorescence (rather than absorbance), has been utilized. Fabrication of PEBBLEs based on the described scheme, in the sub-micron size range, has been demonstrated for K^+, Na^+, and Cl^-. Extension of that procedure to existing ionophores for other analytes, both cationic and anionic, simply involves optimization of the matrix for the required sensing elements. [34] Larger liquid polymer sensing particles have also been recently developed by Bakker and co-workers, utilizing a dispersion method to cast PVC particles and plasticizer-free dodecyl acrylate microspheres for ion sensing. [35, 36]

A batch of poly(decyl methacrylate) PEBBLE sensors typically consists of 210 mg of decyl methacrylate, 180 mg hexanedioldimethacrylate, 300 mg of dioctyl sebacate, with 10-30 mmol/kg each of ionophore, chromoionophore, and ionic additives added after spherical particle synthesis. The spherical particles are prepared by dissolving decyl methacrylate, hexanedioldimethacrylate, and dioctyl sebacate (DOS) in 2 ml of hexane. To a 100 ml round bottom flask, in a water bath on a Corning pc-351 hot plate stirrer, 75 ml of pH 2 HCl is added along with 1,793 mg of PEG 5'000 monomethyl ether and stirred and degassed. The monomer cocktail dissolved in hexane is then added to the reaction flask (under nitrogen), stirred at full speed, and water bath temperature is raised to 80°C over 30-40 minutes. Potassium peroxodisulfate (6 mg) is then added to the reaction and stirring is reduced to medium speed. The temperature is kept at 80°C for two more hours, and then the reaction is allowed to return to room temperature and stir for 8-12 hours. The resulting polymer is suction filtered through a Fisherbrand glass microanalysis vacuum filter holder with a Whatman anodisc filter (0.2 μm pore diameter).

The polymer is rinsed three times with water and three times with ethanol to remove excess PEG and unreacted polymer. Tetrahydrofuran (THF) is then used to leach out the DOS and then the PEBBLEs are again filtered and rinsed. They are allowed to dry in a 70°C oven overnight. Dry polymer is weighed out, and DOS, ionophore, chromoionophore and ionic additive are added to this dry polymer, so that the resulting polymer has 40% DOS, 20 mmole/kg ionophore, 10 mmole/kg chromoionophore, and 10 mmole/kg ionic additive. Enough THF is added to this mixture so as to just wet the PEBBLEs. The PEBBLEs are allowed to swell for eight hours and then the THF is removed by rotary evaporation. The resulting PEBBLE sensors are rinsed with double distilled water and allowed to air dry.

For sizing, the PEBBLEs are suspended in a 50/50 water/ethanol solution. A few drops of this suspension are evaporated on a glass cover slip and sputtered with gold. The SEM images were taken on a Hitachi S-3200N Scanning Electron Microscope. Figure 4.3 clearly illustrates the utility of PEG 5'000 as a steric stabilizer for PEBBLE fabrication.

Figure 4.3. SEM of gold coated poly(decyl methacrylate) nanoshperes produced by emulsion polymerization. *Left.* PEG 5'000 monoethyl ether as surfactant. 48 % of the PEBBLEs have a diameter of 500 nm ± 40 nm; 37% of the PEBBLEs have a diameter of 600 ± 40 nm. *Right:* Poly (decyl methacrylate) PEBBLEs fabricated using the same procedure excluding PEG as a steric stabilizer.

4.2.3. Polyethylene Glycol-Coated Sol-Gel Silica PEBBLEs

Sol-gel silica has also been used as the matrix for the fabrication of PEBBLE nanosensors because of the superior properties it has over organic polymers. Sol-gel glass is a porous, high purity, optically transparent and homogeneous material, [37] thus making it an ideal choice as a sensor matrix for quantitative spectrophotometric measurements. Also, it is chemically inert and thermally stable compared to polymer matrices. The preparation of sol-gel "glasses" is technically simple and tailoring the physico-chemical properties (i.e. pore size or inner-surface hydrophobicity) of sensor materials can be achieved easily by varying the processing conditions and the amount or type of reactants used. This enables the pore sizes to be optimized such that the analyte is able to diffuse easily and interact with the sensing molecules whilst the latter are prevented from leaking out of the matrix (also true for polyacrylamide-based sensors). Furthermore, this "glass" is produced under so-called soft chemical conditions, allowing the inclusion of biomolecules.

A range of sol-gel-based sensor configurations has been described in the literature, including monoliths, thin films, miniaturized probe-tips and powders. [37] Immobilization of the sensing reagent in a supportive matrix is a critical step in the fabrication of optical sensors. It can be achieved by either chemical or physical entrapment of the sensing reagent in the pore structures of the sol-gel network. An important advantage of physical entrapment is the minimal alteration in the spectral and binding properties of the sensing molecules, due to the weak interactions with the supporting matrix. A key advantage of sol-gel sensors is that the "soft" chemical techniques (i.e. room-temperature synthesis) used are ideal for the entrapment of enzymes and proteins. For instance, one can incorporate oxidase enzymes into proven ratiometric oxygen sensing sol-gel PEBBLEs[38] for the sensing of analytes such as glucose (using the glucose oxidase enzyme).

The sol-gel silica PEBBLEs are coated with polyethylene glycol (PEG) polymers. We found that the coating afforded several advantages. In early experiments with sol-gel PEBBLE production, we sometimes found that after collecting the particles as a dry solid, aggregation among particles took place when re-suspending them in an aqueous solution for applications. This is because the suspended silica particles formed agglomerates, probably due to interparticle hydrogen bonding.[39] Surface barriers introduced onto the PEBBLES, by adsorbing or grafting polymers, act as steric hindrances, preventing aggregations among particles in an aqueous suspension. Direct contact between particles is prevented by the polymer chains extending into the medium, and the associated electrical charges. Therefore, either no aggregates of particles are formed or, at least, the rate of coagulation will be decreased, depending on the degree of coverage of polymers onto the particles and the thickness of the polymer layer. [40-42] We chose to use a coating of steric stabilizer by covalently bonding one end of PEG to the particle. The free end of the PEG extends far into the surrounding medium. Thus, in an aqueous suspension, the aggregation among the nanoparticles is prevented by the repulsion force and solvation layer of the PEG surface moiety. [43] Other important advantages of having PEG coatings on the silica PEBBLEs involve their applications to biological systems. PEG is non-toxic and its attachment to silica nanoparticles provides a biocompatible and protective surface. [44] The PEG coatings would reduce protein and cell adsorption onto the particles. It is also known that "PEGylated" materials are cleared by organs and kidneys at a reduced rate; PEG coatings, therefore, effectively increase particle circulation time (plasma residence time) for in vivo applications. [45-47] These properties of the PEG-coating, onto the sol-gel silica PEBBLEs, are crucial to both their intracellular and in-vivo applications.

To produce a monodisperse batch of PEG-coated sol-gel silica PEBBLEs, PEG is attached to the PEBBLE surface or its precursor (starting compound for the sol-gel reaction) tetramethyl orthosilicate (TMOS) through transesterification (reaction [1]). The sol-gel forms from an equilibrium step (reaction [2]) and a condensation step (reaction [3]) involving TMOS. Reactions [2] and [3] predominate over reaction [1], resulting in a silica sol-gel matrix with a PEG coating. These reactions are executed in the presence of a relevant dye, which is encapsulated, to form a functional PEBBLE. [48]

$$...Si - OR + HO(CH_2CH_2O)_nH \leftrightarrow ...Si - O(CH_2CH_2O)_nH + ROH \qquad [1]$$
$$...Si - OR + H_2O \leftrightarrow ...Si - OH + ROH \qquad [2]$$
$$...Si - OR + HO - Si \leftrightarrow ...Si - O - Si + ROH \qquad [3]$$

A typical synthesis procedure for the PEG-coated silica PEBBLEs is as follows: PEG polymers of different molecular weights (usually MW of 5000 or 6000) are dissolved in mixed solution of ammonium hydroxide and methanol. Upon mixing, the solution becomes transparent, and TMOS is added drop-wise to initiate the hydrolysis reaction. The solution is then stirred vigorously at room temperature for a few hours before the reaction is stopped. A typical reaction solution, which gives an average particle size of ~100 nm, consists of PEG MW 5000 monomethyl ether (2 g), methanol (99.9%, 24 ml), ammonium hydroxide (30% wt of ammonia, 6 ml), and TMOS (99.9%, 0.2 ml).

After the reaction is stopped, a liberal amount of ethanol is added to the reaction solution and the mixture was transferred to an Amicon ultrafiltration cell (Millipore Corp., Bedford, MA). A 100 kDa membrane separates the silica particles from the unreacted monomers, PEG, and ammonia, under a pressure of 10 psi. The particles are further rinsed with at least 500 ml distilled water and 200 ml ethanol to ensure that all unreacted components are removed from the silica particles. The silica particle suspension is passed through a suction filtration system (Fisher, Pittsburgh, PA) to collect the particles which are dried to yield a final product of PEGylated silica nanoparticles.

A typical SEM image (left) and the result of static light scattering measurements (right) of sol-gel silica PEBBLEs produced by the above mentioned sol-gel method are shown in Figure 4.4. It should be noted here that the SEM and light scattering results were obtained using the same batch of particle sample. The two results matched quite well, indicating that no or very little aggregation took place during the light scattering measurements when the silica particles were suspended in a buffer solution, due to the steric stabilization effect of the PEG polymer on the particle surface.

 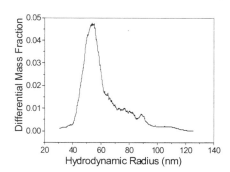

Figure 4.4. *Left:* A typical SEM image of PEG-coated silica nanoparticles. The scale bar is 1 μm. The particles are prepared by using PEG MW 5000 monomethyl ether (2 g), methanol (99.9%, 24 ml), ammonium hydroxide (30% wt of ammonia, 6 ml), and TMOS (99.9%, 0.2 ml). *Right:* AFFF-multi-angle static light scattering result, giving the size distribution of the same batch of silica nanoparticles as depicted in Figure 4.4 *left*.

4.3. CLASSIFICATION AND CHARACTERIZATION OF PEBBLE SENSORS

A variety of PEBBLE sensors based on the three different matrices discussed in the previous section have been fabricated for the real-time monitoring of a variety of analytes (such as biologically important ions and small molecules) in biological systems, i.e. single cells. The performance of these different PEBBLE sensors will be detailed in this section.

4.3.1. Ion Sensors

This section focuses on the PEBBLEs designed to sense ionic moieties in vitro.

4.3.1.1 Ratiometric PEBBLE Sensors for Ions: pH, Calcium, Zinc, and Magnesium

This type of ion sensors is based on the encapsulation of a fluorescent indicator dye and usually a reference dye inside the polyacrylamide matrix. The sensor response is based on the fluorescence emission intensity ratios between the indicator dye and the reference dye.

It should be noted that as PEBBLEs get smaller and smaller, the concentration of dye in the PEBBLE matrix should remain the same, but fewer and fewer dye molecules incorporated per PEBBBLE and are available for the giving of signal or for participating in the equilibrium based thermodynamic sensing schemes. It is known that with currently available intensified CCD (Charge Coupled Device) technology at least 5 dye molecules per PEBBLE are required to image a single PEBBLE. This lower limit has already been exceeded with our polyacrylamide-based PEBBLEs. Calculations based on the dye concentration present in the polyacrylamide emulsion predict that for a polyacrylamide PEBBLE of 20 nm there is less than one dye molecule per PEBBLE on average.

Therefore, we do not attempt to image these particles singly, but rather use an ensemble of polyacrylamide PEBBLEs for imaging. For single polyacrylamide PEBBLE work, we have used 200 nm PEBBLEs, which contain approximately 100-1000 dye molecules. In the case of liquid polymer (PDMA) PEBBLEs, a minimum number of molecules is required for the thermodynamic equilibrium sensing scheme to be useful.

The protection of sensing elements from interference due to protein binding or cellular sequestration allows the calibration of the PEBBLE sensors in solution with confidence that this same calibration will be valid during intracellular applications. It is important to note that the interference due to protein binding or large molecule/heavy metal interactions are eliminated (but this does not improve the selectivity). Most polyacrylamide PEBBLE calibrations are performed on a Fluoro-Max 2 spectrofluorometer (ISA Jobin Yvon-Spex, Edison, NJ, USA) with the excitation and emission slits set at 2-nm bandwidth.

The polyacrylamide PEBBLEs are hydrophilic and water can diffuse freely through them. A dye that has a chromometric response to the analyte is entrapped in the pores of the hydro-gel. Poor extraction of analyte ions into the hydro-gel is not a consideration because water and small ions are allowed to diffuse freely through the hydro-gel. A chromoionophore-analyte complex is formed within the matrix, similar to the response of the "naked" dye in solution. The dynamic range and selectivity of the PEBBLE is dependent on the K_D of the dye with respect to the analyte and any interfering ions. The response mechanism and a description of K_D for calcium sensors are shown in Equations [4] and [5] below:

$$[Ca^{2+}]_{(aq)} + C_{acrylamide} \leftrightarrow [C(Ca^{2+}]_{(acrylamide)} \cdots [4]$$

$$K_D = \frac{[C]_{(acyrlamide)}[Ca^{2+}]_{(aq)}}{[C(Ca)^{2+}]_{(acylamide)}} \cdots\cdots\cdots [5]$$

Many fluorescent dyes with good selectivity for pH and calcium are available. Thus, calcium and pH sensing present a good opportunity to compare PEBBLE response to that of "naked" dye in similar calibration environments. Complete details on these comparisons can be found in the literature. [16, 22, 23] Table 4.1 and 4.2 summarize the results for pH sensitive dyes and calcium dyes, respectively. Both dye solutions and PEBBLEs utilized the inert sulforhodamine 101 dye as an internal standard. It can be seen that, with few exceptions, in PEBBLEs the slope of the calibration decreases, resulting in reduced sensitivity of the measurements, but the linear ranges are not significantly affected by incorporating the dye in the PEBBLE. The polyacrylamide matrix does not adversely affect the reversibility of the indicator dye and can be tailored to enclose the dye for periods long enough for sensing in live cells.

The reversibility of pH PEBBLEs containing the fluorescent pH probe CDMF (5- (and -6-) carboxy-4',5'-dimethylfluorescein and the reference dye sulforhodamine 101(SR) is illustrated in Figure 4.5. The pH of the solution was changed between pH 6.4 and pH 7.0 for several cycles. The CDMF emission maximum was ratioed against the SR emission maximum and plotted, clearly demonstrating full reversibility of the sensors. Assays of the leaching of CDMF and calcium green PEBBLEs (each containing the sulforhodamine 101 reference dye) show that less than 50% of the dye leaches from the

PEBBLEs in a 48 hr time period. On the time scale of the current single cell experiments, at most a couple of hours, the dye loss is acceptable. With comparable leaching rates for both the indicator and reference dyes, the ratiometric sensing scheme used with these PEBBLEs is maintained. [22, 23]

Table 4.1. Calibration of five pH sensitive dyes and the corresponding polyacrylamide PEBBLE sensor. Sulforhodamine 101 was used as an internal reference in all cases.

pH Indcator	Linear Range pH units	slope[a] ± SD	intercept	r^2	n
CNF (dye)	7.0 – 7.7	7.8 ± 0.4	-53	0.98	6
CDMF + SR (dye)	6.2 – 7.4	3.3 ± 0.5	-19	0.96	12
BCPCF + SR (dye)	6.5 – 7.6	1.3 ± 0.07	-7.3	0.99	6
FSA + SR (dye)	5.8 – 7.2	3.1 ± 0.3	-17	0.99	7
SNAFL (dye)	7.2 – 7.7	-0.75 ± 0.01	6.8	0.99	4
CNF (PEBBLEs)	7.0 – 7.7	0.24 ± 0.01	-0.66	0.96	5
CDMF + SR (PEBBLEs)	6.2 – 7.4	0.67 ± 0.05	-3.0	0.99	6
BCPCF + SR (PEBBLEs)	6.2 – 7.2	0.43 ± 0.07	-1.6	0.90	6
FSA + SR (PEBBLEs)	5.8 – 7.0	3.4 ± 0.4	-17	0.99	5
SNAFL (PEBBLEs)	7.2 – 8.0	-0.49 ± 0.02	4.6	0.99	7

[a]Ratio of normalized fluorescence intensity vs. pH.
CNF: 5-(and 6-) carboxynaphthofluorescein.   **CDMF**: 5-(and 6-) carboxy-4',5'-dimethylfluoroscein. **BCPCF**: 2',7'-bis-(2-carboxypropyl)-5-(and 6-) carboxyfluorescein. **FSA**: flouroscein-5-(and 6-) sulfonic acid, **SNAFL**: 5-(and 6-) carboxy SNAFL®-1. **SR**: sulforhodamine 101.

Table 4.2. Calibration of three calcium selective dyes and the corresponding polyacrylamide PEBBLE sensor. Sulforhodamine 101 was used as an internal standard in all cases.

Calcium Indicator	Linear Range µM calcium	slope ± SD	intercept	r^2	n
Calcium Green + SR (dye)	0 – 0.15	30 ± 0.7	1.7	0.94	4
Calcium Orange + SR (dye)	0 – 0.15	1.5 ± 0.03	1.0	0.95	6
Calcium Green 5N + SR (dye)	3 – 30	0.010 ± 0.05	0.99	0.95	7
Calcium Green + SR (PEBBLEs)	0 – 0.15	7.3 ± 0.05	0.97	0.99	6
Calcium Orange + SR (PEBBLEs)	0 – 0.1	1.3 ± 0.05	1.0	0.79	5
Calcium Green 5N + SR (PEBBLEs)	0 – 5	0.022 ± 0.007	1.0	0.99	4

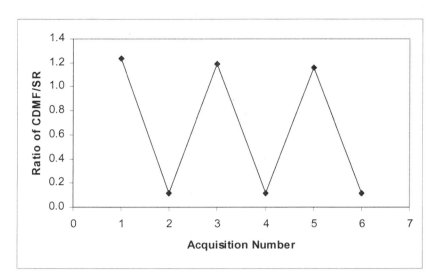

Figure 4.5. Demonstration of polyacrylamide PEBBLE reversibility. The pH of a solution of CDMF/SR PEBBLEs was cycled between pH 6.4 to pH 7.0 for several measurements. The ratios of the two peaks in the spectrum attained from each of the measurements were then plotted, illustrating that the PEBBLE sensors are reversible.

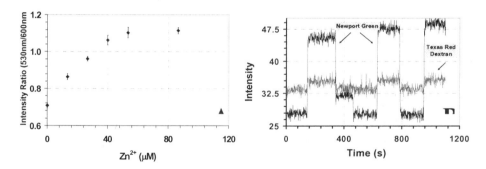

Figure 4.6. (A) 3-mg/mL Newport Green/Texas Red PEBBLE suspension calibrated with zinc nitrate. The samples were excited at 508-nm on a fluorometer, and the emission spectra were collected. From the emission spectra the ratios of intensities were collected. (B) Reversibility test of the 3 mg mL^{-1} PEBBLE suspension using $Zn(NO_3)_2$ and EGTA. Samples were excited at 508 nm and emission was monitored at 530 nm (Newport Green) and 600 nm (Texas Red-dextran).

The calibration result of zinc-sensitive polyacrylamide sensors, containing immobilized Newport Green as the indicator dye for Zn^{2+} sensing and the reference dye, Texas Red-dextran, is shown in Figure 4.6A. The maximum percent increase in the "Intensity Ratio" is 50%, which is less than observed with the Newport Green dye in solution (which has an increase of ~250%). The data indicates a linear range between 10 and 40 μM Zn^{2+}, with a detection limit of 4 μM. Although the sensor has a low affinity for zinc, these PEBBLEs may be useful for neural analysis where zinc levels can reach micromolar concentrations. [25]

The reversibility of the zinc-sensitive PEBBLEs was tested using ethylene glycol-bis (beta-aminoethylether)-N,N,N',N'-tetraacetic acid (EGTA). EGTA, a common metal ion chelator, has a dissociation constant for zinc in the nanomolar range; thus, it preferentially binds to zinc over Newport Green (K_d 1 μM) by orders of magnitude. Figure 4.6B shows that as the zinc ion binds to Newport Green encapsulated in the PEBBLE, the intensity of the dye increases. Conversely, as the zinc ion is removed by the chelator, EGTA, the fluorescence returns to its original state. The data indicates that the sensor configuration is reversible. In addition, it shows that the sensors respond to zinc ion concentrations on a time scale (<4 s) that makes them applicable to real time imaging. In addition to demonstrating the response and reversibility of the system, it also shows that the reference dye, Texas Red–dextran, is insensitive to the analyte ion.

Another kind of polyacrylamide-based ion sensor is the newly developed magnesium-sensitive PEBBLE, [26] which contains immobilized Coumarin 343 (C343) as the indicator dye for Mg^{2+} sensing, and Texas Red-dextran as a reference dye. In a typical calibration curve for these sensors, the C343 and Texas Red-dextran peak intensities were ratioed and plotted against the magnesium concentration. All calibrations were conducted in a 0.1 M MOPS pH 7.19 buffer. The C343/Texas Red-Dextran PEBBLEs have been calibrated with a dynamic range of 0-30 mM and a detection limit of 340 μM. The fluorescence intensity, I_f, of the sample is represented by equation [6], where ϕ_D is the quantum efficiency of the unbound dye, [D], and ϕ_{DMg} is the quantum efficiency of the magnesium bound complex, [DMg]:

$$I_f = \Phi_D[D] + \Phi_{DMg}[DMg].$$ [6]

Equation [7] is the equilibrium reaction that occurs between the dye and magnesium ions:

$$[D] + [Mg] \leftrightarrow [DMg].$$ [7]

Equation [8] is the K_D expression for the reaction:

$$K_D = \frac{[D][Mg]}{[DMg]}.$$ [8]

Equations [9] and [10] are the conservation of mass equations for the total dye, $[D]_T$, and total magnesium, $[Mg]_T$, present in the sample, which are substituted into [8]:

$$[D] + [DMg] = [D]_T,$$ [9]

$$[Mg] + [DMg] = [Mg]_T,$$ [10]

giving

$$K_D = \frac{([D]_T - [DMg])([Mg]_T - [DMg])}{[DMg]}. \qquad [11]$$

Once equations [9] and [11] are expressed in terms of [D] and [D-Mg], respectively, and substituted into equation [6], I_f can then be expressed, in terms of the known constants, ϕ_D, ϕ_{DMg}, $[D]_T$, and $[DMg]_T$, in equation [12]:

$$I_f = \left(\frac{([D]_T + [Mg]_T + K_D) + \sqrt{([D]_T + [Mg]_T + K_D)^2 - 4[D]_T[Mg]_T}}{2} \right)(\phi_{DMg} - \phi_D) + \phi_D[D]_T \qquad [12]$$

This last equation can then be fitted to the experimental points using a sum of least squares method to determine the K_D. The calculated dissociation constant (K_D) of the sensors was 19 mM. The interference from calcium is a major complication of current strategies aimed at the measurement of magnesium concentrations in biological environments. A "same solution interference test" was performed. The calibration of these sensors in the presence of 1 mM calcium background shows very little deviation from the original calibration curve. A second "same solution test" containing 1 mM calcium, 20 mM sodium and 120 mM potassium also shows almost no deviation from the background-free calibration (see Figure 4.7A). This test verifies that these sensors are suitable for any intracellular environment, in which calcium will reach at most micromolar levels.

Separate solution responses to various intracellular cations were also conducted in order to determine any interference with the Mg^{2+} PEBBLEs. The fluorescence intensity of a free dye C343 solution was monitored as the potassium, sodium, and calcium concentrations were increased to biologically relevant levels of 700 mM, 100 mM, and 300 μM, respectively. C343 shows no change in fluorescence properties in response to high concentrations of intracellular ions other than magnesium. Changes in fluorescence intensity due to potassium and sodium were not anticipated; however, most commercially available magnesium indicators exhibit a higher selectivity for calcium than for magnesium. Therefore to reliably monitor magnesium, the sensitivity of C343 to calcium is of primary concern. The above calibrations prove that C343 is insensitive to calcium, demonstrating that the magnesium PEBBLE is a reliable indicator of intracellular magnesium levels.

The reversibility of the magnesium sensitive PEBBLE sensors was tested by two different methods. In the first method, a forward and reverse calibration of the sensors was conducted over a two-hour period. The peak ratio values at each level of magnesium show a perfect overlap between forward and reverse calibration throughout the dynamic range (not shown). The second method involved oscillating the magnesium concentration of the sample. A plot of the peak ratio vs. time (Figure 4.7B) shows the full reversibility of the sensors. It also shows response times, $t_{90\%}$, of less than 4 seconds. The determination of response times was procedure limited. Data points could only be acquired at 4-second intervals in order to allow for addition of analyte. Thus, the value of 4 seconds is an upper limit.

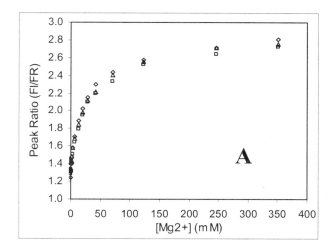

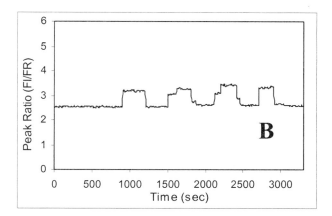

Figure 4.7. (A) Calibration of magnesium selective PEBBLE sensors without background ions (diamond), in the presence of 1 mM Ca^{2+} (square), and in the presence of 1 mM Ca^{2+}, 20 mM Na^+, and 120 mM K^+ (triangle). The samples were excited at 445 nm on the fluorometer and the emission spectra were collected. The peak ratio (F_I/F_R) is plotted against magnesium concentration. All calibrations were buffered at pH 7.19 with 0.1 M MOPS buffer. (B) The peak ratio of a PEBBLE solution was monitored as alternating aliquots of $MgCl_2$ and EDTA were added. The sample was excited at 445 nm and the emission spectrum was collected. After an initial baseline, an aliquot of $MgCl_2$ is added and the peak ratio increased accordingly. After the ratio stabilized an aliquot of EDTA was added to remove free Mg^{2+} from the solution resulting in a decreased peak ratio. This was repeated four times demonstrating the reversibility of the sensors. Experiments were conducted on a fluorometer and the sample was buffered at pH 7.19 in a 0.1M MOPS buffer.

4.3.1.2. Ion Correlation-based PEBBLE Sensors: Potassium, Sodium, and Chloride

This type of ion sensors is fabricated by the immobilization of a selective lipophilic ionophore, a fluoroionophore and an ionic additive inside a PDMA matrix. The general fabrication method of these ion correlation-based sensors was detailed in section 4.2.2. The fluorescence response scheme used to follow analyte binding of non-fluorescent ionophores in PDMA-based PEBBLE sensors closely follows previous work on PVC-based fiber optic sensors selective for potassium and sodium ions. [1, 7, 8] For example, in the cationic sensors, the hydrogen-ion selective chromoionophore competes with the optically silent ionophore as cations enter the liquid polymer matrix. A lipophilic additive maintains ionic strength in the matrix and aids in preventing the co-extraction of anions. It allows for charge neutrality in the membrane without negative counter-ions being brought from the solution into the membrane. [29] "Ion correlation" is a general term that refers to either ion exchange or ion co-extraction.

The work described here takes advantage of an indicator with two fluorescence emission maxima (λ_1, λ_2), giving a relative intensity that changes with the degree of protonation (Π). This degree of protonation, Π, can be evaluated in terms of the fluorescence intensity ratio, $F_{\lambda 2}/F_{\lambda 1}$, given by the protonated chromoionophore intensity $F_{\lambda 2}$ and the deprotonated chromoionophore intensity $F_{\lambda 1}$ (see Figure 4.8 for spectra). [7] The experimentally obtained spectra are normalized to the iso-emmisive point of the dye and Equation [13] is used to determine the degree of protonation of the dye at any given analyte concentration at a known pH. [34] Superscripts P and D denote the completely protonated state and completely deprotonated state of the chromoionophore, respectively, lack of superscript denotes intermediate points.

$$\Pi = \frac{\dfrac{F^{D}_{\lambda 2}}{F^{D}_{\lambda 1}} - \dfrac{F_{\lambda 2}}{F_{\lambda 1}}}{\dfrac{F^{D}_{\lambda 2}}{F^{D}_{\lambda 1}} - \dfrac{F^{P}_{\lambda 2}}{F^{D}_{\lambda 1}} + \dfrac{F_{\lambda 2}}{F_{\lambda 1}}\left(\dfrac{F^{P}_{\lambda 1}}{F^{D}_{\lambda 1}} - 1\right)} \qquad [13]$$

The degree of protonation (Π) of the indicator spectra obtained from the PEBBLE calibration is related to the analyte concentration by using the theoretical treatment of ion-exchange sensors developed by Simon, Bakker and colleagues. [1, 7, 28, 29] For the incorporation of a selective neutral ionophore into a matrix, along with a selective chromoionophore for indirect ion monitoring (ion exchange sensors), the metal ion activity a_i^{v+} in solution (see Equation [14]) is a function of the hydrogen ion activity a_{H+} in solution, the interfering cations a_j^{z+} (where K^{opt}_{ij} is the selectivity coefficient for the interfering ion) and the constants [L_{tot}], [C_{tot}], [R_{tot}], which are total ionophore (ligand) concentration, total chromoionophore concentration, and total lipophilic charge site concentration, in the membrane. Note that [CH] is the protonated chromoionophore concentration and [C] is the free base concentration. It is assumed that all components added during the PEBBLE swelling procedure go into the matrix. The parameter Π has been defined as the relative portion of the protonated chromoionophore, $\Pi = [CH]/[C_{tot}]$. [7]

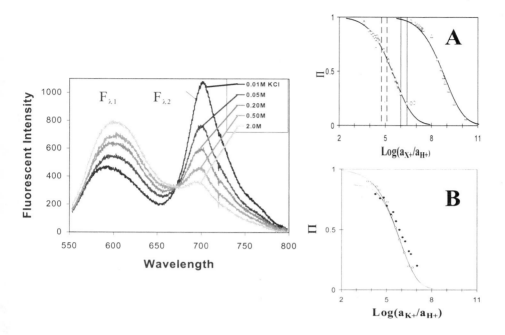

Figure 4.8. *Left*: Normalized emission spectra from suspended K$^+$ PEBBLE sensors using the pH chromoionophore ETH5350 for ion-correlation spectroscopy in tandem with BME-44 for the monitoring of K$^+$ activity. PEBBLEs are excited using a mercury arc lamp and green dichroic mirror. The spectra show response in going from 10 mM KCl to 2.0 M KCl (well beyond saturation of the sensor), all buffered at pH 7.2 with 10 mM Tris buffer. *Right*: **(A)** Response of BME-44-based poly decyl methacrylate PEBBLEs to potassium (o), and sodium (Δ), along with theoretical curves. Theoretical curves are constructed solving equation 5 for a_{Iv+}, (I$^+$ being K$^+$ or Na$^+$ in this case), for a given value of Π. The vertical solid lines delimit values for log (aK$^+$/aH$^+$) typically found in intracellular media and the vertical dashed lines delimit the typical extracellular ratios. **(B)** Response of K$^+$ PEBBLEs to standard additions of KCl in Tris buffer (o) compared to a similar experiment run in a constant background of 0.5M Na$^+$ (●). The theoretical lines are drawn using Equation 5 with log K$^{opt}_{ij}$ = -3.3.

It follows that

$$a_{i^{v+}} + K^{opt}_{ij} \, a_{j^{z+}} = \frac{1}{K_{exch}} \left(\frac{(1-\Pi)\, a_{H+}}{\Pi} \right)^v \left(\frac{[R^-_{tot}] - (\Pi)[C_{tot}]}{v([L_{tot}] - \frac{1}{v}\{[R^-_{tot}] - (\Pi)[C_{tot}]\})} \right) \quad ....[14]$$

which, for primary and interfering ions with charges of 1, simplifies to:

$$ a_{i^{v+}} + K_{ij}^{opt} a_{j^{z+}} = \frac{1}{K_{exch}} (\Pi^{-1} - 1) a_{H+} \left(\frac{[L_{tot}]}{[R_{tot}^-] - \Pi[C_{tot}]} - 1 \right)^{-1} \quad \text{....[15]} . $$

Calibration of a K^+ sensor based on these principles is shown in Figure 4.8 along with normalized spectra (to demonstrate the ratiometric nature of the sensor). The data were taken using an Olympus inverted fluorescence microscope with an Hg lamp (Olympus, Melville, NY) as the light source. For potassium sensing, the chromoionophore is ETH5350, the ionophore is BME-44, and the lipophilic additive is potassium tetrakis-[3,5-bis(trifluoromethyl)phenyl] borate (KTFPB). [8, 34) The data points for potassium and sodium responses are plotted along with corresponding theoretical curves based on Equation [15]. The constant K_{exch} is determined from a line fit to the experimental data. Then the theoretical curve is plotted using the experimentally determined K_{exch} and the constants R_{tot}, C_{tot} and L_{tot} so as to find the expected a_{I+} for a given value of Π. Dashed lines delimit typical extracellular activity ratios and the solid lines delimit the intracellular levels (log (aK^+/aH^+)). [49)

It was found that the actual response matches well with the theory (Figure 4.8A), which is gratifying considering the small size of the systems. The dynamic range at pH 7.2 extends from 0.63mM to 0.63M a_{K+}. The log of the selectivity for potassium vs. sodium, determined by measuring the horizontal separation of the response curves at Π = 0.5, is –3.3. This value, when used to plot the expected response in a sample with a constant 0.5M Na^+ interference, matches the experimental data obtained (see Figure 4.8B). This value indicates a selectivity similar to or better than that obtained for other and larger matrices incorporating BME-44, e.g. -3.1 in PVC-based fiber optic work, and –3.0 in PVC-based microelectrodes. [8, 34) It also exactly matches the value given in the review by Buhlmann, Pretsch, and Bakker[1] for a thin PVC film sensor. This selectivity should be more than sufficient for measurements in intracellular media where potassium concentration is about 100 mM and sodium is about 10 mM. [49)

It can be easily demonstrated that most ionophores available for use in ion-selective electrodes can be used in the development of PEBBLEs. This is illustrated by using the same PEBBLE formulations described except for replacing the BME-44 ionophore with sodium ionophore IV, (DD-16-C-5;2,3:11,12-didecalino-16-crown-5;2,6,13,16,19-Penta-oxapenta-cyclo[18.4.4.4^{7,12}.0^{1,20}.0^{7,12}] dotria-contane), and the chromoionophore ETH5350 with a Chromoionophore II or 9-dimethylamino-5-[4-16-butyl-2,14-dioxoaeicos-yl)phenylimino]-5H-benzo[a]phenoazine (ETH 2439)/ 1,1'-dioctadecyl-3,3,3',3'-tetramethyl-indocarbocyanine perchlorate (DiI) combination (a ratiometric combination based on the inner filter effect). [7) The results are seen in Figure 4.9. The Na^+ PDMA PEBBLEs are formulated to have 30 mM/kg ionophore, 15 mM/Kg chromoionophore, and 15 mM/Kg negative lipophilic additive. Calibrations of the PEBBLE response to NaCl and KCl were carried out in Tris buffer, pH 7.3, and theoretical response curves were constructed using Equation [15], after using Equation [13] to plot the experimental data, (as with the K^+ PEBBLEs). It is shown that these PEBBLEs have a log selectivity of approximately –2.0, or are 100 times more selective for sodium than potassium. This selectivity works well for extracellular applications but intracellular applications will require a more selective ionophore.

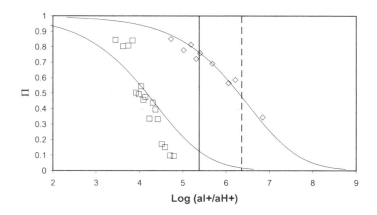

Figure 4.9. Response of sodium ionophore IV-based poly(decyl methacrylate) PEBBLEs to sodium (⊖), and potassium (◊), along with theoretical curves. Theoretical curves are constructed solving Equation 5 for a_{Iv+}, (I^+ being Na^+ or K^+ in this case), for a given value of Π.

We have discussed two kinds of PDMA tandem PEBBLEs designed for cations. Another example showing the versatility of these ion-correlation-based PEBBLEs is a recently developed sensor system for the detection of a biologically important anion, chloride. These PDMA-based Cl^- PEBBLEs are the final sensor systems needed for PDMA PEBBLE-based study of Na^+, K^+, and Cl^- co-transport, a phenomenon that has been reported in almost every animal cell type. [50] Three different methods were used to produce Cl^- sensitive PDMA PEBBLEs. [51] Here we will only focus on one method yielding PEBBLEs that are most suitable for cellular applications. In order to produce a PDMA based Cl^- PEBBLE sensor that was sensitive to Cl^- at biologically relevant concentrations with good reproducibility, a mercury organic compound (Chloride ionophore III) developed by Pretsch and co-workers specifically for anion sensing electrodes[52] was incorporated in the PDMA matrix. The activity of chloride was indirectly followed with the encapsulated fluorescent chromoionophore III (ETH 5350) (reporting the coextraction of H^+, required to maintain charge neutrality) with a small amount of tridodecylmethylammonium chloride (TDDMACl) added to improve the permselectivity of the anions. [52] Figure 4.10 demonstrates the co-extraction response mechanism of these Cl^- PEBBLES. The combination of neutral ionophore with a neutral chromoionophore allows for the coextraction of Cl^- and H^+ while maintaining charge neutrality in the PDMA PEBBLE matrix. Charge neutrality is maintained but the ionic-strength in the membrane changes with changing analyte concentration. Thus, for these sensors empirical calibration without the benefit of a simple theoretical model of the response is relied upon. Typical normalized spectra in different Cl^- concentrations of these PEBBLEs are shown in Figure 4.11 (top). The calibration data of these PEBBLEs for Cl^- and all other anions tested is demonstrated in Figure 4.11 (bottom), and is normalized to the degree of protonation of the chromoionophore by Equation 13.

Figure 4.10. Illustration of the co-extraction response mechanism for the discussed type of Cl⁻ PEBBLEs. It is evident that extraction of the anion-proton changes the ionic strength of the polymer matrix in this sensor configuration.

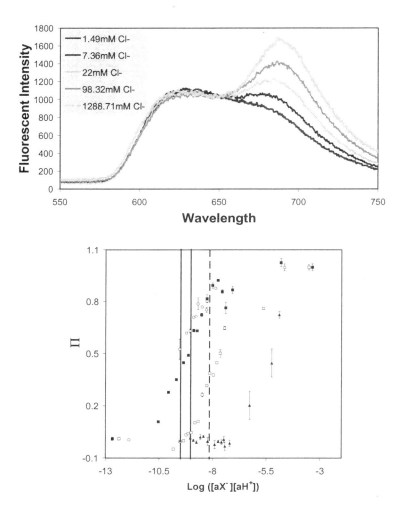

Figure 4.11. *Top:* Normalized fluorescent spectra of chloride ionophore III PEBBLEs in the presence of different chloride concentrations in pH 7.22 HEPES buffer. *Bottom:* Chloride Ionophore III PEBBLE sensor response: (■) chloride; (○) thiocyanate; (□) perchlorate; (▲) nitrite. Solid lines indicate the range of typical intracellular Cl⁻ concentrations, dotted line demarcates typical extracellular Cl⁻ concentrations.

At a pH of 7.2 these PEBBLEs have a limit of detection of 0.2mM Cl⁻ with a linear dynamic range of 0.4mM–190mM Cl⁻ (Π =0.1-0.9, R^2=0.992), which covers the biologically relevant regions of Cl⁻ concentration. The log of the selectivity versus thiocyanate for chloride, perchlorate, and nitrite were determined by measuring the horizontal separation of the linear fits of the experimental data at $\Pi = 0.5$. Figure 4.3 shows the selectivity of the Cl⁻ PEBBLEs over some major interfering anions, as determined by equation 16 where X⁻ is the analyte ion, Y⁻ is:

$$K^{opt}_{X-Y-} = \frac{K^{Y-}_{COEX}}{K^{X-}_{COEX}} \approx \frac{\kappa_{Y-}\beta_{LY-}}{\kappa_{X-}\beta_{LX-}} \qquad [16]$$

the interfering ion, κ is the distribution coefficient of the ion (X⁻ or Y⁻) between the organic phase and the aqueous phase, and β is the stability constant of the ionophore-anion complex. Based on the selectivity result shown in Figure 4.3, these PEBBLEs have good selectivity to Cl⁻ over ClO₄- and NO₂- (almost 100 times and 10,000 times more selective, respectively). It should be noted that the selectivity to Cl⁻ for these sensors is near equal to that of SCN⁻. However, the expected intracellular biological concentrations of Cl⁻ (5-15 mM) are up to almost 4 orders of magnitude greater than for SCN⁻ (0.007-0.017 mM), so the selectivity of these Cl⁻ PEBBLE sensors should be more than sufficient for cellular measurements of Cl⁻.

Table 4.3. Relative selectivity of the Cl⁻ PDMA PEBBLEs.

Ion	Cl⁻ PEBBLEs
SCN-	0
Cl-	-0.4
ClO₄-	-2.0
NO₂-	-4.3

4.3.2. Gas Sensors and Biosensors

This section focuses on sol-gel- and acrylamide-based PEBBLE sensors for the detection of some biologically important gases (NO, and O_2) and small molecules (glucose).

4.3.2.1. Sol-gel-based Nitric oxide PEBBLEs

Nitric oxide sensitive PEBBLEs were fabricated using the typical sol-gel method as described in Section 4.2.3, with minor modifications. [53] This work is an extension of the

NO fiber optic work of Barker. [54] These sensors incorporate into the PEGylated silica matrix Oregon Green-488 carboxylic acid, succinimidyl ester (SE) as the indicator dye for NO sensing, and dextran-linked Texas Red as the reference dye. In the fabrication of sol-gel NO PEBBLEs, a major difference compared to the general sol-gel method described before was that a 10 nm colloidal gold suspension was added to the reaction solution. These gold particles were effectively incorporated into or onto the sensing matrix. This is crucial for the NO sensing mechanism, since when NO binds to the surface of an Oregon Green dye-labeled gold nanosphere there is an observed change in the dye's emitted fluorescence intensity.

During aqueous phase measurements using the NO PEBBLEs, as the nitric oxide concentration in the solution increases there is an observed decrease in the fluorescence intensity from the Oregon Green 488 carboxylic acid-SE as shown in Figure 4.12. These spectra were taken on a FluoroMax-2 and a FluoroMax-3 fluorometer (ISA Jobin Yvon-Spex, Edison, NJ), with slits set to 2 nm for both the emission and excitation. It can be seen that, as a reference for the ratiometric measurements, the emission from Texas red stays relatively constant with the addition of NO-saturated phosphate buffer solution.

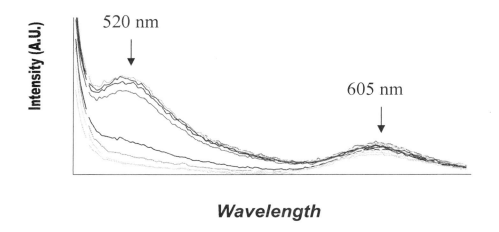

Figure 4.12. The solution phase measurements of sensor emission spectra in response to increasing (top to bottom) concentrations of nitric oxide. The peak on the right is the emission from the reference dye and the peak on the left is the emission from the indicator dye. The fluorescence intensity ratio used in calibration is the ratio of $I520/I605$.

The calibration data of the NO PEBBLEs in aqueous phase (10 mM phosphate buffer, pH 7.4) was obtained by adding aliquots of NO-saturated buffer to a PEBBLE buffer solution, and a typical result is shown in Figure 4.13. The fluorescence intensity ratio (R_0/R) is used to construct the calibration curve for nitric oxide concentration, where R_0 stands for the ratio between the fluorescence intensities of the indictor and reference dye at the absence of NO and R stands for the same ratio at a given concentration of NO.

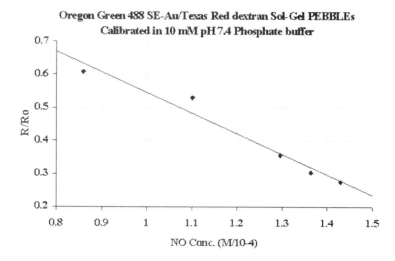

Figure 4.13. Calibration of nanosensors in 10 mM phosphate buffer solution over a region from 85 to 145 uM nitric oxide using nitric oxide saturated phosphate buffer solution.

Testing the reversibility of the sensor response to NO was accomplished by allowing particles to settle and adhere to a silanized glass cover slip. A piece of this cover slip was placed inside a sealed glass chamber and alternating aliquots of buffer solution and nitric oxide-saturated buffer solutions were added to the sealed glass chamber. Between each measurement the chamber and cover slip were rinsed in de-ionized water and purged with argon. The fluorescence intensity ratios for a series of measurements are shown in Figure 4.14, and the sensor response was found to be reversible.

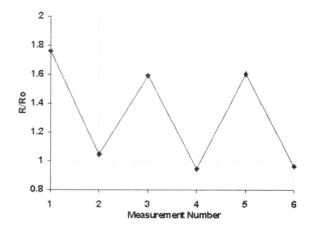

Figure 4.14. Reversibility of NO PEBBLE sensors for a series of measurements taken in alternating solutions of 10 mM pH 7.4 phosphate buffer and 10 mM pH 7.4 phosphate buffer saturated with nitric oxide.

4.3.2.2. *Sol-gel-based Oxygen PEBBLEs*

The sol-gel-based oxygen sensitive PEBBLEs are fabricated using the typical formulations for producing the PEG-coated silica nanoparticles as described in section 4.2.3. These sensors incorporate an oxygen sensitive fluorescent indicator, Ru(II)-tris(4,7-diphenyl-1,10-phenanthroline) chloride ($[Ru(dpp)_3]^{2+}$), and an oxygen insensitive fluorescent dye, Oregon Green 488-dextran, as a reference for the purpose of ratiometric intensity measurements.

The oxygen sol-gel PEBBLEs are based on the quenching of luminescence of the ruthenium dye by the presence of oxygen. The oxygen quenching process is ideally described by the linear Stern-Volmer equation:

$$I_0/I = 1 + K_{SV} \, p[O_2] \qquad\qquad\qquad [17]$$

where I_0 and I are the luminescence intensities in the absence and presence of oxygen at a partial pressure of $p[O_2]$, respectively, and K_{SV} is the Stern-Volmer constant, which depends directly upon the diffusion constant of oxygen, the solubility of oxygen, and the quenching efficiency and lifetime of the excited-state of the fluorophore. [55] Sol-gel PEBBLE sensor response is determined from the ratio (R) of the fluorescence intensities of $[Ru(dpp)_3]^{2+}$ to Oregon Green 488-dextran. The overall gas phase quenching response, Q_G, is given by:

$$Q_G = (I_{N2} - I_{O2}) \, / \, I_{N2} \qquad\qquad\qquad [18]$$

where I_{N2} and I_{O2} denote intensities in 100% N_2 and 100% O_2, respectively. The measured value of Q_G for the sol-gel PEBBLEs is ~92%.

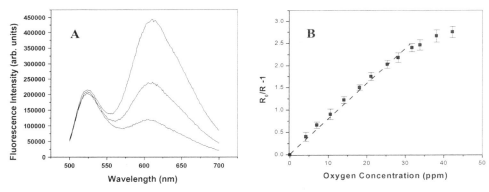

Figure 4.15. A. Aqueous phase emission spectra of sol-gel oxygen PEBBLEs excited at 488 nm: top line: PEBBLE solution purged with N_2; middle line: PEBBLE solution purged with air; bottom line: PEBBLE solution purged with O_2. B. Stern-Volmer plot of relative fluorescence intensity ratios for ratiometric sol-gel oxygen PEBBLEs in the aqueous phase.

While sol-gel-based oxygen PEBBLEs have retained the sensing properties of thin film sol-gel sensor for gas phase sensing, for biological applications the main interest is in solution-based oxygen sensing. [38] Figure 4.15A shows the response of the TEOS-based sol-gel PEBBLEs to dissolved oxygen (DO). The data were taken on a FluoroMax-2 spectrofluorometer (ISA Jobin Yvon-Spex, Edison, NJ), with slits set to 2 nm for both the emission and excitation. The Q_{DO} for the sol-gel oxygen PEBBLEs is ~80%. (Quenching response to dissolved oxygen, Q_{DO}, is defined in a similar way to Q_G, where I_{N2} (intensity at 100% N2) and I_{O2} (Intensity at 100% O2) are replaced by I in fully deoxygenated water and I in fully oxygenated water, respectively). This value represents a reduction in performance vis-à-vis gaseous oxygen; however it is a great improvement with regards to TEOS-based sol-gel films. [56] These sol-gel films had an excellent response to oxygen in the gas phase (Q_G = 90%), but a poor quenching response to dissolved oxygen (Q_{DO} = 20%). MacCraith and co-workers have subsequently reported significant improvements to the Q_{DO} ratio by preparing organically modified sol-gel (Ormosil) films using methyltriethoxysilane (MTEOS) and ethyltriethoxysilane (ETEOS) as the precursors. The success of the Ormosil films in raising the Q_{DO} ratio to 70-80% is largely attributed to the increased hydrophobicity of the film which reduced the water solubility in the film and enhanced the partitioning of oxygen out of solution and into the film. [57] It is thought that the Q_{DO} response of the TEOS sol-gel PEBBLEs (similar to Ormosil films) might be caused by the PEG content of the sensing matrix playing a role analogous to the Ormosil precursors and thus partitioning the oxygen preferentially into the sol-gel PEBBLEs. It is well known that oxygen has a higher solubility in organic liquids than in water, [58] so it should dissolve much better in an organic phase compared to an aqueous phase. In summary, doping the sol-gel PEBBLEs with PEG adds organic components to the sensing matrix, thus encouraging the partitioning of oxygen into the matrix and increases the accessibility of oxygen to the entrapped indicator dye molecules.

The Stern-Volmer plot of fluorescence intensity ratios to oxygen concentrations is shown in Figure 4.15B. The data were taken using an Olympus inverted fluorescence microscope with an Hg lamp (Olympus, Melville, NY) as the light source. Although the performance of the sol-gel PEBBLEs is reduced in the aqueous phase, as opposed to the gas phase, the sensors still demonstrate good reversibility and reproducibility. [38] The dashed line in Figure 4.15B shows the extent of the biologically relevant regime of oxygen concentrations. We note that in this regime, the Stern-Volmer plot is quasi-linear (r^2 = 0.988). Also, as demonstrated in Figure 4.16, the reversibility of the oxygen PEBBLEs was tested and the sensors showed at least 95% recovery each time that the sensing environments were changed among air-, O_2- or N_2-saturated sensor solutions. [38]

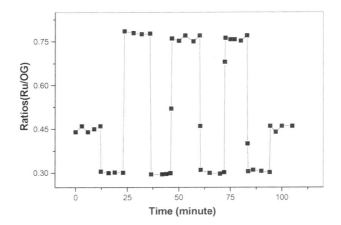

Figure 4.16. Reversibility of PEBBLE sensor response to dissolved oxygen. Alternating measurements were taken in air saturated, O_2 saturated and N_2 saturated PEBBLE sensor solutions.

4.3.2.3. *Polyacrylamide-based Glucose PEBBLE Sensor*

Among the different kinds of PEBBLE sensors developed to date, the newly developed polyacrylamide-based glucose PEBBLE sensor is the first bio-sensor that has been produced by the PEBBLE technique in the sense that its fabrication and sensing mechanism involve the encapsulation of biological molecules (the glucose oxidase enzyme) within the sensing matrix. The glucose sensitive PEBBLEs were made using the general fabrication method for polyacrylamide PEBBLEs described in section 4.2.1. The sensors incorporate glucose oxidase (GO_x), an oxygen sensitive fluorescent indicator $(Ru[dpp(SO_3Na)_2]_3)Cl_2$, and an oxygen insensitive fluorescent dye, Oregon Green 488-dextran, as a reference for the purpose of ratiometric intensity measurements. The enzymatic oxidation of glucose to gluconic acid results in the local depletion of oxygen, which is measured by the oxygen sensitive ruthenium dye. The acrylamide PEBBLE matrix enables the synergistic approach in which there is a steady state of oxygen consumption due to enzymatic oxidation of glucose. This cannot be achieved by separately introducing free enzyme and fluorescent dyes into a cell.

The typical emission spectra of ratiometric glucose PEBBLEs are similar to what is shown in Figure 4.15A for the sol-gel-based oxygen PEBBLEs, since essentially the same fluorescent dyes were used for both kinds of the sensors. The common glucose-sensing scheme involves the employment of glucose oxidase, which catalyzes the oxidation of glucose according to Equation 19 and 20. The measurement of the reduced oxygen level, when glucose is being oxidized, serves as an indirect indication of the glucose concentration. The oxygen sensitive ruthenium dye reports the change in oxygen concentration via a change in fluorescence. The reference dye is present as a marker to ensure that fluctuations in illumination intensity do not cause erroneous results to be recorded.

$$\text{D-glucose} + O_2 \xrightarrow{\text{GOx}} \text{D-gluconolactone} + H_2O_2 \qquad [19]$$

$$\text{D-gluconolactone} + H_2O \rightarrow \text{D-gluconic acid} \qquad [20]$$

Shown in Figure 4.17A are calibration data for the glucose sensitive PEBBLEs, which were taken on a FluoroMax-2 spectrofluorometer (ISA Jobin Yvon-Spex, Edison, NJ), with slits set to 2 nm for both the emission and excitation. The analytical parameter on the Y-axis is the increase in the PEBBLE fluorescence intensity ratios between $(\text{Ru}[\text{dpp}(SO_3Na)_2]_3)Cl_2$ and Oregon Green 488-dextran, and it is plotted against the glucose concentrations. This increase is defined as $(R_G - R_B)/R_B$, where R_B and R_G denote the fluorescence intensity ratio (indicator dye over reference dye) of the glucose PEBBLEs in air-equilibrated plain buffer and that of glucose PEBBLEs in air-equilibrated buffer containing glucose in a certain concentration, respectively. As can be seen, the dynamic range (range within which 95% of the total signal change occurs) of the glucose polyacrylamide PEBBLEs is ~0.3-8 mM, with a linear range between 0.3 and 5 mM, which has a correlation coefficient of 0.997. A deviation from linearity is observed at glucose concentrations higher than ~5 mM. This could be due to either enzyme saturation or a depletion of oxygen within the sensing region. The nature of the calibration curve in Figure 4.17A supports the diffusion-limited condition of the oxygen-based enzymatic reaction, since the non-linearity starts at relatively low substrate concentrations. [59]

The reversibility of the polyacrylamide glucose nanosensors is illustrated in Figure 4.17B; the sensing environment was changed between 0.26 and 3.92 mM glucose. In this experiment, initially a glucose solution was added to a glucose-free, air-saturated PEBBLE buffer solution (pH=7.2, in a glass chamber), to obtain a starting glucose concentration of 0.26 mM. More glucose solution was added to the PEBBLE solution to bring the concentration up to 3.92 mM. The solution was then diluted with buffer to a glucose concentration of 0.26 mM again. After the sensor response stabilized, enough glucose solution was added to bring the concentration of the solution back to 3.92 mM. As shown in Figure 4.17B, the sensors were found to be effectively reversible in this process, each time the sensing environment was changed between 0.26 and 3.92 mM. It should be noted that, due to the limited volume of the glass chamber and restrictions of the experimental setup, it was not possible to further dilute the PEBBLE solution to get more oscillations between 0.26 and 3.92 mM glucose, for this reversibility test. Figure 4.17B also indicates the sensor response time to increasing and decreasing glucose concentrations. As can be seen, in going from 0.26 to 3.92 mM glucose the response times (t_{100}) were ~150-200 s (a slightly longer response time was observed after the dilution of the sensor solution with the addition of buffer). In the reverse direction, the response time was ~100 s. This response time is comparable to other glucose biosensors based on optical oxygen transduction. [60]

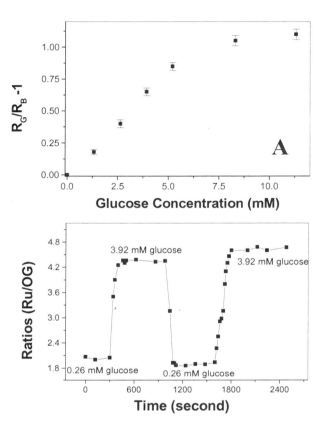

Figure 4.17. (A) Calibration curve for the glucose nanosensors. All measurements were performed in air-saturated phosphate buffers at pH 7.2 and at room temperature. R_B and R_G denote the fluorescence intensity ratio (indicator dye over reference dye) of the glucose PEBBLEs in air-equilibrated plain buffer and that of glucose PEBBLEs in air-equilibrated buffer containing glucose in a certain concentration, respectively. (B) Reversibility and response time of the glucose PEBBLEs.

4.4. PEBBLE SENSORS AND CHEMICAL IMAGING INSIDE LIVE CELLS

In this section, we introduce four different methods that have been used for the delivery of PEBBLE sensors into cells and discuss several examples of intracellular applications,using PEBBLEs based on all three different matrices (polyacrylamide, PDMA, and sol-gel silica).

4.4.1. PEBBLE Delivery Methods

One of the most important considerations when applying PEBBLE nano-sensors to single cell studies is the delivery of the PEBBLEs into the cell. The many delivery methods that have been explored include gene gun, picoinjection, liposomal delivery, and sequestration (phagocytosis and pinocytosis) into macrophages. All of these methods will be detailed in this section and are summarized in Figure 4.18.

Up to now, the gene-gun delivery method has been most commonly used for successful PEBBLE delivery into cells, and it can best be thought of as a shotgun method (Figure 4.18A). Sample preparation for the particle delivery system requires dispersion of the PEBBLEs in ethanol or water and the careful application of a thin layer of PEBBLEs onto the carrier (delivery) disk. This disk is set below a rupture disk. Helium pressure is then built up behind the rupture disk, which ruptures at a specific helium pressure and propels the PEBBLEs from the carrier disk into adherent cells in a culture dish (sitting at a certain distance below the carrier disk) in a shotgun fashion. The gene gun can be used to deliver one to thousands of PEBBLEs per cell into a large number of cells very quickly (dependent on the concentration of PEBBLEs on the delivery disk). [21-23, 34, 38] Cell viability is excellent, 98% viability compared to control cells, [22] for small numbers of PEBBLEs, and hinges directly on the number of PEBBLEs delivered, the delivery pressure, and the chamber vacuum. In gene gun delivery, it is not easy to direct the sensors to a particular region of the cell since the PEBBLEs are shot in a random pattern. However, changing the firing pressure or the distance between the carrier disk and the cells will result in PEBBLEs with different momenta. By controlling this momentum, PEBBLEs will selectively stay either in the nucleus or in the cytosol of the cells.

Picoinjection is used to inject picoliter (pl) volumes of PEBBLE containing solution into single cells (Figure 4.18B). This method of delivery is dependent on the fabrication of pulled capillary "needles", through the use of a pipette-puller and a micro-forge. The smallest volume deliverable is 10 pl and the most concentrated PEBBLE solution to work in the pulled capillary syringe is 5 mg/ml PEBBLEs. The maximum number of PEBBLEs one can put in is dependent on the volume of solution that can be injected without damaging the cell. Picoinjection can give a wide range of PEBBLE concentrations in the cell, and cell viability is good (if done by an expert), but because each cell must be individually injected, the method is time consuming and tedious. [22]

Commercially available liposomes can also be used to deliver PEBBLEs into cells (Figure 4.18C). The liposomes are prepared in a solution of PEBBLEs and then placed in the cell culture where the liposomes fuse with the cell membranes and empty their contents (the PEBBLE containing solution) into the cell cytosol. Three factors play a key role in determining the number of PEBBLEs delivered to each cell with this method: the original concentration of the PEBBLEs, the concentration of liposomes placed in the cell culture, and the length of time the liposomes are left with the cells. [16, 22, 23] The parameters must be tailored for each cell line used in order to obtain the desired concentration of PEBBLEs in the cells. While it would be difficult to deliver a single PEBBLE to each cell with this method, it does seem that a low end of between 10–50 PEBBLEs per cell would be possible, with the high end being the maximum number of PEBBLEs the cell could take without losing viability. Liposomal delivery is useful for delivering PEBBLEs to a lot of cell cytosols simultaneously. The challenge is in tailoring the delivery, for the concentrations desirable and for the cell line being used. Cell viability is excellent. Obviously, the PEBBLE size needs to be small enough for this method, and delivery is essentially limited to the cell cytoplasm.

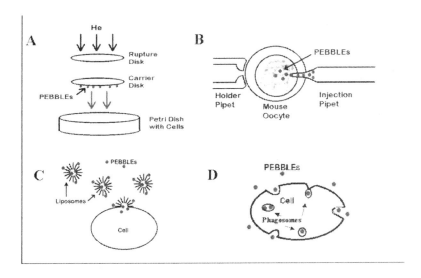

Figure 4.18. Schematic representation of PEBBLE delivery methods. A) Gene gun delivery, B) Picoinjection, C) Liposomal delivery and D) Phagocytosis.

Macrophages, a specialized immune system cell, take up PEBBLEs automatically (Figure 4.18D). The number of PEBBLEs that each macrophage takes up is dependent on the concentration of the PEBBLE solution and the amount of time the macrophages are allowed to stay in the PEBBLE solution. The advantage of this delivery method is that one can easily deliver varying concentrations of PEBBLEs to macrophages. The disadvantages are that it is mainly useful for macrophages (which are hard to culture) and that PEBBLEs are only internalized into certain cell regions. This method also provides excellent cell viability. [22]

4.4.2. Typical Examples of Biological Applications of PEBBLE Nanosensors

The ability of PEBBLE sensors to measure intracellular analytes has been demonstrated in mouse oocytes and rat alveolar macrophage, as well as in neuroblastoma, myometrial and glioma cells. [21-23, 26, 34, 38, 51, 61] All currently applied PEBBLE technologies have relied on fluorescence emission ratios for signal transduction. Cell imaging was accomplished with either an Olympus FluoView 300 scanning confocal microscope system equipped with an Ar-Kr and He-Ne (or similar system) or an Olympus inverted fluorescence microscope, IMT-II (Lake Success, NY), using Nikon 50mm f/1.8 camera lenses to project the image into an Acton 150mm spectrograph (Acton, MA) with spectra read on a Princeton Instruments, liquid nitrogen-cooled, 1024 X 256 CCD array (Trenton, NJ). The Olympus fluorescent microscope (or similar system with an upright microscope) was used to obtain PEBBLE spectra. Following are examples of real time cell imaging, using standard fluorescent microscopy techniques.

4.4.2.1. Polyacrylamide-based Calcium PEBBLEs

The first PEBBLEs produced were polyacrylamide-based, and were first shown successful in intracellular measurements with macrophages. Alveolar macrophages were obtained from rats' lung lavage with Krebs-Henseleit buffer. Macrophages were maintained in a 5% CO_2, 37°C incubator in Dulbecco's Modified Eagle Medium (DMEM) containing 10% fetal bovine serum and 0.3% penicillin, streptomycin and neomycin. PEBBLE suspensions ranging from 0.3-1.0 mg/ml were prepared in DMEM and incubated with alveolar macrophage overnight. Macrophage images were then taken on a confocal microscope and spectra of the same cells were obtained on the fluorescence microscope. Polyacrylamide PEBBLEs selective for calcium (containing Calcium Crimson) were used in order to monitor calcium in phagosomes within rat alveolar macrophage, because of the ease in which macrophage phagocytose particles. This method for delivering the PEBBLEs into cells provided a simple, yet important, test of the PEBBLE sensors in a challenging (acidic) intracellular environment. Macrophage that had phagocytosed 20 nm calcium-selective PEBBLE sensors (Figure 4.19A-B) were challenged with a mitogen, Concanavalin A (Con A), inducing a slow increase in intracellular calcium, which was monitored over a period of 20 minutes (Figure 4.19C). PEBBLE clusters confined to the phagosome enabled correlation of ionic fluxes with stimulation of this organelle.

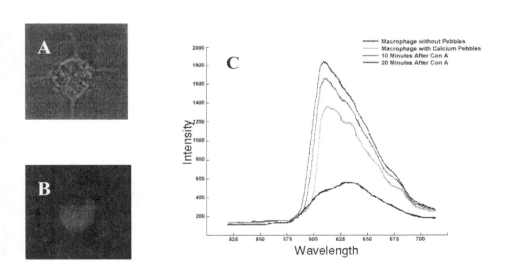

Figure 4.19. (Please see Color Inserts Section) A rat alveolar macrophage with calcium PEBBLE sensors (60x). A) Nomarski illumination and B) Fluorescence illumination C) The increasing intracellular calcium, monitored by calcium PEBBLEs in alveolar macrophage following stimulation with 30 µg/ml concanavalin A.

The calcium PEBBLEs in the macrophage experiment clearly demonstrate a time resolved observation of a biological phenomenon within a single, viable cell. One can clearly obtain relevant time domain data with a fluorescence microscope, spectrograph and CCD. With a confocal microscope system and the appropriate dye/filter sets, one can attain both temporal and spatial resolution, as demonstrated below.

Calcium PEBBLEs have been used, containing "Calcium Green-1" (Molecular Probes) dye, in combination with sulforhodamine dye, as sensing components. Calcium Green fluorescence increases in intensity with increasing calcium concentrations while the sulforhodamine fluorescence intensity remains unchanged, regardless of biologically relevant concentration of ions, pH, or other cellular component. Thus, the ratio of the Calcium Green/sulforhodamine intensity gives a good indication of cellular calcium levels regardless of dye, PEBBLE concentration, or fluctuations of light source intensity. Figure 4.20 shows a confocal microscope image of human C6 glioma cells containing calcium green/sulforhodamine PEBBLEs. The PEBBLEs were delivered by liposomes to the cytoplasm of the cells. In the image, the sulforhodamine fluorescence shows up red (reference peak) and calcium green (green fluorescence) increases with increasing calcium intensity (both dyes confined in PEBBLEs). The toxin, m-dinitrobenzene (DNB), was introduced to the left side of the image and allowed to diffuse to the right, which caused calcium PEBBLEs inside different cells to 'light up' from left to right in time. As a result, high resolution in both the spatial and temporal domains was obtained. The effect of DNB is to cause a disruption of the mitochondrial function, resulting in the uncontrolled release of calcium, which is associated with the onset of a mitochondrial permeability transition (MPT). [23] Calcium PEBBLEs were also used to determine that the half-maximal rate of calcium release (EC_{50}) occurred at a 10-fold lower concentration of m-DNB in human SY5Y neuroblastoma cells, compared to human C6 glioma cells. [23]

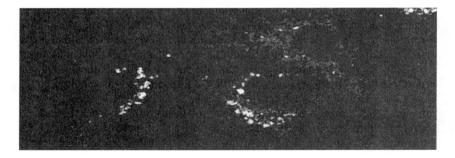

Figure 4.20. (Please see Color Inserts Section) Confocal microscope image of human C6 glioma cells containing Calcium Green/sulfarhodamine PEBBLEs (DNB moving left to right).

4.4.2.2. *PDMA-based Potassium PEBBLEs*

PDMA PEBBLEs were delivered into rat C6-glioma cells, using a BioRad (Hercules, CA) Biolistic PDS-1000/He gene gun system, with a firing pressure of 650 psi, and a vacuum of 15 Torr applied to the system. Immediately following PEBBLE delivery, cells were placed on an inverted fluorescent microscope. The gating software for the CCD was set to take continuous spectra at 1.3 s intervals. After 20 s, and after 60 s, 50 µl of 0.4 mg/ml kainic acid was injected into the microscope cell. Kainic acid is known to stimulate cells by causing the opening of ion channels. Confocal microscopy was used to determine the localization of the PEBBLE sensors after gene gun delivery. [34] Figure 4.21 (inset) shows the confocal fluorescent image of the PEBBLEs, overlaid with a Nomarski differential interference contrast image of the cells. The image indicates that the PEBBLE sensors are localized in the cytoplasm of the glioma cells. Figure 4.21 (graph) shows the PEBBLE sensors inside the cells responding to kainic acid addition to the cell medium, after 20 and 60 seconds. One can see that log (aK$^+$/aH$^+$) increases, indicating either an increase in K$^+$ concentration or a decrease in H$^+$ concentration (increase in pH).

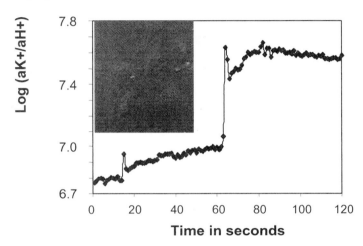

Figure 4.21. Ratio data of poly(decyl methacrylate) K$^+$ PEBBLEs. In C6-glioma cells during the addition of kainic acid (50 mL of 0.4 mg/ml) at 20 s and 60 s. Ratios were converted to log (aK$^+$/aH$^+$) using solution calibration of the PEBBLEs. Log (aK$^+$/aH$^+$) increases after kainic acid addition (and subsequent K$^+$ channel openings). *Inset.* Confocal image of poly(decyl methacrylate) K$^+$ PEBBLE fluorescence overlaid with Nomarski image of rat C6-glioma cells. 488 nm excitation, 580 nm long pass filter.

The amount of kainic acid added is not known to affect the pH of cells in culture and kainic acid by itself has no effect on the sensors. Therefore, the change is likely due to increasing intracellular concentration of K$^+$, which is the expected trend. The membrane of C6-glioma cells can initiate an inward rectifying K$^+$ current, induced by specific K$^+$ channels, a documented role in the control of extracellular potassium. [62] Thus, when stimulated with a channel opening agonist, the K$^+$ concentration within the glioma cells is expected to increase.

4.4.2.3. *Sol-gel-based Oxygen PEBBLEs*

Sol-gel silica, a relatively new PEBBLE matrix, provides the flexibility of being able to be tailored to the properties of the matrix, to accept either hydrophilic or hydrophobic dyes. It has also been proven as a matrix compatible with the use of protein-based sensors. [37] Using the gene gun, ratiometric sol-gel PEBBLEs, with $[Ru(dpp)_3]^{2+}$ as oxygen-sensitive dye and Oregon Green 488-dextran as reference dye, were inserted, under conditions similar to the PDMA PEBBLEs delivery protocol, into rat C6-glioma cells, in order to monitor intracellular oxygen. Figure 4.22 shows the confocal images of C6 glioma cells containing sol-gel PEBBLEs under Nomarski illumination overlaid with: (A) The green fluorescence of Oregon Green 488-dextran and (B) the red fluorescence of $[Ru(dpp)_3]^{2+}$. It can be seen that the cells still maintained their morphology after the gene gun injection of PEBBLEs and showed no sign for cell death. The dyes were excited, respectively, by reflecting the 488 nm Ar-Kr and the 543 nm He-Ne laser lines onto the specimen, using a double dichroic mirror. The Oregon Green fluorescence from the PEBBLEs inside the cells (Figure 4.22A) was detected by passage through a 510 nm long-pass and a 530 nm short-pass filter, and the fluorescence of $[Ru(dpp)_3]^{2+}$ (Figure 4.22 B) through a 605 nm (45 nm band-pass) barrier filter. A 40X, 1.4 NA oil immersion objective was used to image the Oregon Green and $[Ru(dpp)_3]^{2+}$ fluorescence. The distribution of PEBBLEs shown in the overlaid images demonstrates that the green and red fluorescence in Figure 4.22A and Figure 4.22B were truly from PEBBLEs inside cells. It should be noted that most of the PEBBLEs were loaded into the cytoplasm, but there were also some in the nucleus.

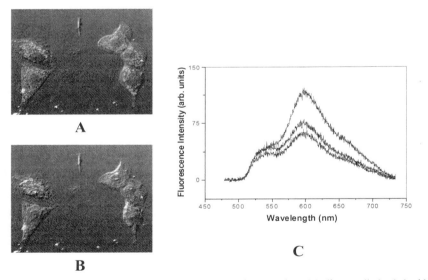

Figure 4.22. (Please see Color Inserts Section) Confocal images of rat C6 glioma cells loaded with sol-gel oxygen PEBBLEs by gene-gun injection. (A) Nomarski illumination overlaid with Oregon Green fluorescence of PEBBLEs inside cell. (B) Nomarski illumination overlaid with $[Ru(dpp)_3]^{2+}$ fluorescence of PEBBLEs inside cells. (C) Fluorescence spectra of a typical ratiometric sensor measurement of molecular oxygen inside rat C6-glioma cells; bottom line: cells (loaded with sol-gel PEBBLEs) in air-saturated DPBS; middle line: cells in N_2-saturated DPBS, 25 seconds after replacing the air-saturated DPBS; top line: cells in N_2-saturated DPBS, after 2 minutes.

The spectra in Figure 4.22C shows the response of oxygen sensitive sol-gel PEBBLEs, inserted inside rat C6 glioma cells, to changing intracellular oxygen concentrations. After gene gun injection, the cells were immersed in DPBS (Dulbecco's Phosphate Buffered Saline) and a spectrum was taken of these cells, using 480 ± 10 nm excitation light. The air-saturated DPBS was then replaced by nitrogen-saturated DPBS, to cause a decrease in the intracellular oxygen concentration, and the response of the oxygen PEBBLE sensors inside the cells was monitored during a time period of 2 minutes. As can be seen, the fluorescence intensity of $[Ru(dpp)_3]^{2+}$ went up successively as the oxygen level inside the cells decreased. Average intracellular oxygen concentrations were determined on the basis of a Stern-Volmer calibration curve, obtained using the fluorescence microscope-Acton spectrometer system, and are summarized in Figure 4.4. The comparatively large errors are due to the low resolution of the spectrometer. We note that our measured intracellular oxygen value (when cells were in air saturated DPBS) is comparable with the previously reported value of ~7.1 ppm measured inside islets of Langerhans. [63] These results show that the PEBBLE sensors are responsive when loaded into cells and that they retain their spectral characteristics, enabling a ratiometric measurement to be made. [38]

Table 4.4. Real time measurements of intracellular oxygen

Average intracellular oxygen concentrations (ppm)	
Cells in air-saturated buffer	7.9 ± 2.1
Cells in N_2-saturated buffer (after 25 sec)	6.5 ± 1.7
Cells in N_2-saturated buffer (after 120 sec)	≤ 1.5
Air saturated buffer solution	8.8 ± 0.8

4.5. ADVANTAGES AND LIMITATIONS OF PEBBLE SENSORS

As can be seen, all PEBBLE sensors described in this chapter are designed to be ratiometric in the emission mode. Up to now, most of the fluorescent dye-based optical sensors rely on measurements of a single fluorescence peak intensity. These sensors are known to be problematic in most practical applications due to signal fluctuations not directly caused by changes in the analyte concentration, e.g. light scattering by the sample and/or excitation source fluctuations. [64] Ratiometric PEBBLE sensors have been one answer to the problems posed by intensity measurements[2-4, 65], since they compensate for the effect of these factors by taking the ratio of the indicator peak intensity over the reference peak intensity. Another solution has been the measurement of fluorescence lifetimes[54, 65-67]; however, ratiometric methods are experimentally simpler than lifetime measurements.

As mentioned above, the advantage of using ratiometric sensors is the exclusion of factors such as light scattering by the sample and excitation source fluctuations. A typical example is given here by performing a ratiometric check on a polyacrylamide-based glucose PEBBLE suspension (with or without glucose in the solution) in a spectrofluorometer. In this experiment, the excitation slit width on the fluorometer was varied between 0.5-2 nm to obtain changes in the illumination intensities while the emission slit width remained constant, and the resulting fluorescence intensities of the indicator and reference dyes were recorded. During the experiment, it was found that the absolute intensity of fluorescence emission of each dye decreased (by a factor of ~50 eventually) with decreasing illumination intensity (due to decreasing width of the excitation slit), but the ratios between peak intensities of these two dyes remained essentially constant, as shown in Figure 4.23. This result demonstrates that the ratiometric method is substantially more reliable than single-intensity-based measurements, because it can eliminate the effects resulting from excitation power fluctuations, which cannot be avoided in single-intensity-based measurements.

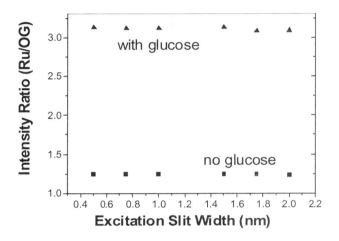

Figure 4.23. A 5 mg/ml glucose PEBBLE suspension in 10 mM phosphate buffer (with and without glucose) was monitored on a fluorometer. The width of the excitation silt was varied to obtain different excitation intensities. From the emission spectra, the fluorescence intensity ratios were calculated, and were found to remain essentially constant under dramatic changes in the excitation powers.

As mentioned, important advantages of the PEBBLE sensor include the separation of the cell and sensing elements, which provides protection for both from perturbation by the other, and the ability to combine components to accomplish complex sensing schemes. The greatest advantage, from a sensing standpoint, is that all of this is accomplished in a sensor with nanometer dimensions.

The advantage gained by protecting the cell from the sensing elements is self evident, especially when the selective sequestration of some dyes into cellular organelles is considered. What may be less intuitive is the interference of cellular components, especially protein, with the function of sensing dyes. [11] The comparison of the function of free dyes to both polyacrylamide and sol-gel PEBBLEs utilizing the dyes as sensing elements clearly demonstrates the advantage of preventing perturbations by macro-molecules on sensing components. The matrix protects the entrapped dyes from the intracellular environment, preventing interference with the fluorescent properties of the dyes. Without this shielding of a dye, its fluorescence would behave unpredictably inside a given cell, making calibration of even ratiometric dyes difficult or impossible.

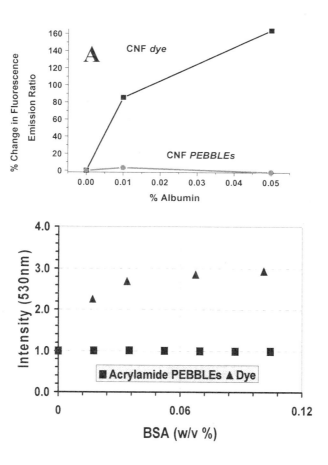

Figure 4.24. Effect of protein on sensors. **A)** Adding as little as 0.01% albumin to a solution of CNF dye molecules causes almost a 90% change in the fluorescence intensity ratio of this pH sensitive dye, even though the pH of the solution remains constant. Under the same conditions the PEBBLEs containing the CNF dye are not affected by the addition of albumin. **B)** Peak emission intensity of Newport Green, 530-nm, monitored on a fluorometer. Spectra are acquired after each successive aliquot of a 10% (w/v) bovine serum albumin solution. The BSA concentrations are plotted versus the peak intensities. As little as 0.02% BSA causes over a 200% increase in Newport Green dye intensity, but the intensity of the Newport Green embedded in polyacrylamide PEBBLE remains unchanged, even at BSA concentrations above 0.10%.

One example of the effect of protein binding is demonstrated with 5-(and 6-) carboxynaphthofluorescein (CNF) pH dye. CNF is a highly photostable, inexpensive and internally ratiometric dye for pH, which is not used for intracellular applications because of the error induced by macromolecule binding. However, protected in a polyacrylamide matrix, CNF becomes a viable tool for intracellular pH study. Figure 4.24A illustrates the benefit of entrapping CNF in an acrylamide matrix (PEBBLE). Incubation of the free dye with as little as 0.01% albumin induces alterations in emission ratios of almost 90% (pH was maintained constant as measured with a standard pH electrode) which is an error equivalent to 1 pH unit. The same dye, protected in the PEBBLE, shows minimal perturbation by albumin; the resulting error is equivalent to about 0.01 pH units. [16]

Another example of the benefit due to sensing element protection by the polyacrylamide matrix is found when comparing the zinc PEBBLE (based on Newport Green) to naked Newport Green in bovine serum albumin (BSA) solution. Aliquots of the BSA solution were added to either a 3 mg/mL PEBBLE suspension or a 125 nM Newport Green dye solution and the resulting fluorescence was monitored. Although Newport Green has good selectivity over intracellular ions, the dye itself is prone to artifacts resulting from non-specific binding of proteins, such as (BSA), as shown in Figure 4.24B. Monitoring the peak of Newport Green at 530 nm, there is a substantial increase in the peak intensity with each successive addition of BSA. In addition to the increase of intensity, there is also a 4-nm shift in peak wavelength. The PEBBLEs containing the Newport Green dye, however, are unaffected by the additions of BSA, which is confirmed by the peak's wavelength and intensity both remaining the same. As little as 0.02% BSA causes a major change in the Newport Green dye intensity (i.e. >200% increase) but the intensity of the Newport Green embedded in the sensor remains unchanged, even at BSA concentrations above 0.10%. [25]

As expected, it has been found that the sol-gel matrix of PEBBLE sensors also prevents macromolecules, such as proteins, from diffusing through the matrix. As with the polyacrylamide PEBBLEs, the effects of non-specific protein binding have also been investigated by the addition of BSA, and sol-gel oxygen PEBBLEs were used as an example. The results show that adding as little as 0.14% BSA to a solution containing $[Ru(dpp)_3]^{2+}$ (indicator dye for oxygen) and Oregon Green 488-dextran dye (reference dye), with the same molar ratio in solution as inside the sol-gel oxygen PEBBLEs, changes the fluorescence intensity ratio of the two dyes by a factor of more than 2.3 (i.e. an increase of over 130%). This change is mostly due to the change in the fluorescence intensity of $[Ru(dpp)_3]^{2+}$ after the addition of BSA. However, under the same conditions, the PEBBLE sensors containing these two dyes are not affected by the addition of BSA and a change in fluorescence intensity ratio of at most 4% is observed when even an increased concentration of BSA (0.23%) is added, demonstrating the same protection as given to dyes in the polyacrylamide matrix. The susceptibility of the sol-gel PEBBLEs to perturbation by heavy metal ions (Hg^{2+} and Ag^+) and to one of the notorious collisional quenchers (I^-) were also examined. $Hg(NO_3)_2$, $Ag(NO)_3$ and KI were added to a PEBBLE solution and to a free dye solution of $[Ru(dpp)_3]^{2+}$ (same concentrations as in PEBBLEs) up to a concentration of about 200 ppm. There was a 5-10% decrease in the fluorescence intensity of the $[Ru(dpp)_3]^{2+}$ free dye each time, while for the PEBBLEs no measurable effect could be observed. [38]

There is not a similar "naked" dye/PEBBLE protected dye comparison for liquid polymer PDMA-based PEBBLEs. This is because the hydrophobic components cannot even be used "naked" in aqueous solution. The separate phase (the liquid polymer) is essential for the PDMA matrix based sensing mechanism and allows for the complex sensing schemes that are based on the ion-exchange or co-extraction mechanisms. [1]

The nanometer dimensions of the PEBBLEs of all three matrices give a useful advantage over traditional, monolithic, optodes in terms of response time. In order to follow biological perturbations in real time, a fast response is required from the PEBBLE sensors. Most PEBBLE sensors depend on bulk-equilibrium, between sensor and solution phase (the oxygen sensor depends on steady-state). The diffusion rate of analyte inside polyacrylamide and sol-gel matrix is expected to be similar to that in aqueous solution phase, while the diffusion of analyte in the hydrophobic PDMA matrix should be much slower. However, in all cases, the small size of the PEBBLE sensors gives a rapid response time, despite the need for bulk equilibrium. Figure 4.25 shows the response times of the various matrices.

The response time of polyacrylamide sensors to Ca^{2+} was determined reliably using an Olympus IX50 inverted microscope equipped with a mercury arc lamp and a PMT. Calcium selective probes were premixed with a caged calcium ion, and this solution was inserted into a quartz capillary. The calcium was uncaged with a pulse of UV light from a Quanta-Ray 10 ns Nd/Yag laser (Quanta-Ray, Mountainview, CA) equipped with a frequency tripler and coupled into an optical fiber positioned over the capillary. In Figure 4.25A, the response time of the PEBBLE sensors was compared to that of the free dye (no polymer matrix), so as to separate the diffusion time through solution from the diffusion time through the matrix. As can be seen from Figure 4.25A, the 90 % response time of the PEBBLE sensor to the increase in free calcium is on the order of one millisecond or less. Theoretically, with an approximate diffusion constant of 10^{-6} cm^2/sec, the average diffusion time should be about ten microseconds for a 100 nm radius sensor, and 100 nanoseconds for a 10 nm radius sensor. [16]

For PDMA based K^+ sensors, using BME-44 as ionophore, ETH5350 as the chromoionophore and KTFPB as the ionic additive, the ratio of the protonated chromoionophore to free base was analyzed vs. time in response time measurements. As shown in Figure 4.25B, in going from log aK^+/aH^+ = 3.6 to 5.7 the response time (10%-90% signal change) is about 0.5s (for a concentration change of over 2 decades). In the reverse direction, the response time is about 0.8s. This fast, sub-second response time of the PEBBLEs is a direct result of their small size. Diffusion in PDMA is in the range of 10^{-8} cm^2/s, with small variations that depend on cross-linker content. [68, 69] Thus, for a PEBBLE radius of about 300 nm, one expects a diffusion time of about 10^{-3} s. This is consistent with the experimental values, which are again upper limit values due to the solution mixing times. [34]

The measured transition times for sol-gel oxygen nanosensors (Figure 4.16) are on the order of 20 to 30s, but these times are much longer than the intrinsic response time of the PEBBLEs, due to the significant contribution of the time used to saturate the solution with O_2 or N_2. It is difficult to measure the exact response time, because changing between oxygenated and deoxygenated PEBBLE solutions takes time. The measured transition times (including the time of saturating the solution) are only an upper limit for the response time. The PEBBLE sensors should intrinsically have shorter response times than previously reported thin film and fiber optic sol-gel sensors (on the order of seconds or minutes) simply because of the smaller sizes of these sensors. According to the Einstein diffusion equation, where $X^2=2D\tau$, a shorter diffusion length X (which is directly related to the size of the sensor) results in a much shorter time for oxygen molecules to diffuse through the sensing matrix (which is basically the response time). A lower limit can thus be estimated, using $D\approx2\times10^{-9}$ m^2/sec (diffusion constant of oxygen in water) and $X\approx3\times10^{-7}$ m, giving $\tau\approx20\times10^{-6}$ sec, i.e. a response time in the microsecond range. An upper limit can be estimated considering that the PEBBLE sensor dimensions are 10-100 times smaller than thin film sensors, and have a spherical shape. This should give the PEBBLE sensors a response time in the millisecond range. [38]

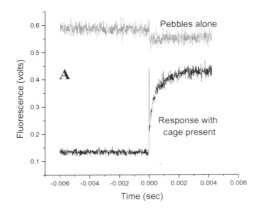

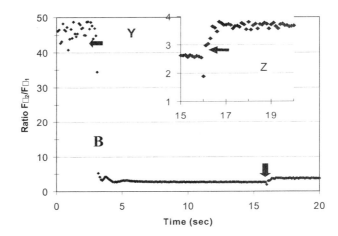

Figure 4.25. (A) Response time of polyacrylamide calcium PEBBLE sensors. Calcium was released using a single 10 ns UV pulse from an Nd/YAG laser, which photolysed the cage, releasing free calcium into a solution of PEBBLEs. The observed response time, less than 1 msec, was indiscernible from that of the corresponding dye not entrapped in a polymer matrix. (B) Response time of K^+ sensitive PEBBLEs to added KCl and added buffer solutions. It can be seen (Y) that response in the forward direction is about 0.5 s (0-40 mM KCl); and in the reverse direction (see inset Z) it is slightly longer 0.8 s (40-20 mM KCl).

Leaching of dye molecules out of the PEBBLE matrix is the greatest limitation of PEBBLE sensors; it is a major concern and is very dependent on the dye and matrix combination. Factors such as the molecular size of the dye (small dyes can more readily diffuse through the pores and leak out of the matrix) and the solubility of the dye in the matrix and in water play a significant role.

PDMA-based potassium sensors have a lifetime of 30 minutes, due to component leaching from the liquid polymer membrane. This is consistent with the lifetime of PVC-based optodes of the same composition. [8] After 30 minutes, the sensor response can deviate up to 7% from the initial calibration data at lower K^+ concentration. After 90 minutes, the deviation is up to 13% at lower K^+ concentrations. The deviations are smaller at larger K^+ concentrations. [34]

Sol-gel silica matrix provides excellent stability with respect to dye leaching. [70] In particular, ruthenium complexes often have excellent stability inside the sol-gel matrix and in agreement with previous reports[57, 67, 70-72] the indicator dye $[Ru(dpp)_3]^{2+}$ shows no significant signs of leaching. For the reference dye, the large size of the dextran molecular backbone to which the Oregon Green dye molecules are bound should greatly reduce leaching. According to the dilution factor, a rough estimate provides an upper limit of 1% for the amount of dye molecules leached out of the sensing matrix over a three-day period. [38]

In summary, what has been determined for three matrices is that the PDMA PEBBLEs have a lifetime of about half an hour, the polyacrylamide PEBBLEs have a lifetime of about 24 hours, while the sol-gel PEBBLEs have a lifetime of more than three days. However, as the PEBBLEs are single use sensors made for quick measurements inside in vitro cells that survive only a short period of time, this is acceptable for most applications. It should also be noted that well documented steps, such as covalently attaching the dye to the matrix polymer backbone or adding cages or lipophilic tails (i.e. using dextran with hydrophilic polymers or adding lipophilic tails to lipophilic dyes), can be used to increase PEBBLE sensor lifetime if and when needed.

4.6. NEW PEBBLE DESIGNS AND FUTURE DIRECTIONS

The next step in PEBBLE sensor technology is to apply the currently used PEBBLE techniques to other biologically interesting analytes. Presently work continues on all three matrices, as well as a new ormosil matrix, [73] to develop sensors for more ions and small molecules of biological relevance. In this section, some newly emerging PEBBLE designs and applications that enhance sensing and combine sensing capabilities with nano-actuators, i.e. nano-sized drug delivery and medical imaging particles are discussed. They serve as typical examples for the potential future directions of PEBBLE techniques. The development of this new class of PEBBLEs is still in progress, but their background and some typical results are given below.

4.6.1. Free Radical Sensors

PEBBLEs are also able to measure free radicals. The response time of PEBBLEs is fast enough to measure the concentrations of transients that would otherwise go undetected due to their short lifetimes.

One successful polyacrylamide-based transient PEBBLE measures the presence of hydroxyl radicals. [74] The hydroxyl radical is one of the most reactive and potentially damaging species found in biological systems. It is primarily formed by reactions involving superoxide and/or hydrogen peroxide and copper or iron ions. A challenge in creating a sensor for the hydroxyl radical is its high reactivity. Because hydroxyl radicals are believed to have an extremely short lifetime, [75, 76] only 1 ns in biological systems, it is extremely difficult to create and calibrate a reliable and precise sensor specific for hydroxyl radical measurements.

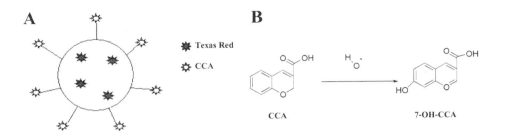

Figure 4.26. A) Hydroxyl radical detecting PEBBLE. B) Reaction of hydroxyl radical with indicator molecule CCA.

As shown in Figure 4.26, the nanoprobe detects the hydroxyl radical when it reacts with a probe molecule specific to ·OH, coumarin-3-carboxylic acid (CCA), which is attached on the outside of the PEBBLE matrix. [74] The product, 7-hydroxy-coumarin –3-carboxylic acid (7-OH-CCA), is highly fluorescent and therefore indicative of hydroxyl radical concentration. In most of the PEBBLEs the dye is entrapped inside the matrix, but for the hydroxyl radical probe, CCA is attached to the outside of the PEBBLE, because the hydroxyl radical is too reactive to enter the polyacrylamide matrix. This direct detection is preferable to another approach that employs a scavenger molecule specific to ·OH. [77, 78] The scavenger reacts with ·OH producing another, longer lived radical. The probe molecule then reacts with the new radical species. The indirect measurement increases response time and is subject to interference if the second radical species is present in situ.

As with most PEBBLEs, a benefit of the nanoprobe ·OH detection system is the inclusion of a reference dye within the sensor matrix. The reference dye, embedded in the PEBBLE matrix and unresponsive to the analyte, provides a means by which the fluctuations of excitation light, PEBBLE concentrations, optical alignment, and eventually cellular interference are easily accounted for. In this particular PEBBLE system, the reference dye is Texas Red. The inclusion of the reference dye in the polymer matrix also protects them from potential destruction by the free radicals.

Increased surface to volume ratios are another advantage of creating a nano-sized probe for hydroxyl measurements. CCA is immobilized on the exterior of the probe; more surface area allows for greater concentration of probe molecules per sensor, allowing for more accurate and reliable results. Development of polyacrylamide PEBBLEs for other transient species is in progress.

4.6.2. MOONs, Tweezers, and Targeting

As discussed, fluorescent PEBBLEs have been developed for various sensing applications. The ability to manipulate the nanosensors, especially in living cells, using noninvasive modulation methods such as light and magnetic fields, greatly enhances their versatility. The modulation allows for exclusion of background signal, controlled sensing of a specific area and other desirable benefits. It is important to discuss briefly the methods of manipulation for these nanoprobes to have a complete view of the potential biomedical and therapeutic implications of PEBBLEs.

4.6.2.1. MOONs

With all the advantages of fluorescent nanoprobes, background fluorescence from samples and instrument optics is still a common problem. In such strong autofluorescence backgrounds, small changes in probe fluorescence tend to be washed out and become nearly undetectable. However, if the probe fluorescence is modulated with respect to background fluorescence, probe fluorescence can be distinguished and separated. This simple procedure increases the signal-to-noise (S/N) or, strictly, signal-to-background ratios by several orders of magnitude.

MOdulated Optical Nanoprobes (MOONs) can be both magnetic[79, 80] (MagMOONs) and Brownian[81] in nature. The magnetically induced periodic motion of the MOON (or random thermal motion in the case of Brownian MOONs) modulates the fluorescence signal, enabling the separation of signal from background. This technique enhances signal-to-noise (S/N) ratios and expands the breadth of applications of PEBBLEs to include samples with highly fluorescent backgrounds or experiments with several fluorescent probes.

4.6.2.1a. MagMOONs.

MagMOONs consist of a polymer or sol-gel silica sphere, metal-coated and with incorporated ferromagnetic material (e.g. Fe_2O_3). The fluorescent, magnetic spheres are placed on a microscope slide and then coated with metal, using vapor-deposited aluminum or sputtered gold, so that one hemisphere is optically coated (Figure 4.27). The metal-coated spheres are then magnetized such that the south side is uncoated. After the particles are suspended in solution and exposed to a magnetic field, they align themselves; if the field is reversed they change direction accordingly. By monitoring the fluorescence from a stationary viewpoint, the MagMOONs appear to be blink "on" and "off" synchronously. It is important to note that while there is some variation in the size of the MagMOONs, the rotation rate is independent of particle size, assuming constant viscosity and magnetic susceptibility. Quenching of the fluorescence by the metallic coating is not an issue, as the coating is not in close proximity to the fluorophores (<10nm).

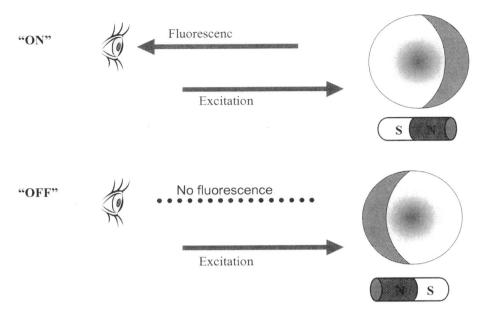

Figure 4.27. MagMOONs align with the magnetic field as the direction of the field is switched, essentially turning the fluorescence "on" and "off" as viewed from a stationary point.

Excitation of the MagMOONs is analogous to that of other PEBBLEs and the emission is collected using a simple CCD or photodiode. A solenoid or permanent magnet is used to modulate the external magnetic field by 180°. For enhanced clarity of the signal modulation, it is preferable for the coating to be much thicker than the skin penetration depth of the excitation and emission light; however, this is not essential. Modulation is still satisfactory with thinner layers.

A variation of MagMOONs, aspherical MagMOONs, behave similarly to the spherical moieties, except that magnetic anisotropy due to the aspherical shape causes the particles to emit differently from the ends than from the sides (a result of self-absorption and internal reflection), as demonstrated in Figure 4.28. Hence, rather than the 180° modulation with the spherical MagMOONs, the aspherical variety have a 90° modulation and are easily distinguished in a mixture of both types, resulting in an even higher degree of signal enhancement by multiplexing.

The aspherical MagMOONs come in two varieties, [79] chain (Figure 4.28A) and pancake MagMOONs (Figure 4.28B). The chain variety is created by linking together several spherical MagMOONs into a longer chain. This can either happen spontaneously under a magnetic field (superparamagnetic particles link together under a magnetic field and then disperse when the field is removed) or by thermally bonding the particles while they are self-aligned under a magnetic field. Thermal bonding by heating the beads over their glass transition temperature creates a more permanent linkage of the MagMOON chains. Other linking methods include: chemical linkage, melting microspheres while in a grooved mold, and briefly heating microspheres as they pass through a microfluidic device.

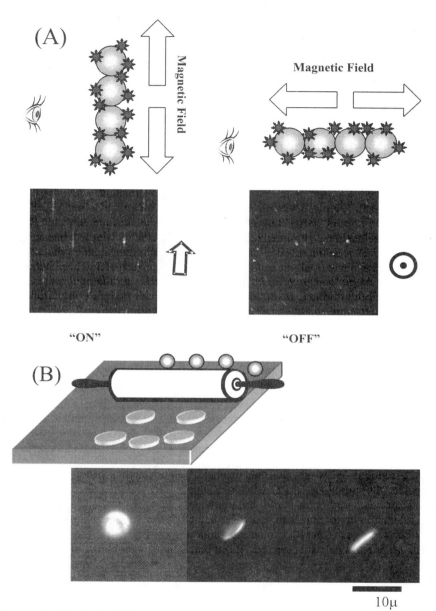

Figure 4.28. A) Chain MOONs. These aspherical MagMOONs are created by linking several spherical particles. Note the difference in fluorescence in the "on" and "off" positions. B)Pancake Moons. Aspherical MagMOONs created by rolling tubing over spherical MOONs. Note the asymmetrical fluorescence as the MOON rotates.

Pancake shaped MagMOONs are created by physical deformation of microspheres. Rolling a piece of glass tubing over the particles, much like using a rolling pin, flattens the spheres. At this time, ferromagnetic "nano-crumbs" can be physically embedded into the particles. "Roll-shaped" MagMOONs can also be created in a similar manner; particles are rubbed between two microscope slides, creating a cylindrical shaped particle. As with the pancake MagMOONs, the "nano-crumbs" are embedded at the time of deformation.

The possibility of creating MOONs with a continuous magnetic modulation, rather than only on/off positions, is on the horizon.

4.6.2.1b. Brownian MOONs.

Brownian MOONs, like MagMOONs, use the principle of signal modulation; however, instead of magnetic modulation used by MagMOONs, the sensors are naturally modulated by random thermal motion, or Brownian motion. [81]

As with MagMOONs, Brownian MOONs are made by coating one side of the sensor sphere with a metal coating, usually ~100 nm thickness of aluminum. This provides the "off" position for the MOONs. During sensing, the coated spheres follow the laws of Brownian motion, tumbling and turning erratically. Consequently, the signal from each sensor appears to blink on and off randomly. The signal can be distinguished and deconvoluted from a constant background of fluorescence, either from the sample or from the autofluoresence of the optics.

The simplicity provided by the Brownian approach gives to the MOON sensors is an enormous incentive for its development. The Brownian MOONs do not require any magnetic fields for sensor modulation, making it a simple system for use in biological applications. Once employed, Brownian MOONs require less complex instrumentation and experimental control.

In measurements using Brownian MOONs, the signal is extracted from sources of noise, autofluorescence and scattering, by using principal components analysis (PCA), a multivariate data analysis technique. More thorough discussions of PCA can be found elsewhere. [82] It should suffice to know that PCA transforms a data set in such a way that any correlation between the signal and background is destroyed; the irrelevant information is discarded, and only the fluorescence correlated to Brownian motion is considered. This deconvolution of fluorescence sources gives a very significant advantage to Brownian MOONs, because it opens the possibility for simultaneous measurements, using two or more sensors with different characteristics (i.e. particle shape or size, differing emission spectra); the signals are simply separated using PCA.

4.6.2.1c. Practical Examples of MOONs in Motion.

Experiments involving both MagMOONs and Brownian MOONs have proved illuminating. As the theory suggests, the background signal is easily eliminated, and the exclusion of background fluorescence simplifies data acquisition.

To illustrate the convenience of MagMOONs in distinguishing between background fluorescence and actual signal, MagMOONs were placed in a common ivy leaf[80] as shown in Figure 4.29. The leaf had a strong constant fluorescent background. Without modulation consideration, the change in overall fluorescence is hardly detectable. However, once the modulation of the MagMOONs is accounted for, the signal from the microscopic MagMOONs is evident enough that an individual MagMOON can be visualized.

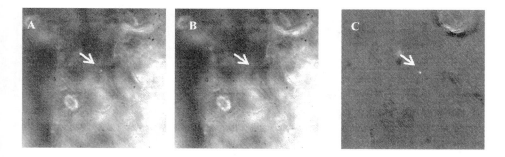

Figure 4.29. MagMOON with fluorescent leaf background. A. MagMOON in "on" position. B. MagMOON in "off" position. C. "On" signal minus "off" signal eliminates background fluorescence.

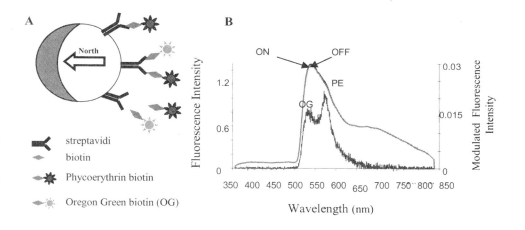

Figure 4.30. Immunoassay using MagMOONs. A) MagMOONs are coated with streptavidin and selectively bind fluorescent labeled biotin. B) Spectra of immunoassay in "on" and "off" positions, which are almost indistinguishable. The difference spectrum (black) shows binding of both Oregon Green biotin and Phycoerythrin biotin.

Modification of the surface with several types of antibodies enable MOON based immunoassays. [79] The ability to separate different fluorescent signals from within the same solution eliminates the need for tedious washing steps often associated with conventional immunoassays. A demonstration of this particular application with the MagMOONs shows that the immunoassay is effective and potentially highly useful for biomedical applications. For the assay, MagMOONs are coated with the antibody streptavidin and then are introduced to a solution containing two types of fluorescent labeled biotin (Figure 4.30A). Biotin is a protein with a particular affinity for streptavidin, and in this example is labeled with either phycoerythrin (PE) or Oregon Green (OG), both fluorescent dyes. Neither label interferes significantly with the binding affinity of biotin. The MagMOONs are monitored in the "on" and "off" positions and the resulting signal is demonstrative of the bound biotin. In the difference spectrum, "on" spectrum minus "off" spectrum, peaks are observed for both PE and OG (Figure 4.30B).

The novelty of the MagMOON immunoassay lies in the fact that no washing step is required. The modulation of the MagMOONs, and by extension, of the bound biotin signal, enables one to easily distinguish the bound protein from any unbound lurking in the background. The MOON modulations also account for other inhomogeneities in the excitation light or other interferences. This approach to immunoassays will allow for several analytes to be detected simultaneously in situ without tedious and time consuming rinsing steps.

Unlike the MagMOONs, the rotation of the Brownian MOONs is dependent on the dimensions of the particles, as well as the local viscosity of the surrounding fluid. By using a well characterized particle, Brownian MOONs can easily be used to measure the local viscosity of intracellular fluids, [81] a highly promising contribution to the field of microrheology, especially inside cells.

4.6.2.2. *Optical Tweezers*

Being able to select and manipulate PEBBLEs or other nanoprobes in different environments is an invaluable asset to nanoprobe technology. By moving a particle to a predetermined detection area, specific locations can be studied exclusively.

Optical tweezers use a focused laser to spatially trap and manipulate particles in the micrometer to nanometer size range. The photons in the focused laser beam impart momentum to the particle as the light reflects and refracts as a result of interaction with the particle. The reflection (or refraction) of the light results in a change in momentum in the propagating laser light and, according to the law of conservation of momentum, some of the momentum must therefore be transferred to the particle.

The imparted momentum results in exerted forces on the particle. The gradient force is a resultant of the Gaussian intensity profile of the beam at the point of focus. The less intense light near the edges of the beam pushes the particle toward the center. Due to the symmetric nature of the beam intensity profile, this acts as a restoring force pulling the particle to the center of the beam. The scattering force pushes the particle in the direction of laser propagation due to the refracted rays. The reflected rays, however, create a scattering force in the opposite direction. When the force incurred by the reflected rays overcomes the force from the refracted rays, the particle becomes "trapped".

For dielectric particles smaller than the wavelength of light, optical trapping becomes difficult and must be enhanced to achieve acceptable trapping strengths. Optical trapping at or near particle resonance frequencies have been shown to enhance the trapping strength by nearly 50 times. [83]

Another advantage of near-resonance optical tweezing is that by choosing a resonance frequency one can selectively trap particles of a particular size. This becomes highly useful in the manipulation of particles in cellular environments. Nanoprobes, in particular, can be tweezed and placed in a specific area of interest within the cell. For example, when monitoring ion-channel activity, it is much more informative to have

PEBBLEs concentrated near the ion channels rather than dispersed uniformly throughout the cell.

4.6.2.3. Targeting

As certain PEBBLEs were primarily designed for use in vivo, ultimately within human patients, spatial control of the PEBBLEs without extraordinarily invasive measures is extremely important. Intelligent placement of PEBBLES is integral in the selective measurement (or therapy) of tissue. A brain tumor, for example, would be best studied by PEBBLEs selectively congregated near the malignancy, rather than by PEBBLEs uniformly distributed throughout the brain. Ideally, a PEBBLE would be designed to inherently gravitate toward the site of interest rather than requiring placement by intrusive means (i.e. surgical implantation).

Differentiated cells have unique membrane markers that distinguish them from other cell types. Usually these cell-surface receptors, essential for normal cell function, are designed to selectively bind free extracellular proteins. Once the natural binding moiety for a receptor is identified, it can be mimicked in the laboratory. When a peptide chain similar to the binding protein is attached to a PEBBLE, it can act as a homing device, binding the PEBBLE to the tissue of interest.

A most promising targeting PEBBLE is the RGD-modified PEBBLE. RGD is a tripeptide (Arg-Gly-Asp) commonly found in adhesive proteins such as fibronectin. When in a cyclic conformation, as in the CDCRGDCFC (RGD-4C), it binds selectively with cells exhibiting the integrins[§] $\alpha v \beta 3$ and $\alpha v \beta 5$. [84, 85] Typically not found in healthy adults, these integrins are present in vasculature undergoing high levels of angiogenesis[86, 87] (blood vessel formation). The elevated growth rate of tumor cells obviously requires angiogenic vasculature, making $\alpha v \beta 3$ and $\alpha v \beta 5$ indicative of a tumor tissue and ideal targets for cancer-related probes.

In the RGD-modified PEBBLE, the amino acid sequence is attached to the exterior of the nanoprobe using a biotin-streptavidin-biotin sandwich. This complex provides a stable base for the RGD as well as a degree of mobility for the sequence; the end chain is free to move appropriately to allow binding with the membrane protein.

Targeting cancer cells in this manner could not only allow real-time tumor supervision, but by coupling the RGD-labeled nanoprobes with a therapeutic agent (see section 4.6.3.2), cancer cells could be specifically treated and destroyed. Because the anti-cancer agents would be delivered only to malignant cells (leaving healthy cells almost wholly unaffected), targeted therapy would enhance the efficacy of the treatment and diminish the disruptive side effects associated with more conventionally used chemotherapy and radiation.

Targeting tumor vasculature provides several advantages over direct tumor therapy. The vasculature surrounding tumors is a genetically sound tissue. It is not subject to the unpredictable genetic mutations cancer cells are infamous for and would therefore not be likely to develop the same drug resistances. Also, targeting vasculature provides a natural amplification method; one vascular endothelial cell death has been calculated to result in

---

[§] : Integrins are transmembrane receptor proteins that are designed to bind with components of the extracellular matrix (ECM). They are a part of the communication network for the cell as they interact with both the cell and the ECM.

at least 100 tumor cell deaths. In addition to the direct restriction of blood flow due to endothelial cell death, vessel damage will promote thrombosis at the site. The clots will further restrict blood flow to tumor cells, making survival impossible. Using the vascular system also provides a simple means of probe introduction; a patient ultimately may receive therapeutic RGD-PEBBLEs intravenously. The PEBBLEs would travel throughout the bloodstream, concentrating only in cancer affected areas.

4.6.3 Nano-explorers and Nano-actuators

4.6.3.1. Magnetic Resonance Imaging Enhancement

Magnetic resonance imaging (MRI) is one of the safest and most widely used medical imaging techniques used today. In order to increase contrast in images, gadolinium chelates or iron oxide nanoparticles are often used to stain the tissue of interest. Gadolinium chelates and iron oxide (dextran covered) nanoparticles are FDA approved and widely used in vivo to diagnose metastases. [88] 70-100 nm nanoparticles containing gadolinium chelates (or iron oxide nanoparticles) have successfully enhanced contrast in MRI images of rat brains. As shown in Figure 4.31, the introduction of the gadolinium chelate PEBBLEs dramatically enhances the contrast of the image, making apparent the previously undetected glioma (brain tumor).

By including gadolinium chelates (or iron oxide) in nanoparticles designed to target tumor cells, we can create a breed of nano-explorers that seek out tumors and aid in cancer detection. Coupled with a therapeutic agent, as discussed below, these MRI contrast enhancing PEBBLEs can effectively image the progress of cancer treatments in real-time.

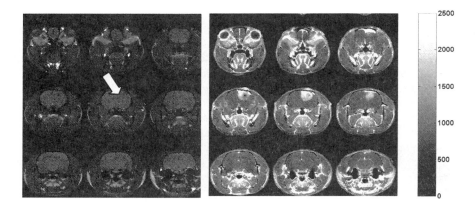

Figure 4.31. Effect of MRI contrast enhancing PEBBLEs. Images on the left are MRI images of a rat brain containing glioma. Images on the right are images of the same rat after the introduction of gadolinium chelate PEBBLEs. Note the tumor enhancement indicated by the arrows.

4.6.3.2. PEBBLE Photodynamic Therapy Nano-actuators

Photodynamic therapy (PDT), a type of light-initiated chemotherapy where a drug is activated by light causing eventual oxidative damage to the cells and resulting in cell death, promises better selectivity and fewer side effects than radiation and chemotherapy. [89, 90] However, like chemotherapy, PDT still suffers from the obstacle of multi-drug resistance (MDR); cancer cells pump introduced drugs back out into the extracellular matrix. Similar to radiotherapy, PDT produces singlet ("killer") oxygen and its oxidizing products, free radicals called reactive oxygen species (ROS), all of which destroy cancer cells. [90] Radiotherapy utilizes x-rays to create ROS with already present oxygen, while PDT relies on non-toxic photosensitizers (PS), dyes (drugs) that produce singlet oxygen (also from already present oxygen) upon irradiation in the visible range. Consequently, PDT requires the tumor to be accessible to light.

In addition to the MDR and light accessibility issues, the delivery of the PS specifically to malignant cells is of concern in PDT. While some degree of control can be achieved by spatially restricting the light therapy, the PS would ideally be delivered only to designated cells. By encapsulating the singlet oxygen producing dyes within a targeted nanoparticle, these concerns can be resolved.

These nano-actuators do not release drugs directly to the cell, as does conventional PDT. Rather, the entire nanoparticle acts as a large PS, delivering a high dose of singlet oxygen near the cell membrane. Active photodynamic dye molecules are not depleted, because they remain intact inside the nanoparticle, in the immediate vicinity of the cell. The problem of MDR is also resolved because only singlet oxygen is actually introduced to the cell. When combined with the targeted delivery technology described above, the overall effect is a high probability of tumor cell kill. Accessibility to light is still an issue with the PDT PEBBLEs; however, brain tissue is essentially transparent to light and an ideal environment for PDT developmental studies.

Nanoparticles containing a water soluble version of 4,7-diphenyl 1,10-phenantroline ruthenium ($Ru(dpp(SO_3)_2)_3$) have demonstrated the ability to produce singlet oxygen effectively while keeping the otherwise toxic dye sequestered from the environment. [91] The details of the nano-actuator synthesis, characterization and singlet oxygen production can be found elsewhere. [91] Another FDA approved drug, photofrin, [92] is also a good source of singlet oxygen and ROS; PEBBLEs containing photofrin have been used in rats with astounding preliminary success, as demonstrated in Figure 4.32.

By combining both PDT and MRI enhancing agents in the same PEBBLE, a tumor can be treated and monitored simultaneously. Increased mobility of water molecules within the tumor is indicative of cell death and can easily be monitored using MRI (NMR). This combination of MRI enhancement and PDT is highly advantageous as the progress of the cancer therapy can be monitored in real-time. In fact, an in vivo study on rat glioma (brain tumor) showed increased water mobility, or tumor dissolution, after only five minutes of therapy with a photofrin nano-actuator. Recently, additional PDT PEBBLEs, with variations in matrix and PS dye components, have also been studied. [93]

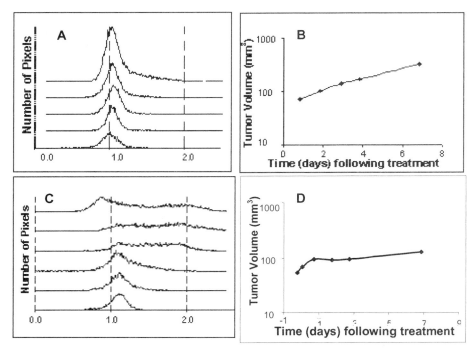

Figure 4.32 Top Panel (A&B): Rat brain glioma were irradiated with laser light with similar exposure time and power in the presence of nanoparticles *lacking* photofrin as a control. Note the uninhibited growth of the tumor. **Bottom Panel (C&D):** Rat Brain glioma were irradiated with laser light for 5 minutes at 740 mW in the presence of the photofrin nanoparticles. C) Water diffusion histogram over several days (voxel is a volumetric unit). The large shift to higher diffusion values indicated dissolution of the tumor. D) Tumor growth post-therapy over 7 days. The therapy has curtailed tumor growth indicating cell death.

4.7. ACKNOWLEDGMENTS

The authors would like to thank the contributions of Rodney Agayan, Jeffery Anker, Dr. Jonathan W. Aylott, Dr. Heather Clark, Dr. Marion Hoyer, Dr. Terry Miller, Dr. Maria J. Moreno, Matthew King, Edwin Park, Dr. Steve Parus, James Sumner, Dr. Ron Tjalkens, and Dr. Fei Yan, as well as University of Michigan Electron Microbeam Analysis Laboratory (funded in part by NSF grant EAR-9628196) for use of the SEM and TEM. We also gratefully acknowledge support from NIH Grants 2R01-GM50300-04A1 (Kopelman) and R01-ES08846 (Philbert), NCI Contract N01-CO-07013, DARPA Biomagnetics Program Grant, and a grant from the Keck Foundation.

4.8. REFERENCES

1. P. Buhlmann, E. Pretsch and E. Bakker, Carrier-based ion-selective electrodes and bulk optodes. 2. Ionophores for potentiometric and optical sensors, Chem. Rev., 98, 1593-1687 (1998).
2. W. H. Tan, Z. Y. Shi and R. Kopelman, Development of Submicron Chemical Fiber Optic Sensors, Anal. Chem., 64, 2985-2990 (1992).
3. W. H. Tan, Z. Y. Shi, S. Smith, D. Birnbaum and R. Kopelman, Submicrometer Intracellular Chemical Optical Fiber Sensors, Science, 258, 778-781 (1992).
4. W. H. Tan, R. Kopelman, S. L. R. Barker and M. T. Miller, Ultrasmall Optical Sensors for Cellular Meaurements, Anal. Chem., 71, 606A-612A (1999).
5. Z. Rosenzweig and R. Kopelman, Development of a Submicrometer Optical-Fiber Oxygen Sensor, Anal. Chem., 67, 2650-2654 (1995).
6. Z. Rosenzweig and R. Kopelman, Analytical Properties and Sensor Size Effects of a Micrometer Sized Optical Fiber Glucose Biosensor, Anal. Chem., 68, 1408-1413 (1996).
7. M. Shortreed, E. Bakker and R. Kopelman, Miniature sodium-selective ion-exchange optode with fluorescent pH chromoionophores and tunable dynamic range, Anal. Chem., 68, 2656-2662 (1996).
8. M. R. Shortreed, S. Dourado and R. Kopelman, Development of a fluorescent optical potassium-selective ion sensor with ratiometric response for intracellular applications, Sens. Actuator B-Chem., 38, 8-12 (1997).
9. S. L. R. Barker, M. R. Shortreed and R. Kopelman, Utilization of lipophilic ionic additives in liquid polymer film optodes for selective anion activity measurements, Anal. Chem., 69, 990-995 (1997).
10. S. L. R. Barker, B. A. Thorsrud and R. Kopelman, Nitrite- and chloride-selective fluorescent nano-optodes and in vitro application to rat conceptuses, Anal. Chem., 70, 100-104 (1998).
11. M. L. Graber, D. C. Dilillo, B. L. Friedman and E. Pastorizamunoz, Characteristics of Fluoroprobes for Measuring Intracellular Ph, Anal. Biochem., 156, 202-212 (1986).
12. M. CohenKashi, M. Deutsch, R. Tirosh, H. Rachmani and A. Weinreb, Carboxyfluorescein as a fluorescent probe for cytoplasmic effects of lymphocyte stimulation, Spectroc. Acta Pt. A-Molec. Biomolec. Spectr., 53, 1655-1661 (1997).
13. C. C. Overly, K. D. Lee, E. Berthiaume and P. J. Hollenbeck, Quantitative Measurement of Intraorganelle Ph in the Endosomal Lysosomal Pathway in Neurons by Using Ratiometric Imaging with Pyranine, Proc. Natl. Acad. Sci. U. S. A., 92, 3156-3160 (1995).
14. B. Morelle, J. M. Salmon, J. Vigo and P. Viallet, Are Intracellular Ionic Concentrations Accessible Using Fluorescent-Probes - the Example of Mag-Indo-1, Cell Biol. Toxicol., 10, 339-344 (1994).
15. W. N. Ross, Calcium on the Level, Biophys. J., 64, 1655-1656 (1993).
16. H. A. Clark, M. Hoyer, M. A. Philbert and R. Kopelman, Optical nanosensors for chemical analysis inside single living cells. 1. Fabrication, characterization, and methods for intracellular delivery of PEBBLE sensors, Anal. Chem., 71, 4831-4836 (1999).
17. H. Xu, J. W. Aylott and R. Kopelman, Fluorescent nano-PEBBLE sensors designed for intracellular glucose imaging, Analyst, 127, 1471-1477 (2002).
18. F. J. Arriagada and K. Osseo-Asare, Synthesis of nanosize silica in a nonionic water-in-oil microemulsion: Effects of the water/surfactant molar ratio and ammonia concentration, J. Colloid Interface Sci., 211, 210-220 (1999).
19. Y. S. Leong and F. Candau, Inverse Micro-Emulsion Polymerization, J. Phys. Chem., 86, 2269-2271 (1982).
20. C. Daubresse, C. Grandfils, R. Jerome and P. Teyssie, Enzyme Immobilization in Nanoparticles Produced by Inverse Microemulsion Polymerization, J. Colloid Interface Sci., 168, 222-229 (1994).
21. H. A. Clark, S. L. R. Barker, M. Brasuel, M. T. Miller, E. Monson, S. Parus, Z. Y. Shi, A. Song, B. Thorsrud, R. Kopelman, A. Ade, W. Meixner, B. Athey, M. Hoyer, D. Hill, R. Lightle and M. A. Philbert, Subcellular optochemical nanobiosensors: probes encapsulated by biologically localised embedding (PEBBLEs), Sens. Actuator B-Chem., 51, 12-16 (1998).
22. H. A. Clark, M. Hoyer, S. Parus, M. A. Philbert and M. Kopelman, Optochemical nanosensors and subcellular applications in living cells, Mikrochim. Acta, 131, 121-128 (1999).
23. H. A. Clark, R. Kopelman, R. Tjalkens and M. A. Philbert, Optical nanosensors for chemical analysis inside single living cells. 2. Sensors for pH and calcium and the intracellular application of PEBBLE sensors, Anal. Chem., 71, 4837-4843 (1999).
24. J. P. Sumner, N. M. Westerberg, A. K. Stoddard, M. Cramer, R. B. Thompson, C. A. Fierke, and R. Kopelman, DsRed Fluorescence Based Copper Ion Nanosensor for Intracellular Imaging, Unpublished, (2003).
25. J. P. Sumner, J. W. Aylott, E. Monson and R. Kopelman, A fluorescent PEBBLE nanosensor for intracellular free zinc, Analyst, 127, 11-16 (2002).

26. E. J. Park, M. Brasuel, C. Behrend, M. A. Philbert and R. Kopelman, Ratiometric optical PEBBLE nanosensors for real-time magnesium ion concentrations inside viable cells, Anal. Chem., 75, 3784-3791 (2003).

27. R. P. Haugland, Molecular Probes Handbook of Fluorescent Probes and Research Chemicals, Molecular Probes, Inc., Eugene, OR (1993).

28. W. E. Morf, K. Seiler, B. Lehmann, C. Behringer, K. Hartman and W. Simon, Carriers for Chemical Sensors - Design-Features of Optical Sensors (Optodes) Based on Selective Chromoionophores, Pure Appl. Chem., 61, 1613-1618 (1989).

29. E. Bakker and W. Simon, Selectivity of Ion-Sensitive Bulk Optodes, Anal. Chem., 64, 1805-1812 (1992).

30. K. Suzuki, H. Ohzora, K. Tohda, K. Miyazaki, K. Watanabe, H. Inoue and T. Shirai, Fiberoptic Potassium-Ion Sensors Based On a Neutral Ionophore and a Novel Lipophilic Anionic Dye, Anal. Chim. Acta, 237, 155-164 (1990).

31. K. Kurihara, M. Ohtsu, T. Yoshida, T. Abe, H. Hisamoto and K. Suzuki, Micrometer-sized sodium ion-selective optodes based on a "tailed" neutral ionophore, Anal. Chem., 71, 3558-3566 (1999).

32. G. J. Mohr, I. Murkovic, F. Lehmann, C. Haider and O. S. Wolfbeis, Application of potential-sensitive fluorescent dyes in anion- and cation-sensitive polymer membranes, Sens. Actuator B-Chem., 39, 239-245 (1997).

33. G. J. Mohr, F. Lehmann, R. Ostereich, I. Murkovic and O. S. Wolfbeis, Investigation of potential sensitive fluorescent dyes for application in nitrate sensitive polymer membranes, Fresenius J. Anal. Chem., 357, 284-291 (1997).

34. M. Brasuel, R. Kopelman, T. J. Miller, R. Tjalkens and M. A. Philbert, Fluorescent nanosensors for intracellular chemical analysis: Decyl methacrylate liquid polymer matrix and ion exchange-based potassium PEBBLE sensors with real-time application to viable rat C6 glioma cells, Anal. Chem., 73, 2221-2228 (2001).

35. S. Peper, I. Tsagkatakis and E. Bakker, Cross-linked dodecyl acrylate microspheres: novel matrices for plasticizer-free optical ion sensing, Anal. Chim. Acta, 442, 25-33 (2001).

36. I. Tsagkatakis, S. Peper and E. Bakker, Spatial and spectral imaging of single micrometer sized solvent cast fluorescent plasticized poly(vinyl chloride) sensing particles, Anal. Chem., 73, 315-320 (2001).

37. D. R. Uhlmann, G. Teowee and J. Boulton, The future of sol-gel science and technology, J. Sol-Gel Sci. Technol., 8, 1083-1091 (1997).

38. H. Xu, J. W. Aylott, R. Kopelman, T. J. Miller and M. A. Philbert, A real-time ratiometric method for the determination of molecular oxygen inside living cells using sol-gel-based spherical optical nanosensors with applications to rat C6 glioma, Anal. Chem., 73, 4124-4133 (2001).

39. G. Tolnai, F. Csempesz, M. Kabai-Faix, E. Kalman, Z. Keresztes, A. L. Kovacs, J. J. Ramsden and Z. Horvolgyi, Preparation and characterization of surface-modified silica-nanoparticles, Langmuir, 17, 2683-2687 (2001).

40. D. H. Napper, Steric Stabilization, J. Colloid Interface Sci., 58, 390-407 (1977).

41. B. Vincent, The Preparation of Colloidal Particles Having (Post-Grafted) Terminally-Attached Polymer-Chains, Chem. Eng. Sci., 48, 429-436 (1993).

42. H. D. Bijsterbosch, M. A. C. Stuart and G. J. Fleer, Effect of block and graft copolymers on the stability of colloidal silica, J. Colloid Interface Sci., 210, 37-42 (1999).

43. K. S. Kim, S. H. Cho and J. S. Shin, Preparation and Characterization of Monodisperse Polyacrylamide Microgels, Polym. J., 27, 508-514 (1995).

44. R. Gref, M. Luck, P. Quellec, M. Marchand, E. Dellacherie, S. Harnisch, T. Blunk and R. H. Muller, 'Stealth' corona-core nanoparticles surface modified by polyethylene glycol (PEG): influences of the corona (PEG chain length and surface density) and of the core composition on phagocytic uptake and plasma protein adsorption, Colloid Surf. B-Biointerfaces, 18, 301-313 (2000).

45. M. Harris, Poly(ethylene Glycol) Chemistry: Biotechnical and Biomedical Applications, Plenum, (1992).

46. P. Kingshott and H. J. Griesser, Surfaces that resist bioadhesion, Curr. Opin. Solid State Mat. Sci., 4, 403-412 (1999).

47. M. Ogris, S. Brunner, S. Schuller, R. Kircheis and E. Wagner, PEGylated DNA/transferrin-PEI complexes: reduced interaction with blood components, extended circulation in blood and potential for systemic gene delivery, Gene Ther., 6, 595-605 (1999).

48. P. Lesot, S. Chapuis, J. P. Bayle, J. Rault, E. Lafontaine, A. Campero and P. Judeinstein, Structural-dynamical relationship in silica PEG hybrid gels, J. Mater. Chem., 8, 147-151 (1998).

49. D. Ammann, Ion-Selective Microelectrodes, Springer, Berlin (1986).

50. J. M. Russell, Sodium-potassium-chloride cotransport, Physiol. Rev., 80, 211-276 (2000).

51. M. G. Brasuel, T. J. Miller, R. Kopelman and M. A. Philbert, Liquid polymer nano-PEBBLEs for Cl– analysis and biological applications, Analyst, 128, (Advance Article) (2003).

52. M. Rothmaier, U. Schaller, W. E. Morf and E. Pretsch, Response mechanism of anion-selective electrodes based on mercury organic compounds as ionophores, Anal. Chim. Acta, 327, 17-28 (1996).
53. C. J. Behrend, and R. Kopelman Unpublished, (2003).
54. S. L. R. Barker, H. A. Clark, S. F. Swallen, R. Kopelman, A. W. Tsang and J. A. Swanson, Ratiometric and fluorescence lifetime-based biosensors incorporating cytochrome c ' and the detection of extra- and intracellular macrophage nitric oxide, Anal. Chem., 71, 1767-1772 (1999).
55. J. N. Demas and B. A. Degraff, Design and Applications of Highly Luminescent Transition-Metal Complexes, Anal. Chem., 63, A829-A837 (1991).
56. A. K. McEvoy, C. M. McDonagh and B. D. MacCraith, Dissolved oxygen sensor based on fluorescence quenching of oxygen-sensitive ruthenium complexes immobilized in sol-gel-derived porous silica coatings, Analyst, 121, 785-788 (1996).
57. C. McDonagh, B. D. MacCraith and A. K. McEvoy, Tailoring of sol-gel films for optical sensing of oxygen in gas and aqueous phase, Anal. Chem., 70, 45-50 (1998).
58. The Merck Index 12th edition, (S. Budavari, ed.), p1195, Merck&Co, NJ (1996).
59. P. C. Pandey, S. Upadhyay and H. C. Pathak, A new glucose sensor based on encapsulated glucose oxidase within organically modified sol-gel glass, Sens. Actuator B-Chem., 60, 83-89 (1999).
60. O. S. Wolfbeis, I. Oehme, N. Papkovskaya and I. Klimant, Sol-gel based glucose biosensors employing optical oxygen transducers, and a method for compensating for variable oxygen background, Biosens. Bioelectron., 15, 69-76 (2000).
61. M. Brasuel, R. Kopelman, I. Kasman, T. J. Miller and M. A. Philbert, Ion Concentrations in Live Cells From Highly Selective Ion Correlation Fluorescent Nano-Sensors for Sodium, Proceedings of IEEE,1, 288-292 (2002).
62. A. Emmi, H. J. Wenzel, P. A. Schwartzkroin, M. Taglialatela, P. Castaldo, L. Bianchi, J. Nerbonne, G. A. Robertson and D. Janigro, Do glia have heart? Expression and functional role for ether-a-go-go currents in hippocampal astrocytes, J. Neurosci., 3915-3925 (2000).
63. S. K. Jung, W. Gorski, C. A. Aspinwall, L. M. Kauri and R. T. Kennedy, Oxygen microsensor and its application to single cells and mouse pancreatic islets, Anal. Chem., 71, 3642-3649 (1999).
64. J. R. Lakowicz, Emerging Biomedical Applications of Time-Resolved Fluorescence Spectroscopy, in Topics in Fluorescence Spectroscopy (J. R. Lakowicz, ed.), Vol. 4, 1-19, Plenum Press, New York (1994).
65. G. Rao, S. B. Bambot, S. C. W. Kwong, H. Szmacinski, J. Sipior, R. Holavanahali and G. Carter, Application of Fluorescence Sensing to Bioreactors, in Topics in Fluorescence Spectroscopy (J. R. Lakowicz, ed.), Vol. 4, 417-448, Plenum Press, New York (1994).
66. Z. Chen-Esterlit, S. F. Peteu, H. A. Clark, W. McDonald and R. Kopleman, A Comparative Study of Optical Fluorescent Nanosensors ("PEBBLEs") and Fiber Optic Microsensors for Oxygen Sensing, Proceedings of SPIE, 3602, 156-163 (1999).
67. I. Klimant, F. Ruckruh, G. Liebsch, C. Stangelmayer and O. S. Wolfbeis, Fast response oxygen micro-optodes based on novel soluble ormosil glasses, Mikrochim. Acta, 131, 35-46 (1999).
68. T. M. Ambrose and M. E. Meyerhoff, Characterization of photopolymerized decyl methacrylate as a membrane matrix for ion-selective electrodes, Electroanalysis, 8, 1095-1100 (1996).
69. T. M. Ambrose and M. E. Meyerhoff, Photo-cross-linked decyl methacrylate films for electrochemical and optical polyion probes, Anal. Chem., 69, 4092-4098 (1997).
70. C. M. Ingersoll and F. V. Bright, Using sol gel-based platforms for chemical sensors, Chemtech, 27, 26-31 (1997).
71. M. T. Murtagh, M. R. Shahriari and M. Krihak, A study of the effects of organic modification and processing technique on the luminescence quenching behavior of sol-gel oxygen sensors based on a Ru(II) complex, Chem. Mat., 10, 3862-3869 (1998).
72. M. L. Bossi, M. E. Daraio and P. F. Aramendia, Luminescence quenching of Ru(II) complexes in polydimethylsiloxane sensors for oxygen, J. Photochem. Photobiol. A-Chem., 120, 15-21 (1999).
73. Y.E. Koo, S. Koo, and R. Kopelman, Unpublished, (2003).
74. M. King and R. Kopelman, Development of a hydroxyl radical ratiometric nanoprobe, Sens. Actuator B-Chem., 90, 76-81 (2003).
75. R. Roots and S. Okada, Estimation of Life Times and Diffusion Distances of Radicals Involved in X-Ray-Induced DNA Strand Breaks or Killing of Mammalian-Cells, Radiat. Res., 64, 306-320 (1975).
76. G. Lubec, The hydroxyl radical: From chemistry to human disease, J. Invest. Med., 44, 324-346 (1996).
77. B. B. Li, P. L. Gutierrez and N. V. Blough, Trace determination of hydroxyl radical using fluorescence detection, Oxidants and Antioxidants, Pt B, 300, 202-216 (1999).
78. X. F. Yang and X. Q. Guo, Study of nitroxide-linked naphthalene as a fluorescence probe for hydroxyl radicals, Analytica Chimica Acta, 434, 169-177 (2001).

79. J. N. Anker, C. Behrend and R. Kopelman, Aspherical magnetically modulated optical nanoprobes (MagMOONs), J. Appl. Phys., 93, 6698-6700 (2003).

80. J. N. Anker and R. Kopelman, Magnetically modulated optical nanoprobes, Appl. Phys. Lett., 82, 1102-1104 (2003).

81. J. N. Anker C. J. Behrend, and R. Kopelman, Brownian Modulated Optical Nanoprobes, Appl. Phys. Lett., Submitted, (2003).

82. A. Hyvarinen, J. Karhunen, and E. Oja, Independent component analysis, J. Wiley, New York (2001).

83. R. R. Agayan, F. Gittes, R. Kopelman and C. F. Schmidt, Optical trapping near resonance absorption, Appl. Optics, 41, 2318-2327 (2002).

84. W. Arap, R. Pasqualini and E. Ruoslahti, Cancer treatment by targeted drug delivery to tumor vasculature in a mouse model, Science, 279, 377-380 (1998).

85. N. Assa-Munt, X. Jia, P. Laakkonen and E. Ruoslahti, Solution structures and integrin binding activities of an RGD peptide with two isomers, Biochemistry, 40, 2373-2378 (2001).

86. B. P. Eliceiri and D. A. Cheresh, The role of alpha v integrins during angiogenesis: insights into potential mechanisms of action and clinical development, J. Clin. Invest., 103, 1227-1230 (1999).

87. S. Kim, K. Bell, S. A. Mousa and J. A. Varner, Regulation of angiogenesis in vivo by ligation of integrin alpha 5 beta 1 with the central cell-binding domain of fibronectin, Am. J. Pathol., 156, 1345-1362 (2000).

88. R. Mathurdevre and M. Lemort, Biophysical Properties and Clinical-Applications of Magnetic- Resonance-Imaging Contrast Agents, Br. J. Radiol., 68, 225-247 (1995).

89. W. Stummer, A. Hassan, O. Kempski and C. Goetz, Photodynamic therapy within edematous brain tissue: Considerations on sensitizer dose and time point of laser irradiation, J. Photochem. Photobiol. B-Biol., 36, 179-181 (1996).

90. T. J. Dougherty, C. J. Gomer, B. W. Henderson, G. Jori, D. Kessel, M. Korbelik, J. Moan and Q. Peng, Photodynamic therapy, J. Natl. Cancer Inst., 90, 889-905 (1998).

91. M. J. Moreno, E. Monson, R. G. Reddy, A. Rehemtulla, B. D. Ross, M. Philbert, R. J. Schneider and R. Kopelman, Production of singlet oxygen by Ru(dpp(SO3)(2))(3) incorporated in polyacrylamide PEBBLES, Sens. Actuator B-Chem., 90, 82-89 (2003).

92. L. C. Penning and T. M. Dubbelman, Fundamentals of Photodynamic Therapy - Cellular and Biochemical Aspects, Anti-Cancer Drugs, 5, 139-146 (1994).

93. F. Yan, and R. Kopelman, Photochem Photobiol, In Press, (2003).

APTAMERS AS EMERGING PROBES FOR MACROMOLECULAR SENSING

Eun Jeong Cho, Manjula Rajendran, and Andrew D. Ellington[*]

5.1. INTRODUCTION

Aptamers, derived from the latin word *aptus* (meaning, "to fit"), are functional nucleic acid binding species that have been selected from combinatorial oligonucleotide libraries by a process known as *in vitro* selection.[1, 2] Since 1990, numerous high-affinity and highly specific aptamers have been selected against a wide variety of target molecules, such as small organics, peptides, proteins, and even supramolecular complexes, such as viruses or cells.[3, 4] Since aptamers have been shown to discriminate between even closely related isomers or different conformational states of the same protein,[5, 6] they are becoming an increasingly popular tool for molecular recognition that may eventually rival antibodies. Their utility has now been demonstrated in a number of analytical applications, such as flow cytometry,[7,8] affinity probe capillary electrophoresis,[9] sandwich assays,[10] capillary electrochromatography,[11,12] affinity chromatography,[13, 14] and more generally as biosensors.[15-18]

In this chapter, we will describe the fundamentals of *in vitro* selection procedures, the characteristics of aptamers compared to antibodies, and our and others' research into adapting aptamers to function as recognition elements for high-throughput screening protocols and multi-analyte biosensors.

5.2. *IN VITRO* SELECTION

Aptamers are selected from random sequence libraries by a process that mimics natural selection. A pool of nucleic acids is sieved for a desired functional property, such as ability to bind to a target molecule. The "winners" (binding species) are selectively

---

[*] Eun Jeong Cho, Manjula Rajendran, and Andrew D. Ellington, Department of Chemistry and Biochemistry, Institute for Cellular and Molecular Biology, The University of Texas at Austin, Austin, TX 78712

recovered and amplified. Over multiple rounds of selection and amplification, the population becomes progressively enriched in binding or functional species.

In greater detail, a pool of DNA sequences generated by chemical synthesis is amplified by the polymerase chain reaction, and converted (via *in vitro* transcription for RNA or strand separation for DNA) into a single-stranded nucleic acid pool that has a random sequence region of 30 to upwards of 200 nucleotides and a complexity of from 10^{13}-10^{16} independent sequences. The single-stranded pool can be DNA, RNA, or modified RNA, and can be sieved for its ability to bind to a given target by any of a variety of means, including column chromatography, alterations in electrophoretic mobility, or co-retention on a filter (**Figure 5.1**).[19] Retained species can be eluted and amplified by some combination of reverse transcription, PCR, and *in vitro* transcription.

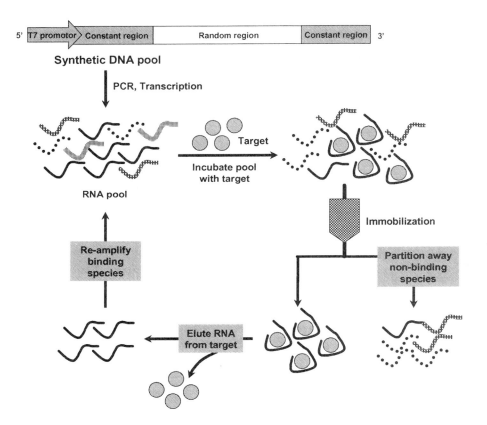

Figure 5.1. An overview of *in vitro* selection. A chemically synthesized, single-stranded DNA pool consisting of a random sequence core flanked by constant regions is PCR amplified and transcribed *in vitro* to generate a RNA or modified RNA pool. Following incubation with a target molecule, sequences that bind to the target are separated from non-binding sequences, re-amplified by reverse transcription, PCR, and *in vitro* transcription, and again selected for binding function. Over iterative rounds of selection with increasingly stringent binding conditions, tight and highly specific aptamers can be isolated.

In the first several rounds of selection, a very small fraction of the population typically binds to the target. However, since these species are preferentially recovered and amplified, the pool progressively becomes less diverse and the functional species eventually dominate the population. The affinity of the selected aptamers can be controlled by varying the stringency of each round of selection, normally by varying the concentration of the target, the buffer conditions employed during the binding reaction, or the number and type of wash steps used during column or filter partitioning. Generally, the population is assayed after every several rounds of selection and amplification, and a plurality of the population will be found to bind a target after 6 to 18 rounds of selection.

While the traditional *in vitro* selection procedure usually requires several weeks to months, recent automation of the methods potentially provides a route from target to novel reagents within only a few days.[20] Individual aptamers can be cloned and sequenced, and sequence comparisons typically reveal one or more families related by common ancestry (mutational variants of a single, original sequence) or by similarity (similar sequence or structural motifs that were selected in parallel, from different, original sequences).

The binding affinities of aptamers are highly target-dependent and range from picomolar (1×10^{-12} M) to high nanomolar (1×10^{-7} M) for various protein targets. When small organics are targeted, the dissociation constants are higher, typically micromolar, as might be expected given the smaller number of interactions that will be formed. In either case, interactions tend to be extremely specific, and aptamers can discriminate between related analytes by over 10,000-fold on the basis of single amino acid changes or even single chemical moieties, such as hydroxyl or methyl groups. The specificities of aptamers can to some extent be controlled during selection; for example, negative selections against related analytes or the matrices used for target immobilization can remove cross-reactive aptamers from a population.[21]

Aptamers have a number of advantages relative to antibodies in analytical and sensor applications. While many antibodies are temperature-sensitive and denature upon contact with surfaces, leading to limited shelf lives and possible compromise of assay integrity, aptamers are extremely stable during storage, can be transported at ambient temperatures, and undergo reversible denaturation. Moreover, unlike antibodies, which must be generated by a living organism, aptamers can be produced chemically with extreme accuracy and reproducibility. Based on sequence, mutational, or deletion analyses the sequence of an individual aptamer can frequently be minimized. Some of the smallest functional aptamers are less than 30 nucleotides in length, and the cost of their production is very competitive relative to antibody overproduction and purification. Finally, as we will see, the ability to chemically synthesize aptamers also allows their site-specific modification with fluorescent reporters or chemical linkers, and the concomitant development of novel analytical reagents that have no antibody counterparts.[22]

5.3. ADAPTATION OF APTAMERS AS SIGNALING TRANSDUCTION REAGENTS FOR MACROMOLECULAR SENSING

In the past, fluorescence studies have yielded extensive knowledge of the structure and dynamics of biological macromolecules. Labeling aptamers with fluorophores is therefore particularly attractive for biosensor applications, and aptamers could potentially serve as simple substitutes for antibodies. Thus, fluoresceinated aptamers have been used in lieu of antibodies in a sandwich ELISA format to detect human vascular endothelial growth factor[10] and as probes in flow cytometry to detect human neutrophil elastase and human CD4.[7, 8] Additionally, fluorescence detection can be combined with a variety of analytical techniques. For example, capillary electrophoresis coupled with laser-induced fluorescence detection (CE-LIF) of fluorescently-labeled aptamers has been used to sensitively detect IgE (Limit of detection, LOD = 46 pM), thrombin (LOD = 40 nM),[9] and HIV-1 RT (reverse transcriptase of the type 1 human immunodeficiency virus, linear detection range up to 50 nM).[23] Aptamers as detection reagents proved to be extremely sensitive and selective (with a mass detection limit of 37 zmol for IgE) and had the added advantage of rapid analysis (<60s for IgE and thrombin, <5min for HIV-1 RT) when compared with other diagnostic assays.

However, labeling aptamers with fluorophores might significantly affect nucleic acid conformation. In this regard, Imanishi and co-workers[24] have shown that labeling an anti-Reactive Green 19 (RG19) aptamer with fluorescein at its 5' end did not significantly affect DNA conformation. The circular dichroism of the aptamer was measured in the presence and absence of RG19 and was found not to change; in addition the affinity of labeled and unlabeled aptamers for immobilized RG19 was similar.

On the other hand, structural studies have also shown that aptamers frequently undergo significant conformational changes upon binding to their cognate ligands.[25-29] Therefore, it seemed reasonable to suppose that it might be possible to incorporate a fluorophore into an aptamer in such a way that it would not perturb the functional structure of the aptamer but would instead report any analyte-dependent conformational changes; for example, changes in the chemical environment of the fluorophore could potentially be read as changes in fluorescence intensity. The binding event might also lead to changes in fluorescence wavelength or anisotropy. The resultant so-called 'signaling aptamers' would be powerful analytical probes that could allow real-time, quantitative determinations of unlabeled analyte samples. In the following section we will discuss various signaling aptamers that have been generated by both rational design and evolutionary engineering strategies.

5.3.1. Signaling Aptamers Based on Fluorescence Intensity Changes

5.3.1a. Designed Signaling Aptamers

In an initial effort to design signaling aptamers, we chose two anti-adenosine (ATP) aptamers, one selected from a RNA pool[30] and one selected from a DNA pool[31] as starting points. The three-dimensional structures of both anti-adenosine aptamers had been determined, and functional residues had been identified by both structural and mutational analyses. We attempted to introduce fluorescent reporters at specific positions such that they would not perturb aptamer structure or function. RNA signaling aptamers were designed and synthesized by incorporating acridine and fluorescein (FAM) in place of residue 13 (ATP-R-Ac13, ATP-R-F13), and FAM at the 5'-end (ATP-R-F1).

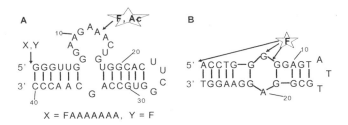

X = FAAAAAAA, Y = F

Figure 5.2. Secondary structure of and sites of dye incorporation for designed ATP signaling aptamers. **(A)** Designed RNA signaling aptamers: Acridine was incorporated in place of residue 13 (ATP-R-Ac13). Fluorescein was incorporated at the 5′ end (ATP-R-F1), at the 5′ end with a heptaadenyl linker (ATP-R-F2), and in place of residue 13 (ATPR-F13). **(B)** Designed DNA signaling aptamers: Fluorescein was incorporated at the 5′ end (DFL0), in place of residue 7 (DFL7), and between residues 7 and 8 (DFL7-8). Residues are numbered from the 5′ ends of the secondary structures that are shown.

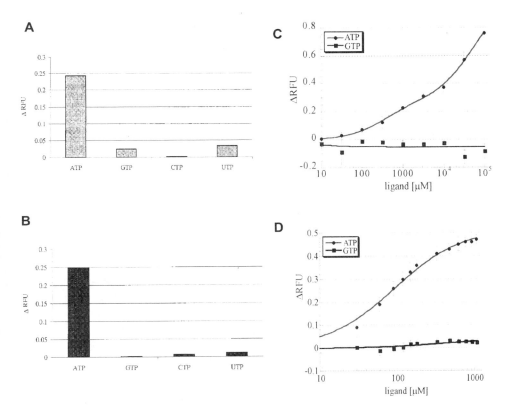

Figure 5.3. Performance of designed ATP signaling aptamers (see also **ref. 32**). Specificities of the signaling aptamers ATP-R-Ac13 **(A)** and DFL7-8 **(B)**. The fractional increase in relative fluorescence units (ΔRFU) was measured in the presence of ATP, GTP, CTP, and UTP (1 mM ligand for ATP-R-Ac13, 200 μM ligand for DFL7-8). Response curves for the signaling aptamers ATP-R-Ac13 **(C)** and DFL7-8 **(D)** at varying concentrations of ATP and GTP.

Similarly, DNA signaling aptamers were prepared by incorporating FAM at the 5'-end (DFL0), in place of residue 7 (DFL7), and between residues 7 and 8 (DFL7-8; **Figures 5.2A** and **5.2B**).[32] We found that ATP-R-Ac13 and DFL7-8 showed significant increases in fluorescence intensity in the presence of 1mM ATP, whereas other designed signaling aptamers showed an insignificant change in fluorescence intensity (5% or less). To assess the specificity of the designed signaling aptamers, changes in fluorescence were measured in the presence of GTP, CTP, and UTP. As shown in **Figure 5.3A and 5.3B**, the designed signaling aptamers can readily discriminate against non-cognate nucleotides, including GTP, UTP, and CTP, just as the original anti-adenosine aptamer does. Fluorescence intensity changes as a function of ATP and GTP concentrations represented a graded increase in fluorescence intensity with ATP, but little or no change in fluorescence intensity with GTP, confirming that designed signaling aptamers can directly quantitate analyte concentrations in solution (**Figure 5.3C and 5.3D**).

Since the NMR structure had shown that anti-adenosine aptamer had two ATP binding sites, K_d values could be inferred based on the following equation:

$$(F - F_0) = \frac{K_1(F_1 - F_0)[L] + K_1K_2(F_2 - F_0)[L]^2}{1 + K_1[L] + K_1K_2[L]^2}$$

where F is the measured fluorescent signal, F_0 is the fluorescence of the uncomplexed signaling aptamer, F_1 is the fluorescence of the aptamer bound to a single ATP molecule, F_2 is the fluorescence of the aptamer bound to two ATP molecules, [L] is the ligand (ATP) concentration, K_1 is the formation constant of the first binding event, and K_2 is the formation constant of the second binding event. This analysis yielded two K_d values: a $K_{d,1}$ of 30 ± 18 μM and a $K_{d,2}$ of 53 ± 30 μM, indicating that the aptamer indeed had two binding sites for ATP even upon the addition of a fluorescent label.

Other fluorophores have also been used to design ATP signaling aptamers. Saito *et.al.*[33] site-specifically introduced bis-pyrene into various positions of the anti-ATP DNA aptamer and found that one of their designed aptamers could report ATP-dependent conformational changes. ATP binding led to a change in the ratio of bis-pyrene excimer emission to monomer fluorescence emission ($K_{d,1}$ of 1.17 ± 0.07 mM and $K_{d,2}$ of 1.64 ± 0.21 mM, slightly less sensitive than the values cited for the fluorescein derivative, above).

Compared to the published K_d value for ATP (6μM)[31], site-specific introduction of the fluorophore had, in general, caused some loss of binding affinity, perhaps by changing the aptamer conformation or by sterically limiting analyte interactions with the aptamer. While these results indicated that it might be possible to use rational methods to readily convert aptamers to aptamer receptors, they also contradicted the assumption that fluorescent labels would report but not perturb binding. The observed loss of binding affinity could potentially limit the utility and dynamic range of signaling aptamers.

5.3.1b. In vitro Selection of Signaling Aptamers

In several instances, the conjugation of fluorescent reporters to aptamers reduced their affinity. In order to better accommodate the fluorescent label, we attempted to introduce it during the selection itself.[34] As an initial attempt, fluorescein was introduced into a random sequence RNA pool by replacing UTP with fluorescein-UTP during *in vitro* transcription.   To avoid a high, intrinsic fluorescence background and enable the sensitive detection of ligand-induced changes in fluorescence intensity, the random sequence region in the pool was skewed so that uridine residues were only sparsely distributed.   Aptamers were then selected that could bind to ATP.   There was of course no assurance that fluorescent aptamers that bound to ATP would also signal in the presence of ATP.   However, we found that one of the RNA aptamers, rafl7s, contained a single, functional uridine residue, could accurately and selectively follow ATP concentration in solution, and could sense ATP concentrations as low as 25 μM with apparent K_d value of 175 ± 5 μM (**Figure 5.4**).   More remarkably, replacement of fluorescein with other fluorophores such as Cascade Blue-7 or Rhodamine Green-5, did not affect adenosine-dependent signaling by the aptamer (K_d = 188 ± 15 μM for Cascade Blue-7 and 571 ± 39 μM for Rhodamine Green-5).   This latter result suggested that the single uridine underwent a strong ligand-induced change in its chemical microenvironment and this change was in turn transferred to the different, pendant fluorescent reporters

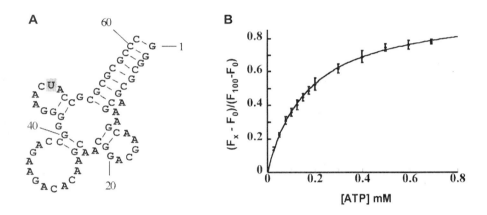

Figure 5.4. *In vitro* selection of an ATP signaling aptamer, rafl7s (see also **ref. 34**). **(A)** Predicted secondary structure of the selected signaling aptamer.   The single uridine residue responsible for signaling has been highlighted. **(B)** Response curve at varying concentrations of ATP. (F_x - F_0) / (F_{100} - F_0) is the relative increase in fluorescence at a given ATP concentration, where F_x is the fluorescence response at a given ATP concentration, F_0 is the fluorescence in the absence of ATP, and F_{100} is the fluorescence at a saturating ATP concentration.

A quantitative comparison of the two general methods for generating signaling aptamers, rational design and evolutionary engineering, is provided in **Table 5.1**.   Each method has its advantages and disadvantages.   Rational design allowed signaling aptamers to be immediately designed and generated, but sometimes required the painstaking synthesis and evaluation of a number of designs.   In addition, the introduction of the fluorescent reporter sometimes led to a decrease in binding sensitivity. In contrast, evolutionary engineering produced signaling aptamers that were optimally adapted to the fluorescent reporter, but otherwise required that new *in vitro* selection experiments be carried out for each target.   For the future, we believe that computational predictions of secondary structure and the energetics of conformational change will increase the throughput and sensitivity of rational design methods, while automation of signaling aptamer selections will reduce the amount of time that must be invested in evolutionary engineering methods.

Table 5.1. Engineered ATP Signaling Aptamers

Engineering method	Fluorophore	Sensing Evidence[a]	K_d	ref
Rational	FAM or Acridine	$+\Delta F$	30 ± 18 µM, 53 ± 30 µM	32
	Bis-pyridine	$+\Delta(F_{excimer}/F_{monomer})$	1.17 ± 0.07 mM, 1.64 ± 0.21 mM	33
Evolutionary	FAM	$+\Delta F$	175 ± 5 µM	34

[a] (+) represents increase in fluorescence intensity in the presence of target and (-) represents decrease in fluorescence intensity in the presence of target.

5.3.2. Aptamer Beacons Based on Fluorescence Resonance Energy Transfer

Fluorescence resonance energy transfer (FRET) involves the transfer of energy from an excited donor fluorophore to a neighboring acceptor molecule. If the acceptor molecule is another fluorophore, then this leads to fluorescence at the acceptor's emission wavelength, whereas if the acceptor is a 'dark quencher' (for example, DABCYL), then it does not subsequently fluoresce. FRET is an exquisitely distance-dependent interaction, and since the lengths (<20-60 Å) over which FRET normally occurs are comparable to the dimensions of biological macromolecules this fluorescence technique has been used with great success in biological studies.

Nucleic acid probes that employ fluorescence quenching, particularly so-called molecular beacons, have been used extensively in molecular biology for both quantitative and qualitative nucleic acid detection. A molecular beacon (MB) is a short synthetic oligonucleotide that possesses a stem-loop structure with a fluorophore and a quencher conjugated to either end of the stem[35-39](**Figure 5.5A**). In the absence of a target sequence, the MB does not fluoresce, because the stem brings the fluorophore and quencher into close proximity.[37-41] However, when the MB encounters a target DNA or RNA molecule that has a sequence complementary to the MB's loop, the loop and target can form a duplex that is longer and more stable than the original hairpin stem. The conformational change from the hairpin stem to the long duplex forces the two arms of the original stem apart and thus permits the fluorophore to fluoresce. Since their development in 1996, molecular beacons have been adapted to many applications, including real-time monitoring of nucleic acid amplification and localization of specific mRNAs.[37-39]

5.3.2a. Designed Aptamer Beacons

It is possible to envision that aptamers can be converted into aptamer beacons in much the same way that they have previously been converted into signaling aptamers. Instead of introducing a single fluorophore into a conformationally labile region of an aptamer, two fluorophores capable of functioning as FRET partners could be introduced into the aptamer structure.

Martin Stanton at Brandeis University hypothesized that aptamers could be engineered to function as molecular beacons by adding sequences that would force the formation of a stable stem-loop structure at the expense of the native, binding structure. Like molecular beacons, the 'aptamer beacon' would exist in a quenched stem-loop structure in the absence of target molecule (**Figure 5.5B**). In the presence of a target molecule, complex formation would shift the equilibrium in favor of the unquenched, native (target-binding) structure, consequently generating a change in the observed fluorescence intensity.

As a model system, our group chose an anti-thrombin aptamer that had been previously selected from a random sequence pool[42] and thoroughly studied.[43-47] According to NMR analysis and x-ray crystallography,[48, 49] the anti-thrombin aptamer formed a chair-form quadruplex with the adjacent 5' and 3' ends in the corners of the quardruplex and two stacked G-quartets linked by TT and TGT loops.[48] The G-quartet structure positioned residues within two connecting TT loops for direct contact,[50] and it

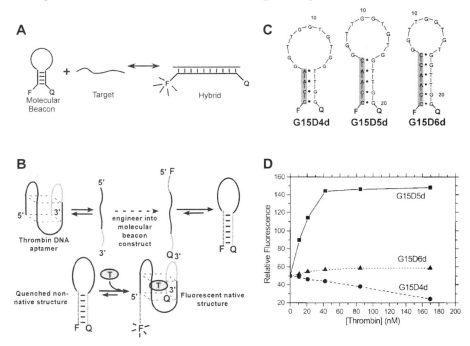

Figure 5.5. Molecular and aptamer beacons. **(A)** Mechanism of molecular beacons: The stem-loop structure in the absence of target contains a fluorophore, **F**, in apposition with a quencher, **Q**. In the presence of a complementary nucleic acid target, formation of an extended helix leads to fluorescence restoration (dequenching). **(B)** Mechanism of aptamer beacon: An aptamer can be forced to form a beacon-like stable stem loop structure by changing its sequence (for example, by adding residues at the 5' end). The additional bases introduced at the 5' end of the aptamer have been highlighted. If a fluorophore and a quencher are appended to the two ends of the new hairpin structure, addition of target will shift the equilibrium in favor of the fluorescent, target-binding native structure. **(C)** Predicted secondary structures of three designed thrombin aptamer beacons. These structures differ primarily in terms of stem length. **(D)** Response curves of the thrombin aptamer beacons at varying concentrations of thrombin (see also **ref. 51**).

therefore seemed likely that derivatization of the 5' or 3' termini should minimally perturb thrombin binding. It was reasoned that the addition of several nucleotides to the 5' end of the anti-thrombin aptamer would cause the aptamer to form an alternate hairpin conformation that would destroy the G-quartet structure. The addition of thrombin to this aptamer would then strongly favor the original thrombin-binding quadruplex (**Figure 5.5B**). Aptamer beacons with different stem lengths were designed and synthesized with a fluorescence quenching pair (5' end, FAM; 3'-end, DABCYL, 4-[[4'-(dimethylamino)phenyl]azo]-benzoic acid; **Figure 5.5C**).[51] The fluorescence responses of the aptamer beacons were evaluated at varying thrombin concentrations (0-120 nM). The aptamer beacon G15D5dMB showed the greatest response, an approximately 2.5-fold increase in fluorescence at saturating thrombin concentrations, and had an apparent K_d of approximately 10 nM (as compared with $K_d \approx 200$nM as determined by affinity chromatography[42]; **Figure 5.5D**). This result implies that the aptamer beacon could potentially be used to directly quantitate protein concentrations in solution.

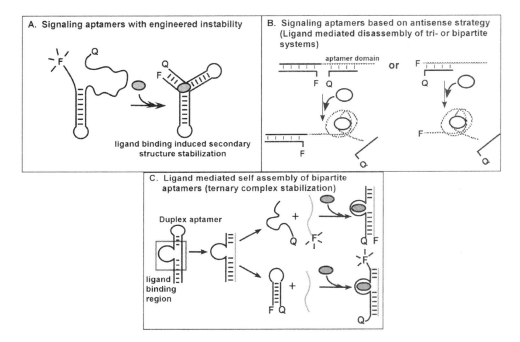

Figure 5.6. Other strategies for designing aptamer beacons. (**A**) Target-dependent stabilization of the native structure relative to a random coil or destabilized secondary structure. (**B**) Target-dependent stabilization of the native structure relative to a duplex formed with an antisense oligonucleotide. (**C**) Target-dependent stabilization of the native structure following division of the aptamer into two halves.

Similarly, conformational changes can be engineered that lead to quenching, rather than dequenching of fluorescence. For example, Landry and his co-workers have destabilized the secondary structures of aptamers so that upon addition of analytes the labeled 5' and 3' ends of the aptamers come together, resulting in quenching (**Figure**

5.6A).[52] In an engineered anti-cocaine aptamer up to 61% of the initial fluorescence was quenched upon addition of cocaine and this quenching aptamer beacon could be used to determine cocaine concentrations ranging from 10-2500 μM. In another example, an anti-platelet-derived growth factor (PDGF) aptamer was converted into an aptamer beacon structure by removing several base pairs from a paired stem and labeling one end of the aptamer with FAM (5'-end) and the other with DABCYL (3'-end).[53] In the absence of PDGF, the aptamer is largely denatured, the fluorophore and quencher are far apart. Upon PDGF binding, the aptamer forms a secondary structure in which the fluorophore and quencher are in apposition and fluorescence is quenched. The PDGF aptamer beacon could detect PDGF concentrations as low as 110 pM in biological samples. One potential problem with quenching aptamer beacons is that there are a variety of ligands or solvents that can interfere with quenching, resulting in a false positive signal.

Tan's group[54] has also taken advantage of conformational changes in the thrombin aptamer. However, instead of engineering conformational changes, as above, these researchers noted that even in the absence of the protein target, an equilibrium existed between the random coil state and the quadraplex states of the thrombin aptamer. Again, target binding shifts the equilibrium in favor of the quadraplex state. The anti-thrombin aptamer was modified to act as an aptamer beacon by incorporating either a fluorophore (fluorescein) and a quencher (DABCYL), or two fluorophores (coumarin (donor) and fluorescein (acceptor)) at the two termini of the aptamer. Thrombin binding was monitored via either quenching of a single fluorophore (fluorescein-DABCYL aptamer, maximal fluorescence decrease $\sim$ 60%) or increasing fluorescence of an acceptor fluorophore (fluorescein-coumarin aptamer, signal enhancement factor = ratio of acceptor to donor intensity before and after binding = ~14 fold). The K_d and limits of detection for the quenching aptamer beacon were 5.20 $\pm$ 0.49 nM and 373 $\pm$ 30 pM, while for the two-fluorophore aptamer beacon these values were 4.87 $\pm$ 0.55 nM and 429 $\pm$ 63 pM. FRET-type aptamer beacons incorporating two fluorophores may be particularly useful for real-time analysis of proteins, and could possibly be used in living specimens in conjunction with ratiometric imaging.

Aptamer beacons can also be designed by using an antisense strategy[55] in which a complementary DNA sequence is used to denature the aptamer (**Figure 5.6B**). Upon addition of ligand, the native aptamer structure is stabilized, and the equilibria is concomitantly shifted away from the denatured, antisense duplex and towards the native structure. The aptamer can be labeled with a fluorophore, and the antisense oligonucleotide with a quencher (denoted QDNA), leading to a quenched fluorescent signal in the absence of target, similar to the unimolecular aptamer beacons described above. Li and his co-workers[55] have demonstrated the potential of this approach by making antisense aptamer beacons that can detect ATP (K_d = 600 μM) and thrombin (K_d = 400 nM). Like the unimolecular aptamer beacons, this strategy requires no foreknowledge of aptamer structure.

In an approach similar to the antisense strategy, aptamers can also be divided to generate quarternary structures that can signal (**Figure 5.6C**). In one example, aptamers were split into two subunits, and each subunit was labeled with a fluorophore and a quencher, respectively. By adding the appropriate target molecule, the two aptamer subunits self-assembled, causing a decrease in fluorescence. Landry and co-workers demonstrated the potential of this approach with an anti-cocaine aptamer (quenched up to

65% of the initial value, detection range of 10-1250 µM) and an anti-ATP aptamer (quenched up to 40% of the initial value, detection range 8-2000 µM).[56] Similarly, Kumar and his co-workers originally selected an anti-Tat aptamer that bound HIV-1 Tat two orders of magnitude (133-fold) better than the natural ligand, TAR, then divided this aptamer to function as a Tat beacon.[57] One half of the anti-Tat aptamer was engineered to fold into a hairpin that contained a fluorophore and a quencher. Upon addition of Tat, the quarternary structure (the complex between the two aptamer halves) was stabilized, resulting in the opening of the hairpin and the generation of a fluorescence signal. The Tat aptamer beacon exhibited a 9-fold fluorescence increase in the presence of 100nM target.

Ligand-dependent conformational transitions can not only influence the optical properties of aptamer beacons with pendant dyes, but can also be exploited for the initial conjugation of fluorescent dyes. Weeks *et.al.*[58] have made use of the differential nucleophilic reactivity of 2' amine substituted nucleotides in flexible versus constrained nucleic acid structures to convert the anti-ATP DNA aptamer into an aptamer beacon.

Table 5.2. Designed Aptamer Beacons

Target	Fluorophore pair	Aptamer Beacon Design Strategy	Sensing Evidence[a]	K_d	ref
Thrombin	FAM/DABCYL	Engineered conformational change	$+\Delta F$	10 nM	51
Cocaine	FAM/DABCYL	Engineered conformational change	$-\Delta F$	10-2500 µM[b]	52
PDGF	FAM/DABCYL	Engineered conformational change	$-\Delta F$	110 pM[c]	53
Thrombin	FAM/DABCYL	Inherent conformational change	$-\Delta F$	5.20 ± 0.49 nM	54
	FAM/coumarin		$+\Delta F$	4.87 ± 0.55 nM	
Thrombin	FAM/DABCYL	Disassembly of aptamer:complement DNA duplex	$+\Delta F$	400 nM	55
ATP				600 µM	
ATP	FAM/DABCYL	Self-assembly of two aptamer subunits	$-\Delta F$	8-2000 µM[b]	56
Cocaine				10-1250 µM[b]	
Tat protein	FAM/DABCYL	Self-assembly of two aptamer subunits	$+\Delta F$	N/A	57
ATP	FCM/Texas-Red	Relative reactivity of FCM to aptamer	$-\Delta F$	160 µM	58

[a] (+) represents increase in fluorescence in the presence of target and (-) represents decrease in fluorescence in the presence of target.
[b] detection range
[c] LOD (limit of detection)

A 2' amine moiety that was site-specifically incorporated into the aptamer could readily form an adduct with fluorescamine (FCM). The fluorescent adduct with FCM could in turn be detected via FRET to a fluorescence acceptor (Texas-Red) that had been incorporated at the 3' end of the molecule. However, in the presence of ATP, adduct formation is considerably reduced. This led to an interesting analytical method in which ATP was detected not only by FRET (K_d = 390 µM), but also by the ability to form the conjugate that leads to FRET in the first place.

The various methods to engineer signaling aptamer beacons are summarized in **Table 5.2**. One of the advantages of these approaches as a whole is that they can potentially be applied to any aptamer, RNA or DNA. In addition, since aptamer beacon design relies upon knowledge of aptamer secondary structural features that can be readily predicted and engineered, their ligand-sensing and -signaling properties can likely be more finely-tuned than was the case for signaling aptamers, which are instead dependent upon small, hard-to-predict conformational changes.

5.3.2b. In vitro Selection of Aptamer Beacons

However, all of the methods that have so far been considered ultimately rely upon engineering known aptamers to generate signals and require a prior knowledge of the detailed primary, secondary, or even tertiary structure of aptamers. If possible it would be much simpler to directly couple selection with signaling.

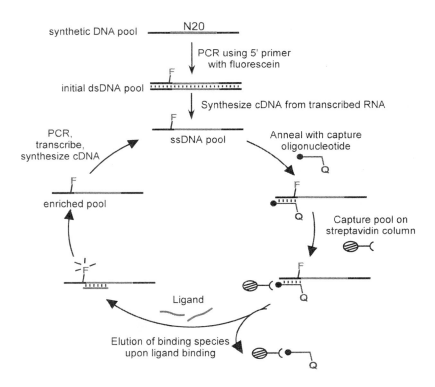

Figure 5.7. *In vitro* selection of molecular beacons. **F** indicates fluorescein, and **Q**, DABCYL. The closed circle at the termini of the capture oligonucleotide represents biotin.

In order to develop a general method for the direct selection of aptamer beacons, our group has first developed a selection method for molecular beacons.[59] This method relies upon ligand-dependent elution of immobilized nucleic acids from an affinity column. As illustrated in **Figure 5.7**, a fluoresceinated, single-stranded DNA pool with 20 randomized positions was annealed with a 2-fold molar excess of a biotinylated capture oligonucleotide which was in turn immobilized on streptavidin agarose beads.

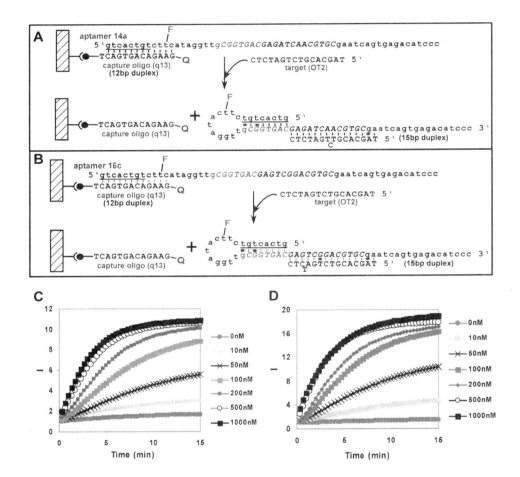

Figure 5.8. Mechanism of elution and fluorescence responsivities of selected molecular beacons (see also **ref. 59**). **(A)** and **(B)** Sequence and proposed signaling mechanism of beacons 14a and 16c. Hybridization of the oligonucleotide target OT2 stabilizes the formation of an internal hairpin stem and disrupts interactions with the capture oligonucleotide. **(C)** and **(D)** Fluorescence response curves of beacons 14a and 16c at varying concentrations of the target oligonucleotide OT2. I is the signal-to-background ratio ($I=(F_{open}-F_{buffer})/(F_{closed}-F_{buffer})$, where F_{open} is the fluorescence of the beacon-capture oligonucleotide complex in the presence of target, F_{closed} is the fluorescence of the beacon-capture oligonucleotide complex in the absence of target, and F_{buffer} is the background fluorescence of the buffer solution alone.

The oligonucleotide affinity column was then developed with an equimolar mixture of two 16-mer oligonucleotide targets (OT1 and OT2). The eluant from the column was amplified (via reverse transcription, PCR, and *in vitro* transcription), purified, and used for the next round of selection. After nine rounds of selection and amplification the population was enriched in variants that could be specifically eluted by target OT2 (but not target OT1). The selected molecular beacons bore a FAM moiety on residue T11 within their 5' constant regions and when hybridized with a capture oligonucleotide containing a DABCYL moiety at its 5' end, their fluorescence was correspondingly quenched (**Figures 5.8A and 5.8B**). Addition of the oligonucleotide target (OT2) resulted in the same conformational change that was originally selected for, the displacement of the capture oligonucleotide bearing the quencher, and thus also resulted in an increase in fluorescence intensity. The predominant selected beacons 14a and 16c showed 9.5-fold and 16.5-fold increases in fluorescence in the presence of a 2-fold of molar excess of OT2, and the fluorescence intensity increased as a function of OT2 concentration (**Figures 5.8C and 5.8D**).

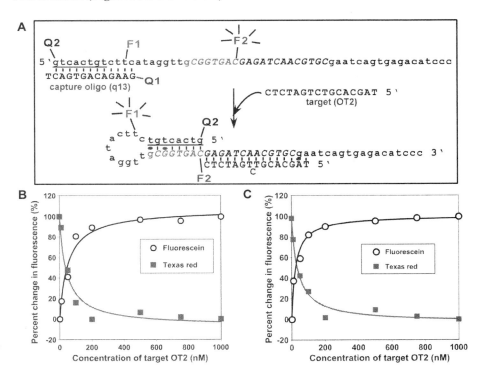

Figure 5.9. Wavelength-shifting selected molecular beacons (see also **ref. 59**). (A) Sequence and predicted mechanism of a wavelength-shifting selected beacon based on 14a. The fluorescein-dT that was labeled **F** in **Figure 5.8(A)** is now labeled **F1**, while a second fluorophore (Texas-Red) is **F2**. There are two DABCYL moieties, at positions **Q1** (as in **Figure 5.8(A)**) and **Q2**. (B) Fluorescence response of wavelength-shifting beacon 14a at different concentrations of OT2. The data for fluorescein was collected by exciting the beacon at 494nm and recording emission at 518nm, while the data for Texas-Red was collected by exciting the beacon at 595nm and recording emission at 615nm. (C) Fluorescence response of a similarly designed wavelength-shifting beacon 16c.

The gain in fluorescence increase was similar to that observed in many designed molecular beacons.[35, 37, 57, 60, 61] The apparent K_ds and LODs for the selected molecular beacons were 37 ± 11 nM and 14 nM for beacon 14a and 34 ± 8 nM and 3.6 nM for beacon 16c, values which were again similar to those previously demonstrated for designed molecular beacons.[62-64] The predicted mechanism for the selected molecular beacons was quite different from that of designed molecular beacons. In selected molecular beacons, a hairpin stem is formed, rather than disrupted, in the presence of the target oligonucleotide, resulting in a loss of the capture oligonucleotide. In order to further exploit the selected conformational changes and thereby expand the potential utility of the selected beacons, a second quencher (DABCYL) was introduced at the 5' end of beacon constructs during chemical synthesis, and a second fluorophore reporter (Texas-Red) was appended to an internal cytidine residue that was predicted to participate in the target-dependent hairpin stem (**Figure 5.9A**). In the presence of the oligonucleotide target, the doubly-labeled beacons should dequench FAM and quench Texas-Red. As shown in **Figures 5.9B and 5.9C,** the 'green' and 'red' fluorescent signals showed inverse changes in intensity as a function of target concentration.

Although the selected molecular beacons have performance characteristics comparable with those of designed molecular beacons, the ability to select beacons may prove useful for identifying available sites on complex targets, such as mRNAs. More importantly, the general method for selection can now be further extended to virtually any target class, potentially yielding selected aptamer beacons.

5.3.3. Signaling Aptamers Based on Fluorescence Anisotropy Changes

Fluorescence anisotropy has been used to study numerous biomolecular phenomena, including conformational changes in proteins and the self-association of peptides and proteins.[41] Like changes in fluorescence intensity, changes in fluorescence anisotropy can be used to monitor binding events. In fact, fluorescence anisotropy may be ideally suited for studying the associations between proteins and appropriately-labeled aptamers. Many target proteins are relatively large molecules compared with aptamers, especially, minimized aptamers. Therefore, protein-binding to an aptamer should bring about a significant change in the overall molecular weight and size of the aptamer and in turn greatly alter the rotational diffusion rate of any fluorophore appended to the aptamer, resulting in a detectable variation in fluorescence anisotropy.

In order to demonstrate that the fluorescence anisotropy of labeled aptamers could be successfully used for sensing interactions with analytes, we again chose the short (15-mer) DNA quadruplex aptamer that binds to the blood-clotting factor thrombin as an example.[65] The 5' end was labeled with FITC (fluorescein isothiocyanate) and 3' end with an alkyl amine which would later be used for surface immobilization. The performance of this aptamer conjugate was first examined in solution. Surprisingly, the labeled aptamer showed a 10-fold higher dissociation constant (1.1 μM) than that of the unlabeled aptamer,[45] but this diminution in sensitivity was reversed following immobilization (see **Section 5.4.3**, below). The labeled aptamer also showed high selectivity for thrombin compared with elastase, another serine protease with an isoelectric point and molecular weight similar to those of thrombin.

Similar fluorescence anisotropy measurements were carried out using fluorescently-labeled anti-PDGF aptamers.[66] The anisotropy change (2-fold) upon PDGF-binding was complete within only a few seconds. As little as 0.22 nM protein could be detected in homogeneous solution in real time. The sensitivity demonstrated by this method should be suitable for the detection of PGDF in serum samples and in biological fluids surrounding tumors (0.4-0.7 nM or higher).

The significance of adapting aptamers for fluorescence anisotropy measurements is that fluorescence anisotropy is a relative (ratiometric) detection technique. As a result, the common problems associated with fluorescence intensity assays, such as bleaching and non uniform emission of the fluorophore during imaging, are not of major concern. Moreover, since fluorescent labels for measuring anisotropy could be introduced at almost any position in an aptamer, it should prove possible to modify many different aptamers for further use as biosensors and to generalize this technique to an even greater extent than was possible with signaling aptamers or aptamer beacons. Finally, since the reporter rather than the analyte is the signal transducing molecule, fluorescence anisotropy should be capable of monitoring real-time changes in analyte concentrations, should require fewer sample preparation steps, and may eventually be useful for *in vivo* measurements.

5.4. APPLICATION OF APTAMERS TO ARRAY FORMATS

A recent trend in the development of biosensors is to use multiple sensors in parallel in an array format to simultaneously detect multiple analytes in a single sample.[67-69] Just as DNA arrays have been used to monitor gene expression patterns, aptamer-based arrays could potentially be used to monitor proteome or metabolome expression patterns. Toward this end, a number of strategies have been developed for adapting aptamer biosensors to aptamer arrays.

5.4.1. Aptamer Chips for High-Throughput Screening

DNA and RNA microarrays[70, 71] are rightly considered the most important methods for characterizing and understanding organismal genomes and transcriptomes. The arrays are basically small chips to which synthetic oligonucleotides or PCR products have been attached, and may contain as many as 40,000 different sequences. A lot of technologies have grown up around microarrays, including arrayers for generating multiple chips and scanners that can readily analyze the multiple fluorescent signals generated by hybridization of labeled cDNAs to the arrays. To the extent that proteins could 'hybridize' to aptamers in the same ways that labeled cDNAs can hybridize to oligonucleotides or PCR products, it would be extremely useful to generate aptamer arrays using some of the same technologies that have proven successful for the generation of gene expression arrays. This is especially true since recent advances in the automation of the *in vitro* selection process[72] should enable the mass production of aptamers as specific receptors for a wide variety of proteins.

As a first step towards the generation of arrays to quantify proteomes, we have generated aptamer arrays for screening selection experiments. The traditional screening method to find the best aptamer at the conclusion of a selection requires cloning, sequencing, and individual binding assays, and is both time-consuming and labor intensive. In addition, the acquisition of sequence data for multiple aptamers from a selection experiment is relatively expensive. If it should prove possible to merely clone and transcribe individual aptamers, and then array them on glass slides, the best aptamer(s) could be quickly identified following incubation with a labeled protein target. Once the best clones had been identified, only their sequences would subsequently be determined.

Our initial efforts centered on screening the affinity of anti-lysozyme aptamers selected via an automated selection process. Merely spotting aptamers onto lysine-coated slides led to inactivation of the aptamers. In order to generate a fixed site for aptamer conjugation to the slide, aptamers were transcribed in the presence of biotin-GpG, which is specifically incorporated at the 5' end of the aptamer. Biotin-labeled aptamers were then spotted onto a glass slide coated at high density with streptavidin; although the streptavidin-biotin interaction is non-covalent, it is extremely robust having a dissociation constant of 10^{-15} M.[73]   Initially, three anti-lysozyme aptamers with a range of affinities (Clones 1, 8, and 12; Round 0 as a negative control) were transcribed with biotin-GPG and spotted onto a streptavidin coated glass slide using a pin printing technique. Cy3-labeled lysozyme solutions were then applied to the aptamer array under a glass cover slip, unbound material was washed away, and the relative affinity of each aptamer for the labeled lysozyme was investigated by using a laser fluorescence scanner GenePix 4000B.

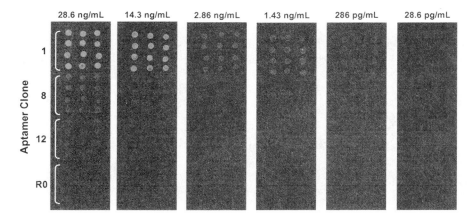

Figure 5.10. An anti-lysozyme aptamer chip array. Three different biotinylated, anti-lysozyme aptamers (Clones 1, 8, 12) and the round 0 unselected pool from which these clones were derived were immobilized on a streptavidin-coated glass slide. A Cy3-labeled lysozyme solution was applied at varying concentrations (as indicated). The slides were rinsed with Tris buffer containing 1M NaCl and 5mM MgCl$_2$, dried, and imaged using a laser fluorescence scanner. Each spot is about 700 μm in diameter.

Figure 5.10 summarizes the false color fluorescence images of aptamer-based microarrays exposed to varying concentration of labeled lysozyme (28.6 pg/mL, 286 pg/mL, 1.43 ng/mL, 2.86ng/mL, 14.3 ng/mL, 28.6 ng/mL, respectively). In general and as expected, more concentrated solutions yielded higher fluorescence intensities. In addition, Clone 1 exhibited the highest affinity to the labeled lysozyme, Clone 8 the next highest, and Clone 12 and Round 0 very little affinity. These results accord with the relative affinities of the aptamers as determined by filter-binding experiments. Indeed, the correlation is quantitative as well as: qualitative binding isotherms constructed from the data revealed that Clone 1 had K_d value of 70 ± 19 ng/mL, which is in fact consistent with the K_d value extracted from filter binding assay (data not shown). In addition, Clone 1 could detect lysozyme in the low pg/mL range.

In addition to showing that it should be possible to generate aptamer arrays to detect labeled protein samples, the development of methods for immobilizing aptamers without loss of function may also lead to the generation of signaling aptamer or aptamer beacons arrays that could directly detect proteins without the need for fluorescent labeling. Such arrays would be especially interesting in that the signaling process would be reversible, and the arrays could potentially be re-used multiple times with different samples.

5.4.2. Sample Processing with Microwell-Based Aptamer Arrays

For all of their potential for screening, chip-based arrays have one major drawback: it is difficult or impossible to perform any processing steps on the array. For example, it is difficult to develop sandwich assay or ELISA methods using chip arrays, since any colorimetric or fluorescent reporters that are generated can diffuse over the entire array. As an alternative, it may be possible to develop microwell arrays that can accomodate the diffusion of reporters, substrates, and analytes.

In collaboration with John McDevitt and Dean Neikirk at the University of Texas at Austin, we have begun to utilize a microwell array format, the so-called 'electronic taste chip' or 'electronic tongue' (ET).[74] In this format microspheres are placed in etched silicon wells. An additional benefit to using microspheres in microwells is that microspheres present a much larger surface area for the display of capture agents relative to planar surfaces, and thus lead to higher sensitivities of detection.[75, 76] This advantage is not unique to microwells, as functionalized microspheres have also been analyzed by flow cytometry[77] and confocal microscopy.[78]

Even though the process of functionalizing microspheres with DNA probes is well-established,[77-83] functionalizing microspheres with aptamers had not been shown until recently, when Walt and co-workers demonstrated the potential of bead-based arrays functionalized with an anti-thrombin aptamer to measure thrombin concentrations in solution.[84] In related applications, microspheres derivatized with target molecules can be analyzed with fluorophore-labeled aptamers via flow cytometry (FACS). Jayasena and co-workers[7] have investigated the ability of an anti - human neutrophil elastase (HNE) aptamer labeled with FAM to bind and facilitate the sorting of HNE-coated beads. In this system, the FAM-labeled anti-HNE aptamer had an affinity for HNE that was similar to that of an anti-HNE antibody, and could be used to detect bead-bound HNE with a K_d of 15 ± 3 nM, a value that was consonant with that obtained in a filter-binding assay (17 nM).[85] In a similar demonstration, a microsphere derivatized with recombinant human CD4 functionalized microsphere[8] could be stained with a modified (2'-fluoropyrimidine) RNA library. Both a selected library and individual aptamers (K_d as low as 0.5 ± 0.05 nM) could detect CD4 via FACS.

In an effort to further exploit the potential of aptamer-functionalized microspheres, our lab has developed methods for adapting aptamers to the electronic tongue platform. Aptamers were derivatized with biotin and immobilized on streptavidin microspheres, which were in turn introduced into the microwells of the ET. For the digital analysis of aptamer-functionalized microspheres, the previously described ET was used with minor modifications, such as a sample injection port (**Figure 5.11A**) before the flow inlet to reduce the sample volume analyzed (200 µL of sample was actually injected) and rectangular microwells instead of square ones in order to accommodate smaller diameter microspheres (diameter 70-100μm, **Figure 5.11B**) and still allow adequate flow without back pressure.

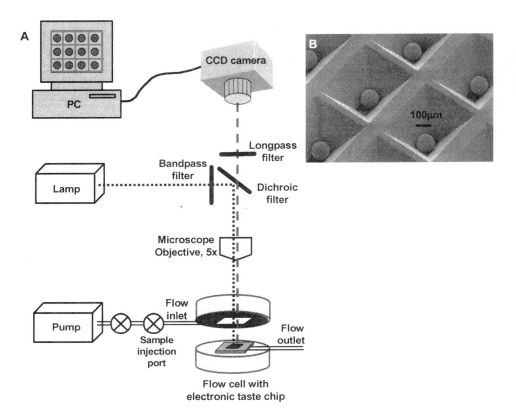

Figure 5.11. Detection system for microwell-based aptamer arrays. **(A)** The "electronic tongue" setup contains a fluid delivery system, fluorescence microscope, CCD camera, flow cell, and computer for data analysis. **(B)** Aptamer-derivatized beads in the micromachined wells of a chip.

Initial analytical determinations with the modified ET focused on the biothreat agent ricin, a class II ribosome-inactivating protein from the castor bean plant *Ricinus communis* that has the potential of being used as a weapon in biological attack.[86-88] Recently, our group was successful in creating RNA aptamers specific for the catalytic ricin A-chain (RTA).[89] The anti-ricin aptamers bear no resemblance to the normal RTA substrate, the sarcin-ricin loop (SRL), and were not depurinated by RTA. The originally selected aptamer consisted of 80 nucleotides but was minimized to 31-nucleotides. The minimal anti-ricin aptamer could recognize ricin with high affinity ($K_d = 7.3$ nM). An advantage of using the anti-ricin aptamer as a test bed for aptamer adaptation to the ET was that any results that were obtained could be directly compared with ricin assays that had previously been developed around a variety of other techniques, such as fiber optic-based immunoassays[90] or noncontact printed microarrays.[91-93]

As a proof-of-principle, Alexa Fluor$_{488}$-labeled ricin was injected into the ET to determine whether and to what extent the aptamer-functionalized microspheres could capture and detect the toxin. **Figure 5.12A** shows the fluorescence images of microsphere beads in the flow path (Bn; blank bead without functionalization, Ra; anti-ricin aptamer functionalized bead). Microspheres functionalized with the anti-ricin aptamer exclusively respond to ricin; as little as 8 μg/mL of ricin could be detected using this configuration (**Figure 5.12B**). The sample incubation, rinse, and measurement procedures were completed in 5 min or less.

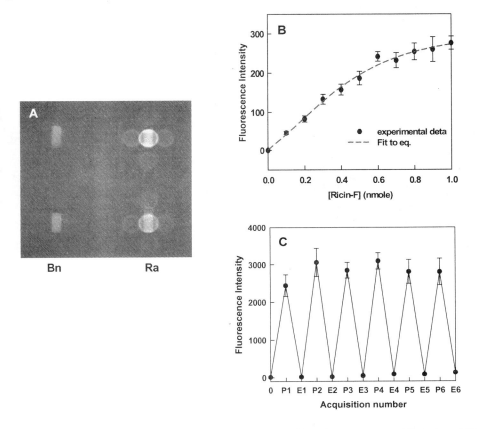

Figure 5.12. Performance of an anti-ricin aptamer in a microwell-based aptamer array. **(A)** An immobilized anti-ricin aptamer binds Alexa-Fluor$_{488}$-labeled ricin. Optical image of anti-ricin functionalized beads (Ra) and blank beads (Bn) inside the chip after incubating with Alexa-Fluor$_{488}$-labeled ricin for 2 min. **(B)** Response curve of the immobilized anti-ricin aptamer as a function of Alexa-Fluor$_{488}$-labeled ricin concentration. Data points are shown as an average of five values with standard deviations. The solid line represents the best fit using the Sigmoidal equation in Sigmaplot software (SPSS Inc., Chicago, IL). **(C)** Reversible response of the immobilized anti-ricin aptamer. Experiments similar to those described in **(A)** were alternated with a 7M urea wash. Points P1 through P6 represent the introduction of Alexa-Fluor$_{488}$-labeled ricin. Points E1 through E6 represent introduction of the urea solution.

We have previously claimed that aptamers are conformationally robust relative to antibodies, and can potentially be reused. To demonstrate the practicality of these assertions, we alternately introduced Alexa Fluor$_{488}$-labeled ricin and 7M urea solution and recorded the changes in fluorescence. As summarized in **Figure 5.12C**, the aptamer-based microspheres showed a highly reproducible response to the labeled ricin. We also investigated the selectivity of the functionalized microspheres by introducing an anti-lysozyme aptamers into the array and probing with labeled lysozyme (**Figure 5.13**); no cross-reactivity between the two aptamers or the two proteins was observed. Although the detection limit for this assay configuration was not as low as an antibody-based immunosensor assay (100 pg/mL),[94] the ET is still an attractive alternative because of this potential for microsphere regeneration coupled with real-time data collection.

As previously stated, though, the microwell and microsphere technology should be most useful for processing samples and therefore for sensing unlabeled proteins. Accordingly, we modified the ricin assay, as described in **Figure 5.14**. In the new configuration, aptamers were again used as capture reagents, but now fluorescent signals were detected following the addition of anti-ricin antibody labeled with Alexa Fluor$_{488}$.

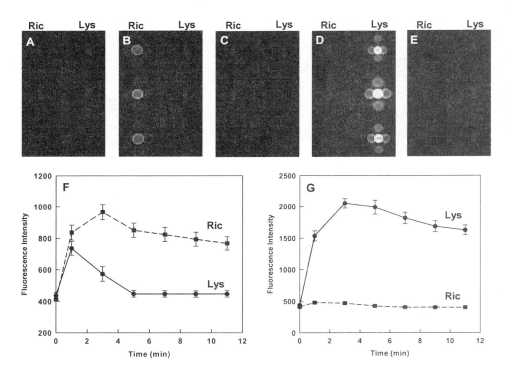

Figure 5.13. Selectivities of immobilized aptamers in a microwell-based aptamer array. **Above.** Anti-ricin (Ric) and anti-lysozyme (Lys) aptamers were immobilized on beads and introduced into the chip. Optical images are shown for the serial introduction of buffer solution **(A)**, Alexa-Fluor$_{488}$-labeled ricin **(B)**, 7M urea **(C)**, Alexa-Fluor$_{488}$-labeled lysozyme solution **(D)**, and 7M urea **(E)**. **Below.** Response curves of immobilized anti-ricin and anti-lysozyme aptamers exposed to Alexa-Fluor$_{488}$-labeled ricin solution **(F)** and Alexa-Fluor$_{488}$-labeled lysozyme solution **(G)**, respectively.

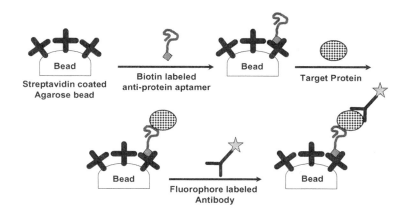

Figure 5.14. Aptamer/protein/antibody sandwich assay. A biotinylated, anti-ricin aptamer was immobilized on a streptavidin agarose bead and used to capture unlabeled ricin. Analyte capture was detected and quantitated with an anti-ricin antibody labeled with Alexa-Fluor$_{488}$.

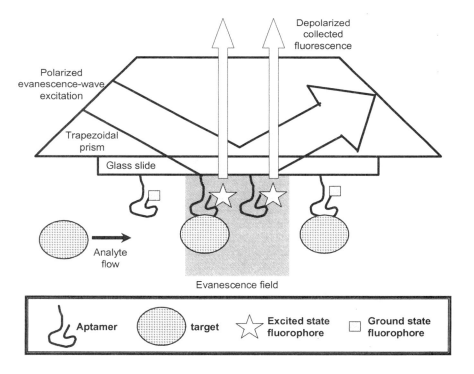

Figure 5.15. Detecting fluorescence anisotropy changes in immobilized aptamer biosensors. A thin film of FITC-labeled anti-thrombin aptamer was immobilized on a glass slide and excited by a polarized evanescent wave. The two polarization components of the depolarized fluorescence emission were then monitored to detect and quantify target-dependent changes in fluorescence anisotropy.

Using a sandwich assay format similar to that which had previously been demonstrated for an antibody-based immunoassay,[95] we were able to detect free ricin concentration as low as 320 ng/mL in near real-time. The sandwich assay format in microwells could also likely be adapted to ELISA-like techniques in which the second antibody was conjugated to an enzyme such as beta-galactosidase or alkaline phosphatase that could turnover colorimetric, fluorescent, or luminescent substrates.

5.4.3. Chip-Based Detection of Changes in Aptamer Anisotropy

If the advantages inherent in the detection of changes in fluorescence anisotropy could be coupled with an array format, this would further potentiate the development of aptamer chips for proteome analysis. To demonstrate that these technologies could be merged, the anti-thrombin DNA aptamer labeled with FITC was immobilized on a microscope cover slip as illustrated in **Figure 5.15**. Next, we again investigated whether the covalent immobilization of aptamer onto a solid substrate altered its sensitivity or selectivity for thrombin. As summarized in **Figure 5.16A**, the labeled aptamer revealed a dynamic range from the nanomolar to micromolar concentration range of thrombin; the detection limit calculated from the slope of the calibration curve was 5 nM. Despite the fact that the fluorescently-labeled aptamer in solution appeared to lose some sensitivity for thrombin (see **Section 5.3.3.**, above), these latter results showed that the labeled, immobilized aptamer had a higher sensitivity (apparent K_d = 47nM) relative to the unlabeled aptamer ($K_d \approx$ 100 nM, as determined by filter-binding assay[45]). As was the case for the aptamer functionalized microsphere, sensing was shown to be reproducible by cycling between PBS buffer and guanidinium hydrochloride solution (**Figure 5.16B**).

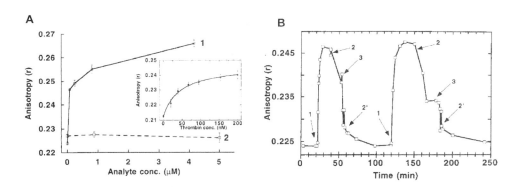

Figuire 5.16. Fluorescence anisotropy of an immobilized anti-thrombin aptamer (see also **ref. 65**). (A) Specificity of the anti-thrombin aptamer: Response curve of the anti-thrombin aptamer labeled with FITC as a function of thrombin (**1**) and elastase (**2**) concentrations. The inset is an expanded view of the data for low target analyte concentrations (0-200 nM). (B) Reversible binding kinetics: Thrombin (**1**, 420nM and 800 nM injected at 20 min and at 120 mM, respectively), a PBS buffer solution (**2, 2'**), and guanidinium hydrochloride (**3**) were applied to an ethanolamine-blocked, aptamer-coated glass slide.

Based on these results, one of the primary difficulties in using fluorescence anisotropy chips would be devising an optical set-up that could quickly move between different sectors on the chip. This problem has recently been tackled by researchers from the company Archemix. By incorporating a moveable mirror into their optical platform they could excite a number of aptamer sensor elements either selectively or all together.[16] The sensor array could be used to specifically detect thrombin, bFGF (basic fibroblast growth factor), IMPDF (inosine monophosphate dehydrogenase), and VEGF (vascular endothelial growth factor) in the presence of complex biological media. The sensitivity of detection for thrombin (K_d = 15.5 nM) was similar to that previously demonstrated for other aptamer biosensors.

5.5. CONCLUSIONS AND FUTURE ASPECTS

Aptamers are the only biomolecules other than antibodies that have universal binding properties. The high sensitivity and specificity of aptamers towards target molecules, and the ease with which they can be chemically synthesized and engineered, make aptamers ideally suited for sensor applications. Fluorescently-labeled aptamers can be engineered either via rational design or *in vitro* selection to function as biosensors for the real time detection of target concentrations in solution. A variety of fluorescent properties, including fluorescence intensity, FRET, or anisotropy changes can be monitored. Aptamers can also be immobilized on arrays and used to detect their cognate ligands. This has enabled the development of aptamer-based chip arrays for high-throughput screening applications.

Because of their exquisite specificity and the ease with which they can be engineered aptamers should prove particularly useful for diagnostic applications. For example, aptamers can be used to distinguish between different isoforms of proteins that are differentially expressed in normal versus cancerous cells,[16] or to detect post-translational modifications such as phosphorylation.[6] Ultimately, the automation of the *in vitro* selection process should enable the high-throughput generation of aptamers against multiple different targets and the consequent development of aptamer-based chip arrays for the acquisition and analysis of organismal proteomes and metabolomes.

5.6. ACKNOWLEDGMENTS

We gratefully acknowledge support by the National Science Foundation, Office of Naval Research, Defense Advanced Research Projects Agency, the Army Research Office, and the Countermeasures to Biological and Chemical Threats Program at the Institute for Advanced Technology (DAAD13-02-C-0079; UTA03-0203).

5.7. REFERENCES

1. Ellington, A. D.; Szostak, J. W. In vitro selection of RNA molecules that bind specific ligands, *Nature* **1990**, *346*, 818-822.
2. Tuerk, C.; Gold, L. Systematic evolution of ligands by exponential enrichment: RNA ligands to bacteriophage T4 DNA polymerase, *Science (Washington, D. C., 1883-)* **1990**, *249*, 505-510.
3. Wilson, D. S.; Szostak, J. W. In vitro selection of functional nucleic acids, *Annu. Rev. Biochem.* **1999**, *68*, 611-647.

4. Famulok, M.; Mayer, G.; Blind, M. Nucleic acid aptamers-from selection in vitro to applications in vivo, *Acc. Chem. Res.* **2000**, *33*, 591-599.

5. Conrad, R.; Keranen, L. M.; Ellington, A. D.; Newton, A. C. Isozyme-specific inhibition of protein kinase C by RNA aptamers, *J. Biol. Chem.* **1994**, *269*, 32051-32054.

6. Seiwert, S. D.; Stines Nahreini, T.; Aigner, S.; Ahn, N. G.; Uhlenbeck, O. C. RNA aptamers as pathway-specific MAP kinase inhibitors, *Chem. Biol.* **2000**, *7*, 833-843.

7. Davis, K. A.; Abrams, B.; Lin, Y.; Jayasena, S. D. Use of a high affinity DNA ligand in flow cytometry, *Nucleic Acids Res.* **1996**, *24*, 702-706.

8. Davis, K. A.; Lin, Y.; Abrams, B.; Jayasena, S. D. Staining of cell surface human CD4 with 2'-F-pyrimidine-containing RNA aptamers for flow cytometry, *Nucleic Acids Res.* **1998**, *26*, 3915-3924.

9. German, I.; Buchanan, D. D.; Kennedy, R. T. Aptamers as ligands in affinity probe capillary electrophoresis, *Anal. Chem.* **1998**, *70*, 4540-4545.

10. Drolet, D. W.; Moon-McDermott, L.; Romig, T. S. An enzyme-linked oligonucleotide assay, *Nat. Biotechnol.* **1996**, *14*, 1021-1025.

11. Kotia, R. B.; Li, L.; McGown, L. B. Separation of nontarget compounds by DNA aptamers, *Anal. Chem.* **2000**, *72*, 827-831.

12. Rehder, M. A.; McGown, L. B. Open-tubular capillary electrochromatography of bovine beta-lactoglobulin variants A and B using an aptamer stationary phase, *Electrophoresis* **2001**, *22*, 3759-3764.

13. Romig, T. S.; Bell, C.; Drolet, D. W. Aptamer affinity chromatography: combinatorial chemistry applied to protein purification, *J. Chromatogr. B Biomed. Sci. Appl.* **1999**, *731*, 275-284.

14. Deng, Q.; German, I.; Buchanan, D.; Kennedy, R. T. Retention and separation of adenosine and analogues by affinity chromatography with an aptamer stationary phase, *Anal. Chem.* **2001**, *73*, 5415-5421.

15. Kleinjung, F.; Klussmann, S.; Erdmann, V. A.; Scheller, F. W.; Fuerste, J. P.; Bier, F. F. Binders in biosensors:high-affinity RNA for small analytes, *Anal. Chem.* **1998**, *70*, 328-331.

16. McCauley, T. G.; Hamaguchi, N.; Stanton, M. Aptamer-based biosensor arrays for detection and quantification of biological macromolecules, *Anal. Biochem.* **2003**, *319*, 244-250.

17. Hesselberth, J.; Robertson, M. P.; Jhaveri, S.; Ellington, A. D. In vitro selection of nucleic acids for diagnostic applications, *J Biotechnol* **2000**, *74*, 15-25.

18. Rajendran, M.; Ellington, A. D. Selecting nucleic acids for biosensor applications, *Comb Chem High Throughput Screen* **2002**, *5*, 263-270.

19. Conrad, R. C.; Giver, L.; Tian, Y.; Ellington, A. D. In vitro selection of nucleic acid aptamers that bind proteins, *Methods Enzymol.* **1996**, *267*, 336-367.

20. Cox, J. C.; Rudolph, P.; Ellington, A. D. Automated RNA selection, *Biotechnol. Prog.* **1998**, *14*, 845-850.

21. Jenison, R. D.; Gill, S. C.; Pardi, A.; Polisky, B. High-resolution molecular discrimination by RNA, *Science* **1994**, *263*, 1425-1429.

22. Jayasena, S. D. Aptamers: an emerging class of molecules that rival antibodies in diagnostics, *Clin. Chem.* **1999**, *45*, 1628-1650.

23. Pavski, V.; Le, X. C. Detection of human immunodeficiency virus type 1 reverse transcriptase using aptamers as probes in affinity capillary electrophoresis, *Anal. Chem.* **2001**, *73*, 6070-6076.

24. Kawazoe, N.; Ito, Y.; Imanishi, Y. Bioassay using a labeled oligonucleotide obtained by in vitro selection, *Biotechnol. Prog.* **1997**, *13*, 873-874.

25. Burgstaller, P.; Kochoyan, M.; Famulok, M. Structural probing and damage selection of citrulline- and arginine-specific RNA aptamers identify base positions required for binding, *Nucleic Acids Res.* **1995**, *23*, 4769-4776.

26. Padmanabhan, K.; Padmanabhan, K. P.; Ferrara, J. D.; Sadler, J. E.; Tulinsky, A. The structure of alpha-thrombin inhibited by a 15-mer single-stranded DNA aptamer, *J. Biol. Chem.* **1993**, *268*, 17651-17654.

27. Hermann, T.; Patel, D. J. Adaptive recognition by nucleic acid aptamers, *Science* **2000**, *287*, 820-825.

28. Ye, X.; Gorin, A.; Frederick, R.; Hu, W.; Majumdar, A.; Xu, W.; McLendon, G.; Ellington, A.; Patel, D. J. RNA architecture dictates the conformations of a bound peptide, *Chem. Biol.* **1999**, *6*, 657-669.

29. Patel, D. J.; Suri, A. K. Structure, recognition and discrimination in RNA aptamer complexes with cofactors, amino acids, drugs and aminoglycoside antibiotics, *Rev. Mol. Biotechnol.* **2000**, *74*, 39-60.

30. Sassanfar, M.; Szostak, J. W. An RNA motif that binds ATP, *Nature* **1993**, *364*, 550-553.

31. Huizenga, D. E.; Szostak, J. W. A DNA aptamer that binds adenosine and ATP, *Biochemistry* **1995**, *34*, 656-665.

32. Jhaveri, S.; Kirby, R.; Conrad, R.; Maglott, E. J.; Bowser, M.; Kennedy, R. T.; Glick, G.; Ellington, A. D. Designed signaling aptamers that transduce molecular recognition to changes in fluorescence intensity, *J. Am. Chem. Soc.* **2000**, *122*, 2469-2473.

33. Yamana, K.; Ohtani, Y.; Nakano, H.; Saito, I. Bis-pyrene labeled DNA aptamer as an intelligent fluorescent biosensor, *Bioorg. Med. Chem. Lett.* **2003**, *13*, 3429-3431.

34. Jhaveri, S.; Rajendran, M.; Ellington, A. D. In vitro selection of signaling aptamers, *Nat. Biotechnol.* **2000**, *18*, 1293-1297.
35. Tyagi, S.; Kramer, F. R. Molecular beacons: probes that fluoresce upon hybridization, *Nat. Biotechnol.* **1996**, *14*, 303-308.
36. Tyagi, S.; Bratu, D. P.; Kramer, F. R. Multicolor molecular beacons for allele discrimination, *Nat. Biotechnol.* **1998**, *16*, 49-53.
37. Tan, W.; Fang, X.; Li, J.; Liu, X. Molecular beacons: a novel DNA probe for nucleic acid and protein studies, *Chemistry* **2000**, *6*, 1107-1111.
38. Fang, X.; Li, J. J.; Perlette, J.; Tan, W.; Wang, K. Molecular beacons: novel fluorescent probes, *Anal. Chem.* **2000**, *72*, 747A-753A.
39. Marras, S. A.; Kramer, F. R.; Tyagi, S. Multiplex detection of single-nucleotide variations using molecular beacons, *Genet. Anal.* **1999**, *14*, 151-156.
40. Lakowicz, J. R. *Principles of Fluorescence Spectroscopy.*, 2 ed.; Kluwer Academic/ Plenum Press: New York, N.Y., 1999.
41. Morrison, L. E. Homogeneous detection of specific DNA sequences by flourescence quenching and energy transfer, *J. Fluorescence* **1999**, *9*, 187-196.
42. Bock, L. C.; Griffin, L. C.; Latham, J. A.; Vermaas, E. H.; Toole, J. J. Selection of single-stranded DNA molecules that bind and inhibit human thrombin, *Nature* **1992**, *355*, 564-566.
43. Wu, Q.; Tsiang, M.; Sadler, J. E. Localization of the single-stranded DNA binding site in the thrombin anion-binding exosite, *J. Biol. Chem.* **1992**, *267*, 24408-24412.
44. Paborsky, L. R.; McCurdy, S. N.; Griffin, L. C.; Toole, J. J.; Leung, L. L. The single-stranded DNA aptamer-binding site of human thrombin, *J. Biol. Chem.* **1993**, *268*, 20808-20811.
45. Macaya, R. F.; Waldron, J. A.; Beutel, B. A.; Gao, H.; Joesten, M. E.; Yang, M.; Patel, R.; Bertelsen, A. H.; Cook, A. F. Structural and functional characterization of potent antithrombotic oligonucleotides possessing both quadruplex and duplex motifs, *Biochemistry* **1995**, *34*, 4478-4492.
46. Tsiang, M.; Jain, A. K.; Dunn, K. E.; Rojas, M. E.; Leung, L. L.; Gibbs, C. S. Functional mapping of the surface residues of human thrombin, *J. Biol. Chem.* **1995**, *270*, 16854-16863.
47. Tasset, D. M.; Kubik, M. F.; Steiner, W. Oligonucleotide inhibitors of human thrombin that bind distinct epitopes, *J. Mol. Biol.* **1997**, *272*, 688-698.
48. Macaya, R. F.; Schultze, P.; Smith, F. W.; Roe, J. A.; Feigon, J. Thrombin-binding DNA aptamer forms a unimolecular quadruplex structure in solution, *Proc. Natl. Acad. Sci. U S A* **1993**, *90*, 3745-3749.
49. Schultze, P.; Macaya, R. F.; Feigon, J. Three-dimensional solution structure of the thrombin-binding DNA aptamer d(GGTTGGTGTGGTTGG), *J. Mol. Biol.* **1994**, *235*, 1532-1547.
50. Kelly, J. A.; Feigon, J.; Yeates, T. O. Reconciliation of the X-ray and NMR structures of the thrombin-binding aptamer d(GGTTGGTGTGGTTGG), *J. Mol. Biol.* **1996**, *256*, 417-422.
51. Hamaguchi, N.; Ellington, A.; Stanton, M. Aptamer beacons for the direct detection of proteins, *Anal. Biochem.* **2001**, *294*, 126-131.
52. Stojanovic, M. N.; de Prada, P.; Landry, D. W. Aptamer-based folding fluorescent sensor for cocaine, *J. Am. Chem. Soc.* **2001**, *123*, 4928-4931.
53. Fang, X.; Sen, A.; Vicens, M.; Tan, W. Synthetic DNA aptamers to detect protein molecular variants in a high-throughput fluorescence quenching assay, *Chembiochem.* **2003**, *4*, 829-834.
54. Li, J. J.; Fang, X.; Tan, W. Molecular aptamer beacons for real-time protein recognition, *Biochem. Biophys. Res. Commun.* **2002**, *292*, 31-40.
55. Nutiu, R.; Li, Y. Structure-switching signaling aptamers, *J. Am. Chem. Soc.* **2003**, *125*, 4771-4778.
56. Stojanovic, M. N.; de Prada, P.; Landry, D. W. Fluorescent sensors based on aptamer self-assembly, *J. Am. Chem. Soc.* **2000**, *122*, 11547-11548.
57. Yamamoto, R.; Baba, T.; Kumar, P. K. Molecular beacon aptamer fluoresces in the presence of Tat protein of HIV-1, *Genes Cells* **2000**, *5*, 389-396.
58. Merino, E. J.; Weeks, K. M. Fluorogenic resolution of ligand binding by a nucleic acid aptamer, *J. Am. Chem. Soc.* **2003**, *125*, 12370-12371.
59. Rajendran, M.; Ellington, A. D. In vitro selection of molecular beacons, *Nucleic Acids Res.* **2003**, *31*, 5700-5713.
60. Poddar, S. K. Detection of adenovirus using PCR and molecular beacon, *J. Virol. Methods* **1999**, *82*, 19-26.
61. Sokol, D. L.; Zhang, X.; Lu, P.; Gewirtz, A. M. Real time detection of DNA.RNA hybridization in living cells, *Proc. Natl. Acad. Sci. U S A* **1998**, *95*, 11538-11543.
62. Liu, X.; Tan, W. A fiber-optic evanescent wave DNA biosensor based on novel molecular beacons, *Anal. Chem.* **1999**, *71*, 5054-5059.
63. Steemers, F. J.; Ferguson, J. A.; Walt, D. R. Screening unlabeled DNA targets with randomly ordered fiber-optic gene arrays, *Nat. Biotechnol.* **2000**, *18*, 91-94.

64. Fang, X.; Li, J. J.; Tan, W. Using molecular beacons to probe molecular interactions between lactate dehydrogenase and single-stranded DNA, *Anal. Chem.* **2000**, *72*, 3280-3285.
65. Potyrailo, R. A.; Conrad, R. C.; Ellington, A. D.; Hieftje, G. M. Adapting selected nucleic acid ligands (aptamers) to biosensors, *Anal. Chem.* **1998**, *70*, 3419-3425.
66. Fang, X.; Cao, Z.; Beck, T.; Tan, W. Molecular aptamer for real-time oncoprotein platelet-derived growth factor monitoring by fluorescence anisotropy, *Anal. Chem.* **2001**, *73*, 5752-5757.
67. Biran, I.; Rissin, D. M.; Ron, E. Z.; Walt, D. R. Optical imaging fiber-based live bacterial cell array biosensor, *Anal. Biochem.* **2003**, *315*, 106-113.
68. D'Orazio, P. Biosensors in clinical chemistry, *Clin. Chim. Acta* **2003**, *334*, 41-69.
69. Sapsford, K. E.; Rasooly, A.; Taitt, C. R.; Ligler, F. S. Detection of Campylobacter and Shigella species in food samples using an array biosensor., *Anal. Chem.* **2004**, *76*, 433-440.
70. Ramsay, G. DNA chips: state-of-the art, *Nat. Biotechnol.* **1998**, *16*, 40-44.
71. Eisen, M. B.; Brown, P. O. DNA arrays for analysis of gene expression, *Methods Enzymol.* **1999**, *303*, 179-205.
72. Cox, J. C.; Hayhurst, A.; Hesselberth, J.; Bayer, T. S.; Georgiou, G.; Ellington, A. D. Automated selection of aptamers against protein targets translated in vitro: from gene to aptamer, *Nucleic Acids Res.* **2002**, *30*, e108.
73. Green, N. M. Avidin, *Adv. Protein Chem.* **1975**, *29*, 85-133.
74. Lavigne, J. J.; Savoy, S.; Clevenger, M. B.; Ritchie, J. E.; McDoniel, B.; Yoo, S. J.; Anslyn, E. V.; McDevitt, J. T.; Shear, J. B.; Neikirk, D. P. Solution-based analysis of multiple analytes by a sensor array: toward the development of an "Electronic tongue". *J. Am. Chem. Soc.* **1998**, *120*, 6429-6430.
75. Buranda, T.; Huang, J.; Perez-Luna, V. H.; Schreyer, B.; Sklar, L. A.; Lopez, G. P. Biomolecular recognition on well-characterized beads packed in microfluidic channels, *Anal. Chem.* **2002**, *74*, 1149-1156.
76. Seong, G. H.; Crooks, R. M. Efficient mixing and reactions within microfluidic channels using microbead-supported catalysts, *J. Am. Chem. Soc.* **2002**, *124*, 13360-13361.
77. Fulton, R. J.; McDade, R. L.; Smith, P. L.; Kienker, L. J.; Kettman, J. R., Jr. Advanced multiplexed analysis with the FlowMetrix system, *Clin. Chem.* **1997**, *43*, 1749-1756.
78. Egner, B. J.; Rana, S.; Smith, H.; Bouloc, N.; Frey, J. G.; Brocklesby, W. S.; Bradley, M. Tagging in combinatorial chemistry: the use of colored and fluorescent beads, *Chem.Comm.* **1997**, *8*, 735-736.
79. Needels, M. C.; Jones, D. G.; Tate, E. H.; Heinkel, G. L.; Kochersperger, L. M.; Dower, W. J.; Barrett, R. W.; Gallop, M. A. Generation and screening of an oligonucleotide-encoded synthetic peptide library, *Proc. Natl. Acad. Sci. U S A* **1993**, *90*, 10700-10704.
80. Hakala, H.; Lonnberg, H. Time-resolved fluorescence detection of oligonucleotide hybridization on a single microparticle: covalent immobilization of oligonucleotides and quantitation of a model system, *Bioconjug. Chem.* **1997**, *8*, 232-237.
81. Hakala, H.; Heinonen, P.; Iitia, A.; Lonnberg, H. Detection of oligonucleotide hybridization on a single microparticle by time-resolved fluorometry: hybridization assays on polymer particles obtained by direct solid phase assembly of the oligonucleotide probes, *Bioconjug. Chem.* **1997**, *8*, 378-384.
82. Van Ness, J.; Kalbfleisch, S.; Petrie, C. R.; Reed, M. W.; Tabone, J. C.; Vermeulen, N. M. A versatile solid support system for oligodeoxynucleotide probe-based hybridization assays, *Nucleic Acids Res.* **1991**, *19*, 3345-3350.
83. Storhoff, J. J.; Elghanian, R.; Mucic, R. C.; Mirkin, C. A.; Letsinger, R. L. One-pot colorimetric differentiation of polynucleotides with single base imperfections using gold nanoparticle probes., *J. Am. Chem. Soc.* **1998**, *120*, 1959-1964.
84. Lee, M.; Walt, D. R. A fiber-optic microarray biosensor using aptamers as receptors, *Anal. Biochem.* **2000**, *282*, 142-146.
85. Lin, Y.; Padmapriya, A.; Morden, K. M.; Jayasena, S. D. Peptide conjugation to an in vitro-selected DNA ligand improves enzyme inhibition, *Proc. Natl. Acad. Sci. U S A* **1995**, *92*, 11044-11048.
86. Knight, B. Ricin--a potent homicidal poison, *Br. Med. J.* **1979**, *1*, 350-351.
87. Lord, J. M.; Roberts, L. M.; Robertus, J. D. Ricin: structure, mode of action, and some current applications, *Faseb J.* **1994**, *8*, 201-208.
88. Vitetta, E. S.; Thorpe, P. E.; Uhr, J. W. Immunotoxins: magic bullets or misguided missiles? , *Trends Pharmacol. Sci.* **1993**, *14*, 148-154.
89. Hesselberth, J. R.; Miller, D.; Robertus, J.; Ellington, A. D. In vitro selection of RNA molecules that inhibit the activity of ricin A-chain, *J. Biol. Chem.* **2000**, *275*, 4937-4942.
90. Narang, U.; Anderson, G. P.; Ligler, F. S.; Burans, J. Fiber optic-based biosensor for ricin, *Biosens. Bioelectron.* **1997**, *12*, 937-945.

91. Rowe-Taitt, C. A.; Hazzard, J. W.; Hoffman, K. E.; Cras, J. J.; Golden, J. P.; Ligler, F. S. Simultaneous detection of six biohazardous agents using a planar waveguide array biosensor, *Biosens. Bioelectron.* **2000**, *15*, 579-589.
92. Taitt, C. R.; Anderson, G. P.; Lingerfelt, B. M.; Feldstein, M. J.; Ligler, F. S. Nine-analyte detection using an array-based biosensor, *Anal. Chem.* **2002**, *74*, 6114-6120.
93. Delehanty, J. B.; Ligler, F. S. A microarray immunoassay for simultaneous detection of proteins and bacteria, *Anal. Chem.* **2002**, *74*, 5681-5687.
94. Poli, M. A.; Rivera, V. R.; Hewetson, J. F.; Merrill, G. A. Detection of ricin by colorimetric and chemiluminescence ELISA, *Toxicon* **1994**, *32*, 1371-1377.
95. Li, Y.; Nath, N.; Reichert, W. M. Parallel comparison of sandwich and direct label assay protocols on cytokine detection protein arrays, *Anal. Chem.* **2003**, *75*, 5274-5281.

MOLECULAR IMPRINTING

Petra Turkewitsch, Robert Massé, and William S. Powell[*]

6.1 INTRODUCTION

Many biological processes are governed by complex interactions between biomolecules. Molecular recognition is, therefore, a fundamental feature of life. Antibody-antigen, ligand-receptor and enzyme-substrate interactions are examples of such biorecognition systems, wherein the recognition domain of the protein binds its ligand or substrate through the cumulative action of multiple intermolecular interactions mostly of a noncovalent nature. These natural molecular recognition molecules offer exquisite selectivity and high affinity, and thus are often incorporated into sensors and used in a variety of bioanalytical techniques, such as ligand binding assays and affinity chromatography.[1-4] However, despite their excellent recognition capabilities, these proteins can lack stability, especially under harsh chemical conditions (e.g. extremes of pH and temperature, organic solvents), can be time-consuming to prepare or costly and difficult to obtain, and their reuse is often limited. For these reasons alternative approaches that use artificial recognition elements with affinities and selectivities similar to their biological counterparts, but with enhanced stability are actively being pursued.[5-7]

Relatively simple artificial receptors that efficiently and selectively bind certain inorganic ions are currently available. By incorporating a fluorophore into or adjacent to the binding domain of the receptor molecule, fluorescent chemosensors have been developed that undergo selective target-induced changes in fluorescence.[6, 7] Many such compounds are now available, and are popular and powerful tools for investigating the intracellular concentrations and localization of calcium and a variety of other ions in real time.[8, 9]

Preparation of an artificial receptor of high affinity and selectivity for a molecule that is larger and more complex than a simple spherical ion is a more daunting task, as it requires the synthesis of a cleft or cavity that has a size and shape to match the analyte, as well as the correct spatial arrangement of functional groups to interact with

* Petra Turkewitsch and Robert Massé, Applied Research and Development, MDS Pharma Services, 2350 Cohen Street, Montreal, QC, Canada. William S. Powell, Meakins-Christie Laboratories, McGill University, 3626 St. Urbain Street, Montreal, QC, Canada.

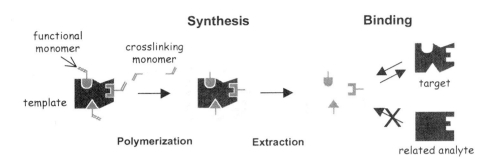

Figure 6.1. Schematic representation of the molecular imprinting process. **Synthesis**: Functional monomers interact with complementary functional groups on the template molecule in solution. Polymerization in the presence of a large amount of cross-linking monomer preserves the orientation and arrangement of the solution template-functional monomer complexes. Extraction of the template molecules from the polymer exposes recognition sites complementary in shape and functional topography to the template. **Binding**: Target molecules possessing the correct shape and arrangement of functional groups selectively rebind to the MIP.

complementary groups on the target molecule.[6] This often involves substantial effort and elaborate organic synthesis. Towards this end, researchers are investigating the technique of molecular imprinting as a means of creating artificial receptors with biological-like binding capabilities (i.e. biomimetic receptors).[5, 10] Molecular imprinting is a process whereby selective recognition sites are created for a target molecule in a polymer by copolymerizing functional and cross-linking monomers in the presence of a template molecule (also referred to as an imprint or print molecule) (Fig. 6.1). The template is normally the target molecule itself, but could also be a close structural analogue. Subsequent removal of the template molecules from the polymer exposes microcavities or imprints with a shape to match the template. These microcavities also contain precisely oriented functional groups complementary to those of the template, held in place through the rigidity of the three-dimensional cross-linked polymer network. Following removal of the template, the molecularly imprinted polymer (MIP) can rebind the target molecule through multipoint attachment to these microcavities, as occurs in enzyme-substrate and receptor-ligand binding, and can exclude other molecules according to the lock-and-key principle originally proposed by Fischer about 100 years ago.[11] This MIP exhibits tailor-made selectivity for the template molecule, and can thus act as an artificial macromolecular receptor.

The concept of molecular imprinting dates back to the 1930's when Polyakov dried silica gel in the presence of various small aromatic molecules, and found that the silica later preferentially adsorbed the aromatic molecule that had been present during drying.[12] In 1949, Frank Dickey,[13] then a student of Linus Pauling, was inspired by Pauling's[14] theory for the production of antibodies *in vitro*, and took the idea of molecular imprinting one step further. He prepared "specific adsorbents" by synthesizing silica gel in the presence of different dye molecules, namely methyl, ethyl, *n*-propyl and *n*-butyl orange. The resulting silica, after washing, displayed "maximum adsorption power for the dye used in the preparation of the adsorbent." However, it wasn't until the 1970's with the work of Wulff and coworkers[15, 16] that molecular imprinting, as we know it today, was born. Wulff prepared MIPs for two optically active templates, D-glyceric acid and D-mannitol, by reacting their diol groups with a polymerizable functional monomer containing a boronic acid group, resulting in the reversible formation of covalent boronate bonds. Each template-functional monomer complex was copolymerized with a cross-linking monomer in the presence of a porogen to produce MIPs, which, after chemically cleaving the templates from the polymer, were selective for the templates. Upon equilibrating each of the MIPs with a racemic mixture of the template, the enantiomer used as the template was preferentially adsorbed, while its antipode remained in solution. Subsequent experiments done in this batch binding mode, combined with chromatographic experiments where the MIPs were used as stationary phases in HPLC, demonstrated that this covalent imprinting approach could be used to create highly selective imprinted polymers that could be used for the resolution of racemates.[10, 17, 18] However, the limited number of types of reversible covalent interactions suitable for molecular imprinting, and the requirement for the synthesis of a polymerizable template-functional monomer complex limit the versatility of this approach. These shortcomings were overcome in the 1980's when Mosbach and coworkers[19, 20] prepared an imprinted polymer using exclusively noncovalent interactions between the template and the functional monomer(s). This noncovalent approach simplified the imprinting procedure and extended the list of molecules amenable to imprinting, as the MIP could be prepared by simply mixing the functional monomer(s) with the template in an organic solvent to allow complementary interactions to form between them.[21-24]

MIPs have now been prepared that are selective for a wide variety of small molecules, including amino acids,[10, 25-28] sugars,[10, 29-33] therapeutic drugs,[34-41] environmental pollutants,[42-49] nucleotide bases and nucleotides,[33, 50-55] steroids,[56-62] and metal ions,[63-68] including copper and zinc. Although selective recognition sites have been prepared for peptides,[34, 69-74] their creation for large biological macromolecules, such as proteins, as well as complex cell structures, has progressed more slowly.[73, 75-77] However, innovative procedures have recently been developed for the imprinting of proteins, yeast and bacteria,[78-83] and more are expected in the future. The applications of MIPs have expanded considerably from their initial use as stationary phases in chiral chromatography to a variety of other separation techniques, including capillary electrophoresis and solid-phase extraction.[26, 84-87] Intensive research is also underway to develop MIPs as substitutes for biological recognition molecules in assays, and for the screening combinatorial libraries of potential drug candidates, as well as artificial enzyme systems for catalytic applications.[88-92] A further goal of molecular imprinting technology is to employ MIPs as the biomimetic recognition component in sensors.[93-96]

MIPs have many potential advantages over biological recognition molecules.[26, 93, 97, 98] They combine the advantages of easy tailor design with physical and chemical stability and durability. They are much more versatile than their biological counterparts as they can be used in extreme pH and temperature environments, and in organic solvents. Furthermore, MIPs can be stored at room temperature in a dry state, and can be reused. They can be efficiently prepared in large quantities, at lower cost than natural antibodies or receptors, and avoid the use of animals or any material of biological origin. In order to raise antibodies against small molecules, one of its functional groups must be coupled to a carrier molecule. This conjugation can be quite laborious, may be difficult to accomplish,

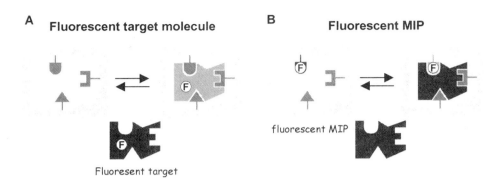

Figure 6.2. Schematic representation of the different approaches used to detect MIP-target interactions using fluorescence. **A. Fluorescent target molecule**: Fluorescence properties of target change upon binding to MIP. **B. Fluorescent MIP**: Fluorescence properties of MIP are altered by target binding.

and could result in conformational changes in the small molecule. Recognition by the antibody will then be based only on the remainder of the small molecule that is not conjugated to the carrier, which could potentially be in an altered conformational state. It may, therefore, be difficult for the antibody to discriminate between the intended target molecule and related structures, especially in that portion of the molecule that was attached to the carrier. In contrast, conjugation is not required for MIPs to be prepared against small molecules, so the imprinted sites are made against the entire molecule. Thus, the possibility of obtaining selectivities that are better than those of antibodies exists. Furthermore, molecular imprints may be made against many different types of target molecules, including compounds for which it is difficult to obtain natural antibodies, such as toxic substances, immunosuppressive drugs, and molecules that are only soluble in organic solvents.

If MIPs are to ultimately be used as biomimetic recognition elements in sensors, a sensitive method that detects the chemical or physical change generated by the binding of the target, and converts it into a physically measurable signal must be employed.[26, 93] Because of the multiple advantages of fluorescence, it has been widely used as a signal transduction mechanism for detecting the binding of target molecules to MIPs.[7, 96, 99-101] This is relatively simple if the target molecule is itself fluorescent, as the fluorescence intensity and/or wavelength is often affected by conformational changes associated with binding to the MIP (Fig. 2). A more versatile approach is to create a fluorescent MIP by incorporating a fluorescent molecule into the polymer network, in which case its fluorescence properties could be altered by interaction with the target. The fluorescent molecule in the MIP could interact directly with the target, although this does not have to be the case.

6.2 SYNTHESIS AND EVALUATION OF MIPS

MIPs are prepared by first allowing monomers containing functional groups, referred to as functional monomers, to interact with complementary groups in the template molecules in a suitable solvent (also known as the porogen or porogenic solvent) (Fig. 6.1). [5, 10, 90] The template can interact with the functional monomer(s) through covalent or noncovalent bonds, metal ion coordination, or a combination of these interactions. The orientation and arrangement of these template-functional monomer complexes in solution are preserved in the resulting polymer through the addition of a large amount of cross-linking monomer. Typically, the polymerization is carried out by mixing the template, solvent, functional monomer, and cross-linking monomer together before adding the polymerization initiator. The mixture is then degassed, and the polymerization is initiated by heat or light. A rigid, insoluble, macroporous polymer block (bulk polymerization) is obtained, which is ground into fine particles for use. Imprinted sites complementary in shape and functionality to the template are exposed when the template is removed from the polymer by extraction or hydrolysis. The most common polymer matrices for preparing MIPs are based on methacrylate, acrylate, or vinyl functional monomers and cross-linkers, although other organic and inorganic matrices (most notably silica, Section 6.2.2.5) have also been used. MIPs are usually prepared as polymer blocks as described above, but they may also be prepared in the form of spherical beads (Section 6.2.2.3) or thin films (Section 6.2.2.2) coating the surface of a substrate.

6.2.1 Components of MIPs

6.2.1.1 Template

Although it is possible to prepare MIPs using templates consisting of only aromatic rings or a single functional group, the majority of MIPs have been directed at targets containing multiple functional groups (e.g. hydroxyl, amino, carboxylic acid, amide). [10, 102] This provides multiple sites for interaction with functional monomers, resulting in a higher likelihood of producing binding sites with high affinity and selectivity. Such high fidelity binding sites are also favored with structurally rigid templates, as they possess fewer possible solution conformations. This will result in the formation of more homogeneous binding sites in the MIP that will recognize the solution conformation of the target, thus requiring a minimal degree of conformational change and entropy loss, promoting high affinity binding.

6.2.1.2 Solvent

The polymerization is usually carried out in the presence of a solvent, which is also referred to as a porogen because it creates the pore structure of the imprinted polymer, thereby permitting the target molecule to access the imprinted sites. [89] The choice of solvent is often dictated by the solubility of the target molecule and the monomers, as it is usually required to dissolve the components of the prepolymerization mixture. For noncovalent molecular imprinting (see section 6.2.1.3.2 below), an organic solvent should be selected that preferably interferes as little as possible with the interactions between the monomer(s) and the template. Since these interactions are often electrostatic and hydrogen bonding in nature, aprotic solvents of low polarity and low hydrogen bonding capacity are favored to maximize the strength of the monomer-template interactions, and thus the production of well-defined high affinity binding sites.

6.2.1.3 Functional Monomer

The functional monomer(s) is of critical importance in the design of an MIP, because it interacts directly with the template, and is therefore responsible for producing specific interaction sites for the target within the binding cavity after polymerization.[84, 102, 103] Functional monomers that interact strongly with the template, for example through covalent interactions (Fig. 6.3), produce stable template-monomer complexes that have a greater likelihood of remaining intact throughout the polymerization process, creating an MIP that has functional groups localized in the binding cavities. On the other hand, if the functional monomers interact more weakly with the template, as may be the case with noncovalent interactions (Fig. 6.3), a large excess of monomer may be required to drive the equilibrium in the prepolymerization solution towards complex formation. This would result in functional monomers being distributed widely throughout the resulting polymer, rather than being restricted principally to the binding cavities, thus increasing the degree

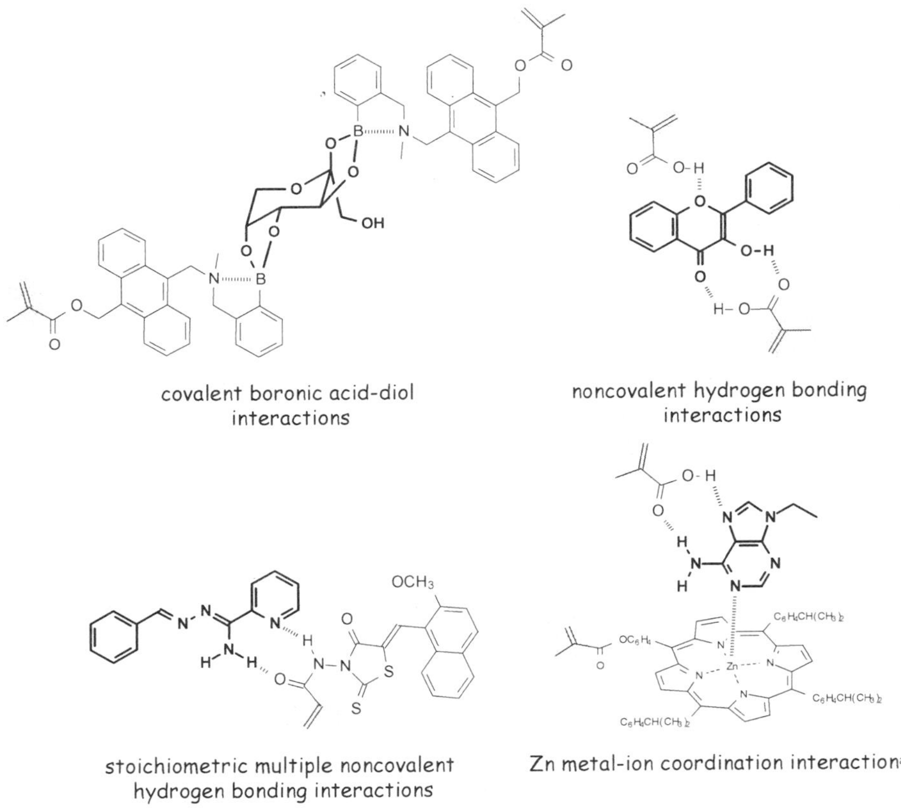

covalent boronic acid-diol
interactions

noncovalent hydrogen bonding
interactions

stoichiometric multiple noncovalent
hydrogen bonding interactions

Zn metal-ion coordination interaction[a]

Figure 6.3. Examples of covalent,[29] noncovalent,[104] stoichiometric noncovalent,[105] and metal-ion coordination[106] interactions between template and functional monomer. Template molecules are depicted using thick bond lines. A shaded double bond indicates the portion of the functional monomer that is incorporated into the polymer matrix. [a] MAA was added for additional hydrogen bonding interactions.

of nonspecific binding. In addition, a range of template-functional monomer complexes, differing in orientation and affinity between the interacting components, is produced when weak interactions exist between the template and functional monomer, resulting in an MIP with a more heterogeneous population of binding sites. Despite these potential problems, functional monomers that interact with the template through noncovalent interactions are most often used because of their flexibility. The choice of functional monomer is ultimately determined by the structure of the template and/or the intended application of the MIP.

The choice of a functional monomer can be facilitated by preliminary studies investigating potential interactions between it and the template. Such interactions can be evaluated using various spectroscopic techniques (^{1}H-NMR, fluorescence, absorption), by chromatography, or molecular modeling studies.[22, 74, 82, 105, 107-109] It would be prudent to examine this interaction under aqueous conditions, if the MIP is to ultimately be used to detect a target molecule in an aqueous environment. There is still a considerable degree of trial-and-error in the choice of a functional monomer, and in this regard, combinatorial techniques are particularly useful because a wide range of MIPs can be synthesized using many possible combinations of monomers, cross-linkers and solvents, and then screened with the target of interest.[110-113]

To permit the detection of a nonfluorescent target binding to an MIP, a fluorescent MIP may be synthesized using a fluorescent functional monomer. This compound may be prepared by adding a fluorescent group to an existing nonfluorescent functional monomer that interacts selectively with the template.[114] Alternatively, a fluorescent functional monomer may be designed in which the fluorescent group participates in the interaction with the template.[53]

6.2.1.3.1 Covalent Interaction with Template. The first MIPs synthesized by Wulff and coworkers utilized the covalent imprinting method, which requires the synthesis of a polymerizable derivative of the template.[10] Popular monomers for covalent imprinting are those containing boronic acid groups, since they form stable, but reversible, boronic esters with templates containing diol groups (Fig. 6.3). The formation of Schiff bases between primary amine groups and aldehyde moieties,[115] as well as ketal formation between diketones and diol-containing compounds have also been explored.[116]

This approach has the advantage that the binding sites are more homogeneous, and the imprinting efficiency is high. Since a more stable interaction of exact stoichiometry exists between the template and functional monomer(s), the functional groups are almost exclusively located in the imprinted polymer cavities, and an excess of functional monomer is not required. This should greatly minimize nonspecific binding sites, which is a problem with the noncovalent approach, in which a large excess of functional monomer is required. However, the flexibility of the covalent approach is limited by the requirement for a reversible covalent interaction between the template and monomer, as a relatively small number of appropriate interactions are available. Another disadvantage of this method is that, after polymerization, the template must be removed from the MIP by chemical cleavage. Furthermore, once the MIP has been prepared, the kinetics of target binding may be slow if covalent bonds have to be formed.

6.2.1.3.2 Noncovalent Interaction with Template. Some of the above problems can be circumvented if the template and functional monomer interact in a noncovalent manner prior to polymerization (Fig. 3).[102, 117] The functional monomers most commonly used are based on methacrylate, acrylate, and vinyl polymerizable groups, because a wide variety of these compounds are commercial available. Figure 4 presents some acidic (e.g. methacrylic acid, MAA), basic (e.g. vinylpyridine, 2-(diethylamino)ethyl methacrylate), hydrogen bonding (e.g. 2-hydroxyethylmethacrylate, acrylamide), and hydrophobic (e.g. styrene) monomers. Their functional groups interact with complementary functional groups of the template, resulting in incorporation of the noncovalently bound template into the polymer, and the formation of binding sites that can subsequently recognize the target molecule.

Because noncovalent interactions are weaker than covalent interactions, to obtain binding sites of high affinity and selectivity it is desirable to use a functional monomer that can interact with several structural features on the template molecule, or if necessary to use more than one functional monomer.[97, 102] Such binding sites function in a manner similar to natural molecular recognition systems, where the binding of a biomolecule occurs through the cooperative action of several weak interactions between the biological binding site and biomolecule.

The ratio of functional monomer to template is very important, as the homogeneity of the binding sites produced in the MIP during the polymerization process will be maximized if the template molecule is fully complexed with the functional monomer. If the ratio is too low, many of the binding sites will contain fewer than the optimal number of functional monomers, resulting in lower affinity and increased heterogeneity. On the other hand, if the ratio is too high, a large number of functional monomer molecules will be incorporated into the polymer at sites other than template binding sites. This will result in high levels of nonspecific binding and reduced selectivity of the MIP.

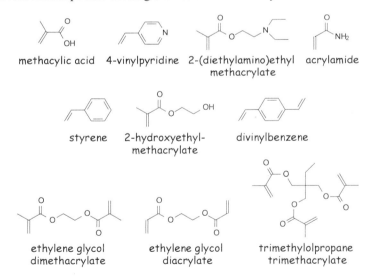

Figure 6.4. Common functional and cross-linking monomers for MIP synthesis.

As a compromise, a ratio of functional monomer to template functional group of about 4:1 is often used.[10] Thus, if the functional monomer can interact with three groups on the template molecule, a ratio of 12:1 would be used. Finally, since the strength of noncovalent interactions is strongly solvent dependent, the choice of solvent must also be considered very carefully. Electrostatic and hydrogen bonding interactions are stronger in nonpolar solvents, whereas hydrophobic interactions are enhanced in aqueous environments or in solvents containing hydroxyl groups.

The preparation of MIPs based on noncovalent template-monomer complexes offers many advantages over other approaches because of its versatility. MIPs prepared in this manner can utilize a variety of distinct interactions to bind the template, including hydrogen bonding, electrostatic and hydrophobic interactions, allowing a large range of molecules to be imprinted. In addition, several different functional monomers that interact simultaneously with the template may be used. The noncovalent method also has the advantage that the bound template molecule can easily be removed from the polymer by mild extraction rather than by chemical cleavage. Because of its simplicity, flexibility, and its similarity to natural systems, where noncovalent molecular interactions are predominant, noncovalent imprinting is now, by far, the more popular method.

Despite its many advantages, the preparation of MIPs based on noncovalent interactions is not without difficulties. The relatively weak nature of these interactions results in considerable heterogeneity among binding sites and increased nonspecific binding compared to the covalent approach. As a result, only a minority of the sites at which the template was present in the initially formed polymer have sufficient affinity to bind the target in analytical applications. One report[23] estimated that less than 1% of the binding sites based on template loading could be reoccupied in binding studies, while in another study[24] a value of 35% was obtained. The choice of solvents may also be more limited in this case, since noncovalent interactions of electrostatic and hydrophobic natures are strongest in nonpolar solvents.

The proportion of high affinity binding sites in the MIP can be maximized, and the degree of nonspecific binding minimized by employing functional monomers that form highly stable complexes with the template. For this reason, functional monomers capable of stoichiometric noncovalent[103, 118] and metal-ion coordination[119, 120] interactions are currently receiving considerable attention. In the case of stoichiometric noncovalent interactions, the functional monomer interacts with the template through multiple noncovalent bonds –typically two or more hydrogen bonds –forming a cyclic hydrogen bonded ion pair (Fig. 6.3).[54, 105, 121-125] This increases the strength of the template-functional monomer complex, and places a high proportion of the functional groups in the imprinted sites upon polymerization. Furthermore, these noncovalent bonds are easily disrupted to release the template, are quickly reformed when template molecules rebind to the unoccupied binding sites, and are less sensitive to the presence of polar protic solvents. A typical example of stoichiometric noncovalent interactions is found in nature in the guanidinium group of the amino acid arginine. This group is involved in protein recognition where it forms multiple hydrogen bonds and electrostatic interactions with the oxygen atoms of phosphate anions. Similar binding patterns are observed with the amidine group, which is able to strongly coordinate both carboxylate and phosphate groups,[125] and with acrylamidopyridine (i.e. 2-acrylamidopyridine and 2,6-bis(acrylamido)pyridine).[54, 121, 123, 124] Another example of stoichiometric noncovalent interactions is the formation of inclusion complexes with cyclodextrins, which are cyclic oligosaccharide structures that can be derivatized to form functional monomers.[69, 126, 127] Since these molecules possess hydrophilic exteriors and relatively hydrophobic interiors of defined sizes, they can bind a variety of substrates in aqueous solution.

Functional monomers containing complexed metal ions provides another means of obtaining high affinity stoichiometric interactions with template molecules (Fig. 6.3).[31, 43, 70, 106, 128-131] Many examples of such interactions occur in biology, including strong interactions of phosphoryl groups with Ni^{2+} and Zn^{2+} ions coordinated to various residues in proteins, and interaction of Cu^{2+} and Fe^{2+} ions with histidine imidazole groups in proteins. These interactions occur rapidly in aqueous media, and are readily reversible under mild conditions, making them especially suitable for targeting biomolecules. Furthermore, there are many metal ions to choose from, allowing the possibility of forming metal complexes with many different functional groups.

6.2.1.3.3 Combined Covalent and Noncovalent Interactions with Template. An alternative strategy for preparing MIPs, which combines certain advantages of both covalent and noncovalent imprinting, is the "sacrificial spacer" method reported by Whitcombe et al. (Fig. 6.5).[62] The stability and exact stoichiometry of covalent bonding between the template and functional monomer(s) is used to prepare the imprinted polymer, while the fast reversible binding kinetics of noncovalent interactions is used when the target rebinds to the free imprinted sites. Thus, the template is covalently bound to the functional monomer with a bridging group (e.g. ester or amide group). Following polymerization, these covalent bonds are cleaved to release the template as well as the bridging group, leaving behind precisely oriented functional groups (e.g. hydroxyl or amino group) that can subsequently bind the target molecule through noncovalent interactions. The bridging group, therefore, acts as a "sacrificial spacer". In an extension of this approach, functional monomers that interact noncovalently with other functional groups on the template can be added to the prepolymerization solution, in addition to the

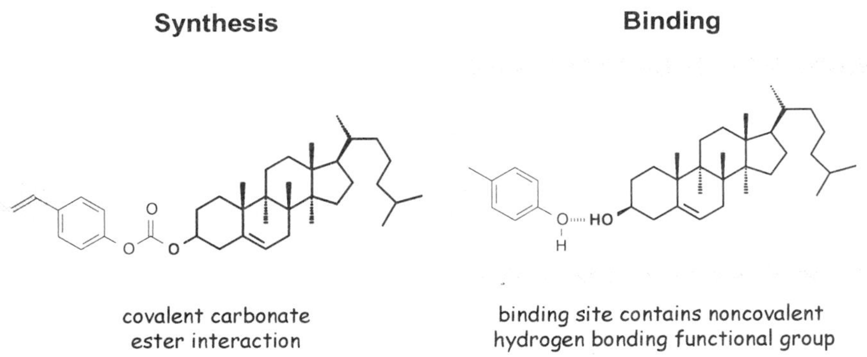

Synthesis   **Binding**

covalent carbonate   binding site contains noncovalent
ester interaction   hydrogen bonding functional group

Figure 6.5. Example of Whitcombe's[62] sacrificial spacer (depicted with thick bond lines) approach for the imprinting of cholesterol. MIP synthesis is performed using the 4-vinylphenyl carbonate ester of cholesterol. Upon removal of the bridging group, rebinding of cholesterol occurs through noncovalent interactions. The shaded double bond indicates the portion of the functional monomer incorporated into the polymer matrix.

covalently bound functional monomer.[132] The sacrificial spacer approach has the advantage that more homogeneous binding sites are created by covalent bonding between the template and monomer without sacrificing the convenience of rapid noncovalent association/dissociation of the target molecule.

6.2.1.4 Cross-linking Monomer

To obtain an imprinted polymer with recognition sites of high affinity and selectivity for the target, the 3-dimensional arrangement of the monomer-template complexes in solution prior to polymerization must be preserved in the final polymer by creating a matrix that is highly cross-linked.[102, 103] To accomplish this, MIPs are typically prepared with a high content of cross-linking monomer (typically ~ 70 to 90 mole % of total monomer).[103, 133] However, a balance must be sought between a polymer that is sufficiently rigid to maintain the correct spatial orientation of complementary functional groups in the binding cavity when the template is removed, yet flexible enough to facilitate rapid release and rebinding of the template from and to the binding site.

Monomers used for cross-linking typically bear di- or tri-vinyl, acrylic or methacrylic groups, such as divinylbenzene (DVB), ethylene glycol diacrylate, ethylene glycol dimethacrylate (EGDMA), and trimethylolpropane trimethacrylate (TRIM) shown in Figure 6.4. It may not be necessary to add a cross-linking monomer if the functional monomer contains more than one polymerizable group, because the functional monomer can then itself act as a cross-linker.

6.2.1.5 Initiator

Free radical initiators are usually used to prepare MIPs based on methacrylate, acrylate and vinyl monomers.[26, 89] 2,2'-Azobisisobutyronitrile (AIBN) and 2,2'-azobis(2,4-dimethylvaleronitrile) (ABDV) are commonly used for this purpose. The polymerization may be initiated either thermally or photochemically (with UV light). Photochemical initiation can be performed at low temperature, and therefore should be considered in cases where a thermally labile template is used, and also when noncovalent imprinting is chosen since noncovalent interactions are stronger at lower temperatures. On the other hand, heat initiation might be preferred with light-sensitive or poorly soluble templates.

6.2.2 Preparation of MIPs

6.2.2.1 Bulk Polymerized MIPs

MIPs are most commonly prepared as polymer blocks, and are processed by grinding and sieving to obtain micrometer-sized particles.[5, 84] Even though the grinding is easy to perform and increases the surface area available for binding, it produces irregularly shaped particles that may not be ideal for all applications (e.g. chromatography). With this polymerization method binding sites are distributed throughout the polymer matrix, so a large number of them remain deeply buried in the polymer matrix and are inaccessible to the target molecule in subsequent binding studies. In addition, the integrity of some of the binding sites may be disrupted as a result of the grinding process. Furthermore, a significant amount of the polymer material is lost during sieving, reducing the yield of "useful" MIP particles are obtained.

Once the MIP has been ground, the template must be removed from the polymer to expose the binding sites. This is a crucial step, because the number of high affinity binding sites available for binding the target depends on the effectiveness of the extraction of template from the MIP.[84] Efficient extraction is also important to prevent subsequent leakage of template buried in less accessible parts of the polymer network, as the released template molecules could compete with target molecules in binding experiments. Extraction involves extensive washing with appropriate solvents, which can be accomplished in a number of ways, including Soxhlet extraction, several steps of stirring and sedimentation, or with the use of a chromatography column. To remove the template molecule as completely as possible, acid or base may have to be added to the extraction solvent. Removal of the template may be assessed by monitoring the washes, for example by HPLC.

Bulk imprinted polymers have also been prepared directly inside chromatographic columns and capillaries by filling them with the imprinting solution, and allowing this solution to polymerize *in situ*.[41, 134-137] Since the imprinted polymer is porous, the template is removed by simply allowing a suitable solvent to flow through the column or capillary. Although this *in situ* polymerization technique dramatically simplifies the synthesis of MIPs, it restricts their use to chromatographic applications.

6.2.2.2 Thin MIP Films

MIPs can also be prepared as thin films coating the surface of a substrate by applying a thin layer of the imprinting solution, containing monomer, cross-linker, template and initiator, directly onto the substrate surface, followed by *in situ* polymerization. This can be accomplished by spin- or dip-coating techniques,[42, 79, 138] or by sandwiching a small amount of imprinting solution between the substrate and another flat surface (e.g. a glass slide).[139, 140] MIPs in the form of uniform thin films allow rapid access of the target to the binding sites. Moreover, thin MIP films provide an important advantage in sensing applications, because the MIP can be brought into close contact with the transducer surface of the sensor.[93]

Many different substrates have be used for the preparation of thin-film MIPs, including polypropylene membranes,[48, 141] silicon wafers,[139, 140] glass filters,[142] and mass-sensitive transducers (e.g. quartz crystal microbalances and surface acoustic wave resonators).[79, 143-146] Electrode surfaces coated with thin imprinted polymeric films by *in situ* electropolymerization have also been reported.[28, 33, 147, 148] The bottom surfaces of vials[111] and microplate wells[82, 112, 149] have also been coated with thin layers of an MIP, which is particularly useful for high-throughput assays. Quartz plates,[143] glass slides,[138] and optical fibers[42, 43] are popular for fluorescence detection of target binding.

Although imprinting matrices based on methacrylate, acrylate and vinyl monomers have been the most widely used to prepare thin films, they can be difficult to prepare reproducibly, and can display a brittleness that is undesirable for sensor applications.[150, 151] Nevertheless, this is a very promising approach, because of its applicability to a wide variety of biomolecules, its potential use in sensors, and its capability for high-throughput screening.

6.2.2.3 Spherical MIP Beads

Problems related to the grinding of bulk polymers, such as the quality of the binding sites, the loss of polymer material, and the irregularly-shaped particles, may be addressed by preparing MIPs in the form of spherical beads. Such beads may be prepared in various ways, including suspension, precipitation, and emulsion polymerization,[52, 61, 67, 68, 152-156] or by coating preformed polymer beads with a thin layer of bulk imprinted polymer.[157, 158] Imprinted beads should be more amenable to mass-production than bulk imprinting, since they may be prepared with greater reproducibility and higher yield.[156, 159] Imprinted microspheres may be used in the same applications as ground polymers, including binding assays, chromatographic separations, and solid-phase extractions.[87, 152, 153, 160-162] Small monodisperse spherical imprinted beads are especially advantageous in chromatographic applications where they provide better flow properties. To facilitate the separation of the polymer fraction from the liquid phase in binding assays, magnetic imprinted beads have been produced using magnetic iron oxide.[163]

Similar to the production of fluorescent imprinted polymers using target-selective fluorescent functional monomers or nonselective fluorescent monomers, one can envision the production of fluorescent imprinted micro- or nano-spheres capable of changing their fluorescence properties in the presence of a target.[61] If sufficiently small, it is conceivable that such fluorescent imprinted spheres may be able to be microinjected into living cells to allow the fluorescence measurement of a variety of small intracellular molecules (e.g. adenosine 3':5'-cyclic monophosphate, cAMP).

Although spherical imprinted beads offer advantages, their synthesis requires a skillful hand, because their preparation is more challenging and not as straightforward as bulk polymers. The methods are also not generally applicable, and in some cases require the synthesis of special functional surfactants.

6.2.2.4 Surface Imprints

To circumvent the problem of inaccessible buried binding sites, MIPs in which binding sites are restricted to the surface of the polymer have been synthesized. Such surface imprints provide binding sites that are easily accessible, resulting in faster mass transfer and binding kinetics. One approach[164] to create surface imprints in a bulk polymer was to covalently attach a template molecule to silica gel. Sufficient cross-linking and functional monomers were then added to fill the pores of the silica gel, and the mixture was polymerized. The silica was then removed by chemical dissolution, leaving behind a porous polymer network in which the binding sites were limited to the surface. As this approach requires that the template have a reactive functional group, analogous to that required for conjugating a hapten to raise antibodies, it may be necessary to first derivatize the template.

Another approach is to use surfactants to restrict the distribution of imprinted sites to the surface of the MIP. Beads with surface imprints have been prepared in this way by emulsion polymerization.[61] Small droplets consisting of monomer and cross-linker were stabilized by a surfactant in water (referred to as an oil-in-water emulsion), and polymerized to yield small near-monodisperse spherical beads. A surfactant is used that strongly binds the template molecule, in this case cholesterol, through covalent bonds or metal-ion coordination. Since the surfactant is amphiphilic, it orients itself at the droplet-water interface, limiting the formation of binding sites to the surface of the droplet following polymerization. A variation of the above approach was used to prepare surface-imprinted bulk polymers.[52, 68, 165] In this case small water droplets containing the template and a surfactant are dispersed within the oil phase containing the monomer and cross-linker. The small water droplets then create a series of cavities in the polymer network after polymerization. Again, because of the presence of the surfactant, binding sites are situated at the interface between the polymer network and the small water droplets, and are fully exposed by grinding the polymer into small particles.

Surface imprints are particularly advantageous for large target molecules (e.g. proteins) and complex structures (e.g. cells) that cannot easily penetrate into the pores of the polymer.[77-80, 83, 166] Recognition sites for blood proteins (e.g. albumin, immunoglobulin G, fibrinogen) lined with disaccharide molecules were created in thin fluoropolymer films by first allowing the protein to adsorb onto a flat mica surface.[83] A thin layer of disaccharide was then applied, covering the protein-mica surface, and forming multiple hydrogen bonds with polar functional groups of the proteins. A fluoropolymer layer, which binds covalently to the disaccharide functional groups, was deposited onto the disaccharide coat, and then attached to a solid support to allow the mica surface to be removed. This exposed the proteins, which were then removed, leaving imprints that selectively adsorbed the protein used as the template over other related proteins in competitive radiolabeled binding studies.

An interesting technique, called bacteria-mediated lithography, was presented to create "imprints" for bacteria on the surface of polymer beads that could potentially be used for cell separation or for the analysis of microorganisms.[78, 81] Once again, an aqueous-organic two-phase system was used, but this time organic droplets, containing monomer and initiator, were suspended in an aqueous solution of bacteria. Water soluble poly(allylamine) molecules were then added to the system, which polymerized around the organic droplets to form a polyamide wall encapsulating the droplets. As this occurred, the bacteria spontaneously partitioned to the interface, and became embedded in the polyamide layer, creating an "imprint" in the surface of the microcapsules. Solid polymer beads were then obtained by polymerizing the inner core of the microcapsules. To create "imprints" for the bacteria that exclusively contained amino groups, exposed amino groups on the surface of the beads not protected by the attached bacteria were reacted with a perfluoropolyether. Upon removing the cells, the "imprints" were visualized using a confocal laser microscope by reacting the unmodified polyamine surfaces beneath the bacteria with a fluorescent molecule. In a brief evaluation of selectivity, the larger rod-shaped Listeria monocytogenes bacteria were twice as likely to bind to their "imprints" over "imprints" created for the smaller round-shaped Staphylococcus aureus bacteria. "Imprints" for Staphylococcus, on the other hand, preferred Staphylococcus to Listeria by a factor of eight.

A stamping procedure was adopted by Dickert et al. to create surface imprints for yeast cells on a QCM.[79] A stamp composed of an array of yeast cells was pressed into a thin layer of partially polymerized polyurethane solution coating the surface of a QCM until the solution was completely polymerized. After removing the stamp from the polyurethane layer and washing off the cells, a honeycomb-like polymer surface was exposed that was able to detect cell concentrations spanning five orders of magnitude with a detection limit of approximately 10^4 cells/mL, and was selective for the particular type of yeast cell imprinted.

6.2.2.5 Silica-based MIPs

Inorganic silica polymer matrices are synthesized by acid- or base-catalyzed hydrolysis and condensation of silanes, known as the sol-gel technique.[138, 167-170] Typical silanes used to prepare imprinted silica materials are tetraethoxysilane, which acts as a cross-linker during polymerization, and various functionalized triethoxysilanes or trimethoxysilanes including phenyl trimethoxysilane, 3-[N,N-bis(9-anthrylmethyl)amino]-propyl triethoxysilane, and 3-[Bis(2-hydroxyethyl)amino]propyl triethoxysilane, which interact with the template through covalent or noncovalent bonds (Fig. 6.6). The functionalized triethoxy- or trimethoxysilanes can also act as cross-linkers, because they still contain free ethoxy or methoxy groups capable of undergoing hydrolysis and condensation. Production of bulk imprinted silica materials that contain binding sites throughout the matrix usually involves mixing a silane functional monomer and cross-linker (typically the main sol-gel component, forming the bulk of the silica matrix) together with the template in an aqueous solution (sometimes containing alcohol) and adjusting the pH to catalyze the polymerization. The polymer can be processed by grinding it into particles and extracting the template, and then evaluated in batch binding and chromatographic experiments, as described below. Alternatively, a thin layer of silica-imprinted polymer may be coated onto the surface of preformed silica particles,[167, 169] or coated onto a flat surface (e.g. a glass slide).[138] Target-selective fluorescent functional monomers,[168] nonselective fluorescent monomers that do not interact directly with the template,[138] and fluorescent templates[167, 169] have all been used to produce silica-based MIPs. Sol-gel materials are attractive for use in sensors, because they are rigid, have a high surface area and porosity, are chemically inert, and, of particular interest for optical sensors, they are often optically transparent. Furthermore, they offer an advantage over organic methacrylate, acrylate, and vinyl imprinted polymers in that they can be prepared in aqueous media, facilitating the preparation of MIPs directed at polar templates, such as biological molecules and drugs.

6.2.3 Evaluation of MIPs

6.2.3.1 Evaluating Target Binding to MIPs

The most common way to evaluate the binding of the target molecule and potential competitors to bulk imprinted polymer particles are by batch binding, chromatography, or a combination of the two. In batch binding experiments, a solution of the target molecule is incubated with the imprinted polymer particles in a vial or tube, and the amount of bound target can be determined using optical methods (e.g. absorbance or fluorescence spectroscopy) or scintillation counting. Chromatographic evaluation is performed by packing the particles into an HPLC column.

In batch binding experiments with fluorescent target molecules, the amount of target bound to the imprinted polymer may be determined by separating the liquid phase from the polymer fraction (i.e. by centrifugation, sedimentation, or filtration), and measuring the fluorescence of either fraction.[171] In some cases it is also possible to measure the

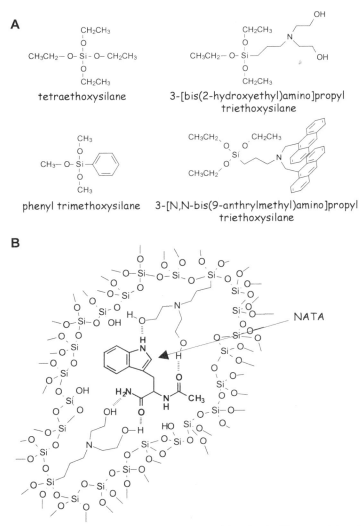

Figure 6.6. **A**. Common silanes used for the synthesis of imprinted silica materials. **B**. Schematic representation of a silica-based imprinted site for NATA (N-acetyltryptophanamide; depicted using thick bond lines).[167]

fluorescence of the polymer suspension directly without separating the two phases. If the liquid phase is analyzed, the amount of bound target is determined by subtracting the amount in the liquid phase (i.e. free target) from the total amount of target. However, the results obtained with this approach are subject to considerable experimental variability if the proportion of bound target is small, as relatively small reductions in free target against high background values must be detected. Alternatively, the polymer fraction containing bound fluorescent target can be resuspended in a suitable solvent and the fluorescence measured.[172] It is also possible to wash the polymer fraction and measure its fluorescence due to the bound target in the dry state.[167, 169] If binding of the fluorescent target molecule by the MIP induces a shift in the wavelength of its maximum fluorescence, the amount bound can easily be determined by measuring the fluorescence of the polymer suspension directly, obviating the need to separate the liquid phase from the polymer.[173]

Assessment of target binding is much simpler if the MIP, rather than the target, is fluorescent. In this case, binding of the target often induces either an increase or decrease in fluorescence intensity, or a shift in the emission or excitation wavelength. It is therefore not necessary to separate the polymer from the solvent, and the fluorescence of the polymer suspension can be measured directly to determine the amount of target bound.[29, 53, 114, 168]

The solvent used in evaluating binding to an MIP also plays a very important role, because the type and strength of the interaction between the target molecule and the MIP depends on solvent polarity. As water molecules strongly interfere with polar interactions, such as hydrogen bonding, recognition in an aqueous milieu will be dominated by hydrophobic interactions. On the other hand, nonpolar organic solvents favor hydrogen bonding and ionic interactions. When organic solvents are used for imprinting, the best affinity and selectivity for the target are often obtained with the same solvent that was used to prepare the polymer.[50, 84, 100]

6.2.3.2 Determining Target Affinity

The Langmuir binding model is most often used to obtain the affinity of the target for its imprinted polymer in batch binding studies.[174] Using a linear form of the Langmuir equation, the amounts of bound and free target determined at various target concentrations are plotted in a Scatchard plot (bound/free vs. bound).[24, 88] The association constant (K_a) and number of available binding sites are then estimated from the slope and x-intercept of this plot. Alternatively, these values may be obtained by plotting the amount of bound versus free target (i.e. binding curve), and directly fitting this data to the Langmuir equation. From the binding curve, an approximate value for the dissociation constant ($K_d = 1/K_a$) may also obtained by estimating the free target concentration at which the concentration of specifically bound target is 50% of maximal. The existence of a heterogeneous population of binding sites in MIPs, especially those created using noncovalent imprinting, yield nonlinear Scatchard plots, suggesting that the binding sites display a continuous range of affinities for the target, akin to polyclonal antibodies. The range of affinities may be estimated by fitting the Scatchard plots to two or three classes of binding sites, and determining the affinity and concentration for each.

Binding affinities can also be estimated using chromatography. In this case, capacity factors (k') may be calculated from the retention times of the target and other related molecules. The capacity factor is equal to the difference between the retention time (t_R) and the time required by the mobile phase to pass through the column (t_M) divided by t_M ($k' = (t_R-t_M)/t_M$). The capacity factor is proportional to the affinity of the target or other analytes for the polymer. Alternatively, binding constants and numbers of binding sites can be determined by frontal chromatography.[27]

6.2.3.3 Determining MIP Selectivity

To determine whether target binding is due to interaction with specific molecularly imprinted sites within the polymer, target binding to the imprinted polymer is compared to binding to a nonimprinted polymer, prepared in an identical fashion in the absence of template. As the nonimprinted polymer contains the functional monomer, a certain degree of binding is expected, and may even exhibit some degree of selectivity, as the functional monomer is chosen based on its ability to interact with the target molecule. However, under optimal conditions, binding of the target to the nonimprinted polymer should always be less than to the MIP if the imprinting process has been successful.

The selectivity of an MIP for its target over other related compounds may be evaluated by comparing their affinities. If the target is fluorescent, or if a radioactive or fluorescently-labeled analog is available, competition binding studies can be performed to calculate the IC_{50} values of various competitors.[89, 133] This is done by plotting fluorescence intensity, peak area, or radioactivity due to the bound target molecule against the concentration of competitor, and determining the concentration that inhibits target binding by 50% (i.e. IC_{50}), which approximates the K_d value. However, this approach cannot be used with fluorescently-labeled MIPs, as the binding of competing molecules could affect the fluorescence of the MIP in the same way as the target molecule. In this case, the affinities of the target and related molecules must be measured individually by determining their K_a or K_d values on the basis of changes in fluorescence intensity. In many cases, selectivity is assessed only qualitatively by estimating binding at a very limited number of concentrations with a limited number of potential competitors, and a K_a or K_d value is not always given. Insufficient information about the cross-reactivity of an MIP with potentially interfering substances sometimes makes it difficult to predict the utility of the MIP in a real biological environment. It is also possible to evaluate the selectivity of an MIP using chromatography to determine selectivity factors (α), which are calculated from the ratio of the capacity factors of two analytes.

The selectivities of MIPs can be altered depending on whether the liquid phase used to evaluate template binding favors hydrophobic or polar interactions.[34, 35, 88] In organic solvents, target binding is much more dependent on polar interactions, whereas hydrophobic interactions are much more important in an aqueous environment. The strong hydrophobic bonding that occurs in aqueous media increases nonspecific binding to the polymer, making it more difficult to demonstrate selective binding of the target under these conditions. This is a significant problem when MIPs are designed to be employed for the analysis of biomolecules present in biological media. Nonspecific binding in aqueous media can be reduced by the addition of a small amount of ethanol or detergent, but more work in this area is required.

6.3 DETECTION OF MIP-TARGET INTERACTIONS BY FLUORESCENCE

6.3.1 Nonfluorescent MIPs

If the target molecule is itself fluorescent, changes in its fluorescence properties due to binding to an MIP can be exploited to determine its concentration (Fig. 6.2A).

Alternatively, a nonfluorescent molecule can be derivatized with a fluorescent tag, but in this case the MIP must be prepared against the derivatized template, otherwise the selectivity of the MIP could be compromised. Another approach for nonfluorescent target molecules is to use competition binding in which the concentration of the target molecule is determined on the basis of its ability to compete with a structurally related fluorescent molecule for binding to the MIP.

Table 6.1. Association constants (K_a) and useful concentration ranges for the binding of various fluorescent target molecules to their MIPs.

Fluorescent target	K_a (M^{-1})	Useful range (μM)	Ref
Dansyl-L-phenylalanine (**1**)	NA^a	~ 25 – 125	171
N-(1-Pyrenyl)maleimidyl-DL-homocysteine (**2**)	$9.3 \times 10^{5\ b}$	~ 0.53 – 2.66	172
(-)-Cinchonidine (**3**)	NA	NA	173
Pyrene (**4**)	9.4×10^7	0.00015 – 0.2	63
Flavonol (**5**)	NA	0.05 – 10	104
β-Estradiol (**6**)	NA	0.1 – 4.0	175
Ricin protein	3.1 and 30×10^6	NA	169
Ricin A polypeptide	9.3×10^6	NA	169
Ricin B polypeptide	$10 \times 10^{6\ c}$	NA	169
NATA (**7**)	4.0 and 77×10^5	NA	167

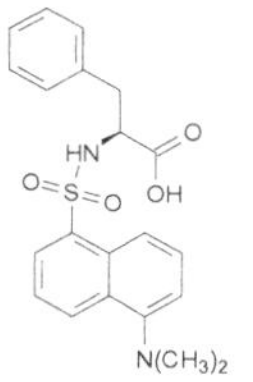

1 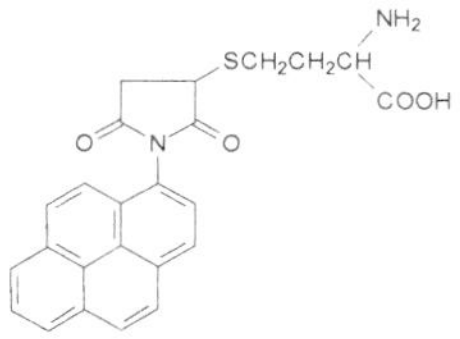   2 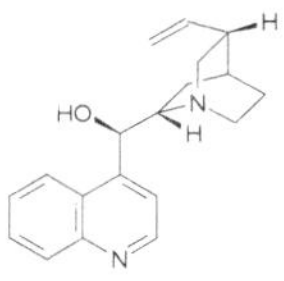   3 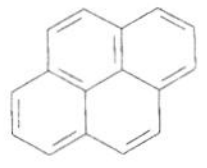   4 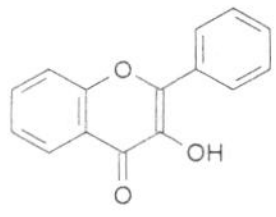   5

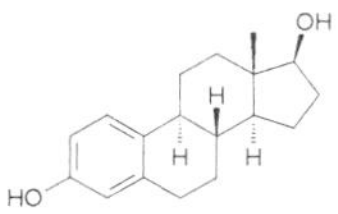

6 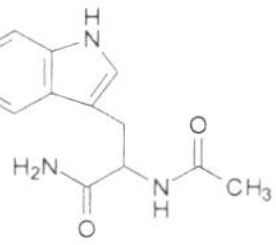   7

^{a} NA = data not available
^{b} site population density = 12 nmol/g polymer
^{c} Ricin B bound to Ricin A MIP with an affinity of 1.1×10^8 M^{-1}, Ricin B bound to Ricin MIP with affinities of 0.40 and 435×10^8 M^{-1}.

6.3.1.1 Fluorescent Target Molecule

Amino acids. The recognition properties of MIPs were first used in combination with fluorescence by Kriz et al.,[171] who constructed a fiber-optic sensing device using an MIP prepared against the fluorescent amino acid derivative, dansyl-L-phenylalanine (Table 6.1). Imprinted polymer particles were brought into close contact with a fiber-optic bundle by securing them at the tip of the fiber, so that the fluorescence of target bound to the polymer could be measured. The fluorescence of the device was proportional to the concentration of the target, and selectivity for the L-enantiomer was observed. Even though the time required to obtain a stable fluorescence response with the analyte was long (~ 4 h), the binding process could be followed in real time. The relatively long response time in this study was most probably due to the large particle size (75-105 µm), and could possibly be reduced by using smaller particles or a thin layer of MIP coated directly onto the optical fiber.

A similar approach was used to prepare an imprinted polymer directed at DL-homocysteine containing a fluorescent pyrene tag (Table 6.1).[172] However, in this case underivatized homocysteine was added directly to the MIP in the presence of a pyrene-based derivatizing reagent, and the derivatization was performed *in situ*. The polymer was then washed, and the fluorescence due to bound pyrene-labeled DL-homocysteine was measured in a cuvette in water. The fluorescence of the imprinted polymer increased linearly with homocysteine concentration between 0.53 and 2.66 µM, similar to the range found in plasma, while that of the nonimprinted control polymer increased only slightly. Scatchard analysis indicated that the average affinity constant for this interaction was 9.3 x 10^5 M^{-1}. Although selectivity was not assessed using the *in situ* derivatization technique, the imprinted polymer was found to be selective for pyrene-labeled homocysteine over pyrene-labeled cysteine. Interestingly, the rate of the reaction between homocysteine and the pyrene reagent was about 5-fold greater in the presence of the imprinted polymer compared to the control polymer, whereas similar reaction rates between cysteine and the pyrene reagent were observed for the imprinted and control polymers. Thus, the MIP also acted as a selective catalyst for the derivatization reaction.

(-)-Cinchonidine. There are also a number of examples of the preparation of MIPs that recognize naturally fluorescent target molecules. The Takeuchi group[173] prepared an MIP against the fluorescent (-)-cinchonidine, a member of the antimalarial alkaloid drug family, using a mixture of MAA and 2-(trifluoromethyl)acrylic acid (TFMAA) as functional monomers (Table 6.1). Upon binding, the fluorescence of (-)-cinchonidine shifted to a lower wavelength, thus avoiding the need to remove the medium containing unbound target, and making this procedure potentially amenable to high-throughput analysis. Towards this end, the same group prepared a library of (-)-cinchonidine-imprinted polymers using different functional monomer (2-sulfoethyl methacrylate) to template ratios in 96-well microplates using an automated liquid handling system.[112] The fluorescence emission intensity of the imprinted polymers following incubation with (-)-cinchonidine was measured using a fluorescence microplate reader without separating the liquid phase from the polymer. The results obtained in this way were validated by measurement of free cinchonidine fluorescence in the liquid phase, and calculation of the bound alkaloid by subtraction. This study demonstrates the feasibility of rapid synthesis and high-throughput screening as a means of optimizing the affinity and selectivity of

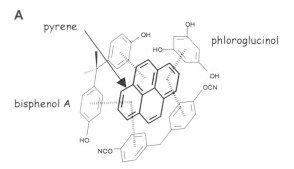

A

pyrene

phloroglucinol

bisphenol A

diphenylmethane 4,4'-diisocyanate

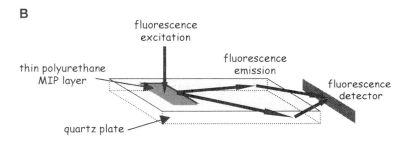

B

fluorescence
excitation

fluorescence
emission

thin polyurethane
MIP layer

fluorescence
detector

quartz plate

Figure 6.7. A. Potential noncovalent aromatic stacking interactions between pyrene (depicted using thick bond lines) and functional monomers. **B**. Quartz plate waveguide arrangement used for fluorescence measurements.[143]

imprinted polymers, and may be widely applicable, especially if the lengthy incubation time (12 h) can be reduced.

Polycyclic aromatic hydrocarbons (PAHs). Polyurethane-based imprints were prepared as thin clear layers on quartz plates and placed in a flow system setup to determine the binding of fluorescent PAHs, including pyrene and acenaphthene (Table 1).[79, 143, 150] Such an approach allows target binding to be followed in real time, and is also amenable to high sample throughput and easy reuse of the MIP. The imprinting solution consisted of the template, as well as bisphenol A, *p,p*'-diisocyanatodiphenylmethane, and phloroglucinol, all of which can act simultaneously as functional monomers and cross-linkers (Fig. 6.7A). Since PAHs lack functional groups, their binding to such an MIP would presumably be through noncovalent aromatic stacking interactions between the phenyl groups of the templates and those of the monomers. A thin layer of the prepolymerization solution was placed on the quartz plate, and polymerized *in situ*. The quartz plate was then placed in a flow cell, and the template was removed by washing, reducing the fluorescence of the MIP by over 99%. The sample containing the target molecule was then injected as a bolus into a flowing stream of solvent, and became concentrated in the flow cell through binding to the MIP. This resulted in increased fluorescence, which was observed perpendicular to the excitation, at the narrow end of the quartz plate, effectively transforming the plate into a planar waveguide (Fig. 6.7B).

Interestingly, the MIP that displayed the greatest fluorescence response to pyrene was not actually obtained using pyrene as a template, but rather the smaller acenaphthene molecule. An imprinted polymer prepared with a 1:3 mixture of acenaphthene and pyrene also bound more pyrene than the pyrene-imprinted polymer.[49] MIPs prepared against anthracene and chrysene behaved similarly, in that molecules larger than the template bound to a greater extent than the template itself. However, when the polymerization was carried out at elevated temperatures or in more dilute solutions, the resulting MIPs displayed greater responses for the larger templates over the smaller molecules. The optimized imprinted polymer for pyrene displayed an association constant of 9.4×10^7 M^{-1} for pyrene in water, and showed a linear increase in fluorescence intensity with increasing pyrene concentrations up to 0.2 μM. A detection limit of 0.15 nM (0.03 ng/mL) was observed for a 3 μm imprinted polymer layer. The response time was only a few minutes. The fluorescence response of the nonimprinted polymer was only 1% that of the imprinted polymer. Common environmental quenchers present in water, such as humic acids, had virtually no effect on the fluorescence response, suggesting that these devices may be suitable for environmental monitoring of PAHs. These imprinted polymers were also combined with mass-sensitive devices, such as a quartz crystal microbalance (QMB), and found to display selectivity patterns similar to those determined by fluorescence measurements, validating the fluorescence results.

The sensitivity, rapidity, and reversibility of this technique are impressive. The high sensitivities obtained are most likely due to the concentration of the target molecules within the MIP in the field of view of the detector. Moreover, the aqueous milieu would favor interaction of the hydrophobic PAH with the MIP. The effects of any unbound interfering substances that may be present in complex mixtures would also be minimized by the selective concentration of the target in the imprinted sensing layer. Polyurethane imprinted polymers have great potential as selective materials in optical sensor applications because they are transparent, and can be reproducibility prepared as ultra-thin films.[150]

The same type of polyurethane-based MIP was also coated on the inside of a quartz glass capillary to create a capillary fiber-optic sensor for anthracene.[176] The sample was placed in the capillary, which was sealed at one end, and the tips of optical fibers to conduct excitation and emitted light were immersed in the sample at the other end. Constructed in this manner, the capillary serves as a waveguide to propagate excited and emitted light, thus delivering excitation light to a high proportion of the surface coated with the MIP. This gives a strong fluorescence signal from a small surface area, and therefore does not require a very intense light source.

Flavonol. A flow setup was also used to detect the fluorescent 3-hydroxyflavone (flavonol) molecule (Table 1, see also Fig. 6.3).[104] MIP particles were packed into a flow cell that was placed in a spectrofluorometer. Injection of flavonol into the system resulted in increased fluorescence, which could be tracked in real time. Under optimal conditions, fluorescence intensity was linear for flavonol concentrations between 50 nM and 10 μM. The response time seemed to vary with concentration, but was generally on the order of minutes. No fluorescence intensity increase was observed with the structurally related molecules, morin and quercitin, suggesting selectivity for the target molecule. The imprinted polymer could be continuously used for up to 2 months with minimal loss in binding. This flow system setup was used to analyze flavonol concentrations in olive oil samples, and the results were validated by conventional liquid chromatography (LC), suggesting its potential usefulness for real sample analysis.

β-Estradiol. Rachkov et al.[175] prepared an MIP for the fluorescent β-estradiol steroid molecule (Table 6.1). The MIP was evaluated by packing it into an HPLC column and monitoring the fluorescence of the eluate. The fluorescence peak area was linear for β-estradiol concentrations between 0.1 to 4 μM. The MIP was selective for β-estradiol, since estrone, in which the 17-OH of β-estradiol is oxidized to a keto group, was barely retained on the MIP column, while testosterone, was only slightly retained.

Glycoproteins. The intrinsic fluorescence of tryptophan residues in peptides and proteins can be used to detect binding to MIPs. Lulka et al.[169] used this approach to determine the binding of the toxic castor bean glycoprotein ricin and its A and B peptide chains to silica-based MIPs prepared by coating the surface of silica gel particles with a thin layer of silica polymer that had been imprinted with these macromolecules (Table 6.1). Following incubation with a solution containing the target, the particles were washed, dried, and glued to a paper card. The card was placed in a spectrofluorometer, and the fluorescence due to tryptophan residues was measured. Scatchard analysis revealed association constants on the order of 10^6 M^{-1} for ricin and it's A and B chains to their respective imprints. This group also prepared silica-based imprints toward the amino acid N-acetyltryptophanamide (NATA) (Table 6.1, see also Fig. 6.6B),[167] but the association constant was an order of magnitude lower than those observed for the protein and peptides. The relatively high binding affinity of the larger molecules is most probably due to multiple-point binding interactions between their polar surface residues and the spatially oriented polar groups in the binding cavities on the imprinted silica particles, compared to small molecules, which would have fewer interaction points. Although the MIPs prepared against ricin and its A chain were selective for their targets, the ricin B chain bound with higher affinity to ricin and ricin A chain imprints than to its own imprint. This may be due to the more flexible random coil structure of the ricin B chain, compared to the extensive α-helix and β-sheet structure of the ricin A chain, which would confer much more rigidity. Thus, the ricin B chain could be sufficiently flexible to bind to the larger imprinted sites for ricin and the ricin A chain, but could have more difficulty binding to its own tightly-fitting sites. Because of the larger size of ricin and the more rigid nature of the ricin A chain, they can bind only to their respective MIPs.

Limitations in the use of fluorescent templates. Although progress has clearly been made in the detection of MIP-target interactions using fluorescent target molecules, this approach may not be ideal, as most analytes are not fluorescent, and thus it is usually necessary to prepare a fluorescent derivative. This can be time-consuming, requires the presence of an appropriate functional group on the target molecule for derivatization, and may compromise selectivity due to the reaction of the derivatizing reagent with other related molecules. Furthermore, the fluorescently-labeled target molecule would have to be stable throughout the polymerization process (i.e. at elevated temperature or to UV irradiation). The sensitivity of this approach may also be compromised by the presence of trapped template molecules in the polymer matrix, which increase the background fluorescence of the MIP.

6.3.1.2 Competition Between Nonfluorescent Target and Fluorescent Reporter Molecule

Another way to detect nonfluorescent target molecules with nonfluorescent MIPs, which avoids some of the above problems, is to use a fluorescent reporter molecule that is either a derivative of the target or a structurally-related molecule.[84, 88, 92] In this case, the MIP is incubated with a mixture of the target and reporter molecules in a competitive binding assay. They can either be added at the same time, or the MIP can be preincubated with the reporter, followed by addition of a solution containing the target. Increasing concentrations of the target will cause corresponding reductions in the amount of bound fluorescent reporter, and increases in its free concentration. The target concentrations can then be determined by measuring fluorescence in either the solid or liquid phases. This approach is analogous to competitive ligand binding assays using antibodies or receptors, and was first applied to an MIP in 1993 by Vlatakis and Mosbach,[37] who used radioactively-labeled target molecules as reporters for the analysis of theophylline and diazepam in human serum. Their MIP-based assay compared favorably with commercially available immunoassays for these compounds. MIP-based assays using fluorescently-, radioactively-, and enzymatically-labeled reporter molecules have now been reported for a wide variety of compounds, including drugs and environmental pollutants.[84, 88] It is an area that is currently under intense investigation.

Herbicides. Piletsky[177] was the first to report the use of an MIP in a competitive binding assay using a fluorescent reporter (Table 6.2). A batch binding assay employing a fluorescein-labeled triazine derivative as the reporter was used to the detect the herbicide triazine. Triazine concentrations spanning four orders of magnitude (0.01 to 100 µM) could be detected, and the MIP was selective for triazine over atrazine and simazine. This approach was extended to MIP-coated microplates,[82] a highly practical format amenable to automation and high-throughput analysis. In this case, the herbicide atrazine was detected using the same fluorescent reporter (Table 2). The fluorescence in the wells from displaced reporter increased with increasing atrazine concentrations. A detection limit of 0.7 µM was obtained. Atraton-D and metribuzin, molecules structurally related to the template, were much less effective than atrazine in displacing the reporter from the MIP, demonstrating its selectivity for the target molecule.

2,4-Dichlorophenoxyacetic acid (2,4-D). Haupt et al.[178, 181] demonstrated that low amounts of the herbicide 2,4-D, could be detected by fluorescence using a 2,4-D-imprinted polymer and 7-carboxymethoxy-4-methylcoumarin (CMMC) as a fluorescent reporter molecule (Table 6.2). Even though the structure of the reporter differs from the target molecule (Fig. 6.8A) and its binding to the imprinted polymer is weaker than 2,4-D, CMMC was still found to be useful, as it competed with radiolabeled 2,4-D for binding to the imprinted sites in both aqueous buffer and acetonitrile solution. Interestingly, 2,4-D labeled with fluorescein isothiocyanate could not be used as a fluorescent reporter, as it was unable to compete with the target for binding. Competitive batch binding experiments were performed by first adding CMMC to MIP particles suspended in aqueous buffer or

Table 6.2. Association constants (K_a) and useful concentration ranges for the binding of various target molecules to their MIPs. Fluorescent competitor molecules used to determine binding are also shown.

Target	Fluorescent competitor	K_a (M^{-1})	Useful range (μM)	Ref
Triazine (**8**)	5-(4,6-dichlorotriazinyl)amino-fluorescein (**13**)	NA	0.01 – 100	177
Atrazine (**9**)	5-(4,6-dichlorotriazinyl)amino-fluorescein (**13**)	NA	~ 0.7 – 100	82
2,4-D (**10**)	7-carboxymethoxy-4-methylcoumarin (**14**)	a	0.1 – 50 (buffer)^b 0.1 – 10 (ACN)^c	178
L-Phenylalaninamide (**11**)	rhodamine B (**15**)	1.7 x 10$^{4\,d}$	~ 50 – 500	179
Chloramphenicol (**12**)	dansylated derivative of 2-amino-1-(4-nitrophenyl)-1,3-propanediol (**16**)	NA	25 – 309	180

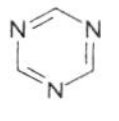

8 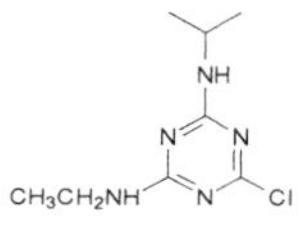   **9** 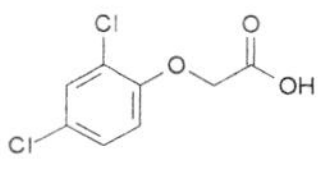   **10** 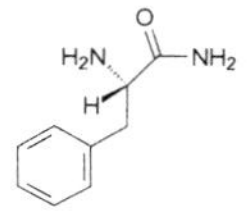   **11** 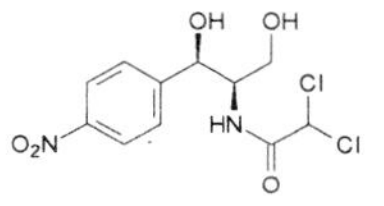   **12**

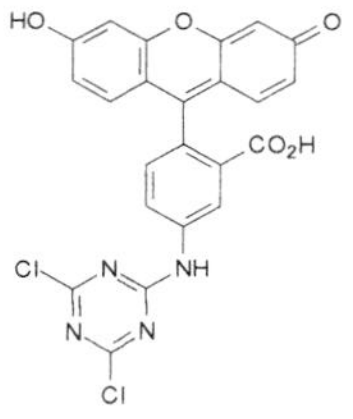

13 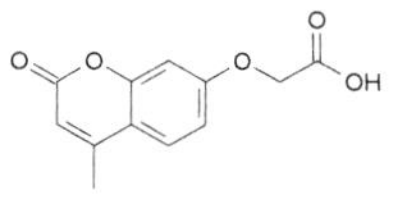   **14** 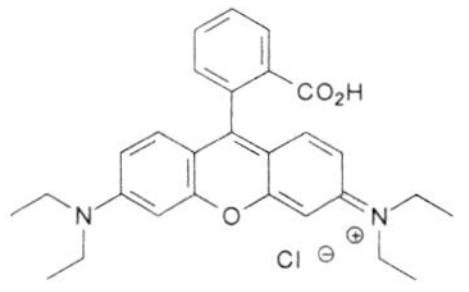   **15** 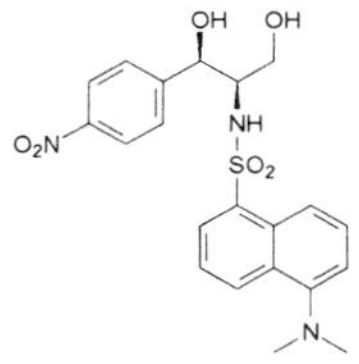   **16**

^{a} IC_{50} ~ 5 μM (buffer); IC_{50} ~ 2 μM (ACN)
^{b} 20 mM phosphate buffer, pH 7, 0.1% Triton X-100
^{c} acetonitrile
^{d} site population density = 1.8 mmol/g polymer

an organic solvent, followed by various amounts of 2,4-D. The amount of CMMC bound to the imprinted polymer was estimated by measuring the fluorescence intensity of CMMC in the liquid phase after a 2 h incubation. A standard curve of % CMMC bound versus 2,4-D concentration indicated that 2,4-D could be detected in the range of 0.1 to 50 μM in buffer, and 0.1 to 10 μM in acetonitrile. These results are almost identical to those obtained using radiolabeled 2,4-D as a reporter molecule.[182]

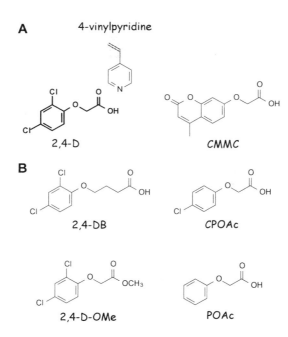

Figure 6.8. **A**. Potential acid-base interaction between 2,4-D (2,4-dichlorophenoxyacetic acid; thick bond lines) and the functional monomer, 4-vinylpyridine. The structure of the fluorescent reporter CMMC (7-carboxymethoxy-4-methylcoumarin) is also shown. The shaded double bond indicates the portion of the functional monomer incorporated into the polymer matrix. **B**. Structures of compounds used to investigate MIP selectivity. 2,4-DB (2,4-dichlorophenoxybutyric acid); 2,4-D-OMe (2,4-dichlorophenoxyacetic acid methyl ester); CPOAc (4-chlorophenoxyacetic acid); POAc (phenoxyacetic acid).[178]

Selectivity studies using the related molecules 2,4-DB (2,4-dichlorophenoxybutyric acid), 2,4-D-OMe (2,4-dichlorophenoxyacetic acid methyl ester), CPOAc (4-chlorophenoxyacetic acid) and POAc (phenoxyacetic acid) revealed significant differences between aqueous and organic media (Fig. 6.8B). In aqueous media, selectivity was in the order 2,4-D (100%) = 2,4-DB (100%) > CPOAc (42%) > POAc (9%) > 2,4-D-OMe (4%). In contrast, in organic solution, selectivity was in the order 2,4-D (100%) >CPOAc (50%) > POAc (14%) > 2,4-DB (2%) > 2,4-D-OMe (1%). The much greater affinity of 2,4-DB for the MIP in aqueous media is presumably due to the additional two methylene groups, which render it more hydrophobic, a characteristic that is much more important for binding in aqueous than in organic solvents. Cross-reactivities determined in buffer were similar to those obtained using radiolabeled 2,4-D as the reporter molecule. Although the sensitivity of the MIP-based assay is less than that of available immunoassays for 2,4-D, the selectivities of the two assays are similar, with the selectivity of one reported immunoassay being in the order 2,4-D (100%) > 2,4-D-OMe (30-160%) > 2,4-DB (1.5-20%) > CPOAc (0.5-2.8%).[183] 2,4-D-OMe, the methyl ester of 2,4-D, displays high cross-reactivity, probably because the carrier used to raise the antibody was coupled to the carboxyl group, making this part of the molecule less recognizable by the antibody.

Other modes of detection have also been used with 2,4-D-imprinted polymers, including electrochemical, chemiluminescence, and enzyme-linked competitive displacement assays, extending MIP-based assays to detection systems widely used in immunoassay technology.[160, 181] Furthermore, 2,4-D detection was also accomplished without the use of a reporter, for example using an integrated electrochemical sensor,[184] and a flow-through sensing system employing FT-IR spectroscopy.[140] The adaptability of MIPs to all of these different detection formats points to their versatility.

L-phenylalaninamide. Fluorescent reporter molecules can also be used to detect target molecules in HPLC assays. Piletsky et al. prepared an MIP using the amino acid derivative, L-phenylalaninamide as the template, and packed it into an HPLC column (Table 6.2).[179] The column was equilibrated with the unrelated fluorescent reporter, Rhodamine B, prior to injecting the target molecule. The area of the peak corresponding to displaced fluorescent reporter was larger for the target than for its D-enantiomer when acetonitrile containing 0.5% acetic acid was used as the mobile phase.

Chloramphenicol (CAP). Fluorescent reporter molecules have also been used to evaluate target binding in flow injection systems. Imprinted polymer particles against the antibiotic CAP were packed into a flow cell, which was placed just upstream of the fluorescence light path of a spectrofluorometer (Table 6.2).[180] The binding sites in the MIP were first saturated with the fluorescent reporter, a dansylated derivative of 2-amino-1-(4-nitrophenyl)-1,3-propanediol (dansyl-ANPD). Injection of CAP resulted in the rapid displacement of dansyl-ANPD, resulting in a fluorescence signal. The detection limit of this assay was 25 µM (8 µg/mL), however the useful range of the assay was very limited, as the upper limit of detection was 309 µM (100 µg/mL). An appealing feature of this setup is the rapid sample turnover of approximately 10 min per sample, compared to other approaches which required analysis times of several hours.

Limitations of the competitive fluorescent reporter approach. Although this approach circumvents the need for a fluorescent template/target, the choice of an appropriate reporter molecule is far from trivial and usually involves a considerable degree of trial and error. The addition of a fluorescent tag to a target molecule would significantly increase its size, which may result in lower affinity and selectivity for the imprinted sites on the MIP, compromising the sensitivity of the assay. A fluorescent reporter that is not a derivative of the template must have some structural features similar to the target molecule, and also cannot be too large and bulky to prevent its binding to the specific binding sites in the imprinted polymer. In addition, reporter molecules must have a high fluorescence quantum yield to maximize the sensitivity of the assay. For these reasons, the competitive binding approach may be more amenable to the use of radioactively-labeled reporter molecules, because of their structural similarity to the target molecules.

6.3.2 Fluorescent MIPs

As most potential target molecules are not fluorescent, and the use of fluorescent reporter molecules may compromise sensitivity and selectivity, there is increasing interest in the development of fluorescent MIPs that change their fluorescence properties upon binding the target molecule (Fig. 6.2B). This approach has the important advantage that it can potentially be applied to all targets, irrespective of their structures. Fluorescent imprinted polymers may be prepared using either fluorescent functional monomers that interact selectively with the template, or, more simply, using a general fluorescent monomer that does not interact selectively with the template. Target-selective fluorescent functional monomers would be incorporated into the binding site, whereas nonselective fluorescent monomers would be situated close to the binding site.

To be successful, target binding must alter the fluorescence of the MIP, permitting the polymer to function as both a recognition element and a signal transduction element. Such fluorescent MIPs would allow the binding event to be followed in real time, and the binding kinetics to be investigated. Furthermore, since a separation step is not required, such polymers would facilitate automation and high-throughput analysis, and would even be ideal for homogeneous binding assays where the amount of bound target may be quantified without separating it from the unbound (free) target in solution.[26]

6.3.2.1 Target-Selective Fluorescent Functional Monomer

The use of target-selective functional monomers in the design of an MIP offers the potential of creating imprinted sites with high sensitivity and selectivity for the target molecule. Two types of target-selective fluorescent functional monomers have been used to create fluorescent MIPs. The first consists of an organic molecule that has been specially selected and/or synthesized for its ability to bind covalently or noncovalently with the template. The second is a fluorescent metal complex that can incorporate a specific type of molecule into its coordination sphere.

6.3.2.1.1 Fluorescent Organic Molecule.
cAMP. We designed and synthesized the fluorescent functional monomer, *trans*-4-[*p*-(*N,N*-dimethylamino)styryl]-*N*-vinylbenzylpyridinium chloride (vb-DMASP), for the purpose of creating a fluorescent molecularly imprinted polymer chemosensor capable of specifically detecting cAMP in aqueous solution (Table 3, Fig. 9A).[53] Previous studies[107, 185] demonstrated that this fluorophore displayed strong solvatochromic behaviour resulting from a large increase (~ 18 D) in its dipole moment upon excitation, indicating that its excited state has considerable charge transfer character. The steady-state fluorescence of vb-DMASP was dramatically enhanced in the presence of the purine nucleotides, cAMP and cGMP (guanosine 3′:5′-cyclic monophosphate), and biological macromolecules[186] in aqueous solution, but was only very slightly affected by pyrimidine nucleotides. The high sensitivity of vb-DMASP fluorescence to its environment, and in particular to purine nucleotides, suggested that it would be an excellent candidate for the functional monomer in the preparation of an MIP directed at cAMP. Given its positive charge and the presence of aromatic units, we anticipated that cAMP (negatively charged) would interact with vb-DMASP through electrostatic and aryl stacking interactions in the prepolymerization mixture, as well as in the imprinted sites in the subsequently formed MIP. To increase the hydrophilicity of the MIP, and to provide additional hydrogen bonding interactions between cAMP and the binding sites, the nonfluorescent 2-hydroxyethyl methacrylate (HEMA) functional monomer was also included. Photochemical polymerization was carried out using trimethylolpropane trimethacrylate

(TRIM) as the cross-linking monomer, and methanol as a porogen. Baseline fluorescence of the MIP and the corresponding nonimprinted control polymer in the absence of cyclic nucleotide (I_o) was measured after prewetting with water overnight, centrifugation, and resuspension in water (Fig. 9B). The polymer particles were then recentrifuged and resuspended in water containing cyclic nucleotides. Fluorescence (I) was measured after 1.5 h (Fig. 9B and Fig. 10). cAMP quenched the MIP fluorescence at concentrations as low as 100 nM, with saturation occurring by about 10-100 µM (Fig. 10). From the fluorescence response curve, an apparent association constant of about 3.5×10^5 M^{-1} (K_d = 2.9 µM) was obtained. In contrast, cAMP had little effect on the fluorescence of the nonimprinted polymer. The MIP was selective for cAMP, since the structurally related molecule cGMP had little effect on the fluorescence of both the imprinted and control polymers at concentrations up to 1000 µM (Fig. 6.10).

The selectivity of the imprinted polymer for cAMP over cGMP suggests that functional groups complementary to the target molecule are exactly placed in the recognition sites of the imprinted polymer to allow cAMP binding through electrostatic interaction of the phosphate group, as well as aryl stacking and/or hydrogen bonding interactions with the aromatic purine base of the nucleotide. The apparent affinity of the imprinted polymer for cAMP is much higher than that of the free vb-DMASP molecule for cAMP in buffered aqueous solution (K_a = 13.8 M^{-1}), suggesting that the size and shape of cAMP was also recognized by the three-dimensional polymer network. Although a distribution of binding sites of higher and lower affinity are often obtained when evaluating the affinity of noncovalent imprinted polymers, we found only one apparent association constant, which presumably represents the average affinity of the population of imprinted sites.

Table 6.3. Association constants (K_a) and useful concentration ranges for the binding of various target molecules to their fluorescent MIPs. Fluorescent monomers used to prepare fluorescent MIPs are also shown.

Target	**Fluorescent monomer**	K_a **(M^{-1})**	**Useful range (μM)**	**Ref**
cAMP (**17**)	*trans*-4-[*p*-(*N,N*-dimethylamino)styryl]-*N*-vinylbenzylpyridinium chloride (**22**)	3.5×10^5	~ 0.1 to 100	53
cGMP (**18**)	2-acrylamidopyridine (**23**)	3.1×10^4	~ 10 to 100	124
N^1-Benzylidene pyridine-2-carboxamidrazone (**19**)	3-acrylamido-5-(2-methoxy-1-naphthylidene)-rhodanine (**24**)	NA	NA	105
L-Tryptophan (**20**)	dansyl derivative of dimethylacrylic acid (**25**)	NA	~ 500 - 10000	114
D-Fructose (**21**)	9-hydroxymethyl-10-[[*N*-methyl-*N*-(*o*-boronobenzyl)amino]methyl]-anthracene-9-methacrylate (**26**)	NA	~ 1000 - 100000	29
2,4-D (**10**)[a]	3-[*N,N*-bis(9-anthrylmethyl)amino]-propyl triethoxysilane[b]	2.2×10^6	~ 100 - 750	168

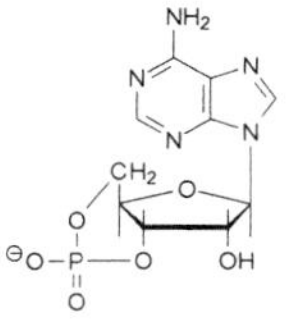
17

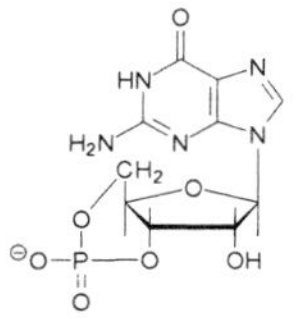
18

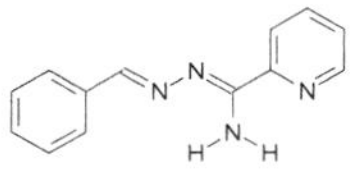
19

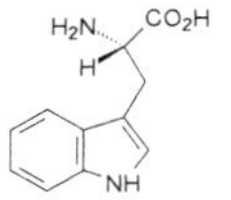
20

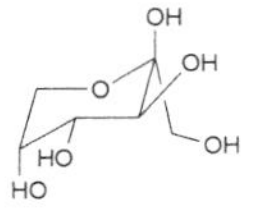
21

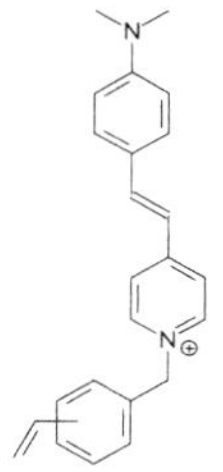
22

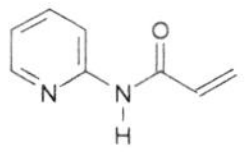
23

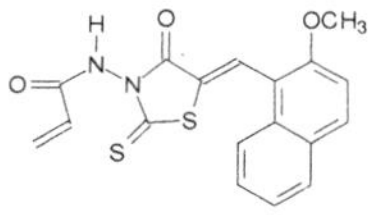
24

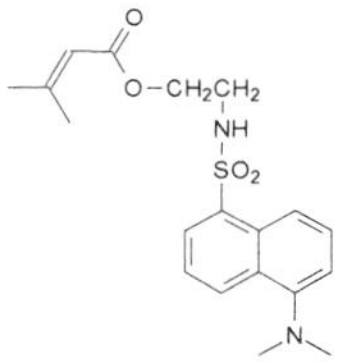
25

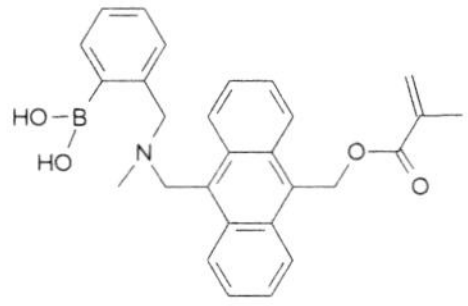
26

[a] see Table 2 for structure
[b] see Figure 6A for structure

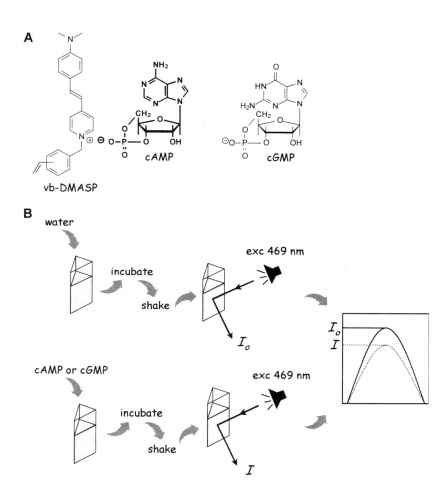

Figure 6.9. A. Potential ionic interaction between cAMP (thick bond lines) and the fluorescent functional monomer, vb-DMASP. The structure of cGMP is also shown. **B**. Method of fluorescence changes of an MIP prepared for cAMP. I_o (solid line; control) and I (dotted line; 1 mM cAMP) show the decrease in fluorescence intensity at 595 nm induced in the MIP by cAMP. The excitation wavelength was 469 nm.[53]

The quenching of the steady-state fluorescence of the imprinted polymer with cAMP is in contrast to the fluorescence enhancement observed for free vb-DMASP with cAMP in phosphate buffer solution. Compared to aqueous solution, vb-DMASP is probably in a different state when immobilized in the restricted polymer matrix. Although the mechanism is difficult to elucidate, it is possible that different interactions between the fluorophore and cAMP are promoted in these two states, resulting in differences in the effect of cAMP on the fluorescence quantum yield and fluorescence lifetime of vb-

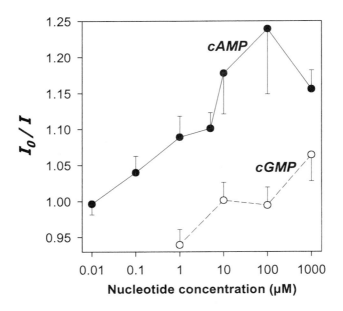

Figure 6.10. Fluorescence of the cAMP-imprinted polymer in the presence of various concentrations of cAMP and cGMP. Excitation wavelength, 469 nm; emission wavelength, 595 nm. See reference 53 for details.

DMASP. However, more experiments are required before a full description of the mechanism can be provided.

Time-resolved fluorescence studies were performed by Wandelt et al.[187, 188] on the cAMP-imprinted and nonimprinted polymers in a manner similar to the steady-state fluorescence experiments presented above. The imprinted and nonimprinted polymers, both in the absence and in the presence of aqueous cAMP, displayed heterogeneous fluorescence decay kinetics that yielded two lifetime parameters upon analysis: a long-lived component (~ 2.6 to 3.0 ns), and a short-lived component (~ 100 to 150 ps). Upon plotting the fluorescence lifetime response of each component (i.e. τ_0/τ, where τ_0 and τ are the fluorescence lifetimes in the absence and presence of cAMP, respectively) versus cAMP concentration, the short- and long-lived components displayed different trends. The lifetime of the short-lived component decreased with increasing cAMP concentrations, similar to the effect of cAMP concentration on the steady-state fluorescence intensity response (I_0/I) of the imprinted polymer. In contrast, no change was observed in the lifetime of the long-lived component. The long-lived component of the control polymer also remained unchanged in the presence of cAMP, but the lifetime of the short-lived component tended to increase. These results suggest that the short-lived component may arise from vb-DMASP molecules in sites that are capable of binding cAMP, whereas the long-lived component may be due to fluorophore molecules buried deep in the polymer matrix that are inaccessible to the analyte.

As observed for steady-state fluorescence studies, there was considerable sample-to-sample variability with the lifetime measurements, and this must be kept in mind when evaluating these results. This variability could arise partly from the grinding process, which would expose binding cavities to different degrees, resulting in a distribution of cavities with different affinities for cAMP, and also from the relatively large size distribution of the polymer particles used. It is also possible that there could be a certain degree of heterogeneity in the polymerization process itself, which could also result in a distribution of binding sites with different affinities.

cGMP. An MIP directed against cGMP was prepared using 2-acrylamidopyridine as the functional monomer, presumably because it is capable of forming two hydrogen bonds with the template, resulting in stoichiometric noncovalent interactions (Table 3).[124, 189] The affinity and specificity of the MIP were investigated by measuring the fluorescence of the imprinted and nonimprinted polymers in 96-well microplates after incubation with various nucleotides in water for 30 min. cGMP quenched the fluorescence of the imprinted polymer, but did not affect the nonimprinted polymer. The maximal response to cGMP occurred at a concentration of 100 μM cGMP, similar to that observed with our cAMP-imprinted polymer discussed above. However, the magnitude of the fluorescence response was greater with the cGMP-imprinted polymer, which exhibited a 50% maximal reduction in fluorescence, compared to a 20% reduction for the cAMP-MIP.[53] The affinity of the cGMP-MIP is about an order of magnitude lower than for our cAMP-imprinted polymer. The fluorescence of the cGMP-imprinted polymer was not quenched in the presence of cAMP, and thus appears to be selective for its target. As we observed with our cAMP-imprinted polymer, there was considerable sample-to-sample variability, as reflected by the relatively large error bars.

Carboxamidrazones. Rathbone et al.[105, 190] created fluorescent MIPs for a series of N[1]-benzylidene pyridine-2-carboxamidrazone antimicrobial agents with the goal of developing a high-throughput fluorescence-based assay to pre-screen large libraries of potential drug candidates, such as those generated by combinatorial chemistry. It was anticipated that the abilities of these antibiotics to bind to the imprinted sites of the MIP would mimic their abilities to bind to their putative target on microorganisms, and that this could therefore be used to predict their potential biological activities (Table 3, see also Fig. 3). Molecular modeling studies suggested that the carboxamidrazones would interact with the fluorescent functional monomer, 3-acrylamido-5-(2-methoxy-1-naphthylidene)-rhodanine, through stoichiometric noncovalent interactions involving two hydrogen bonds. MIPs, prepared by polymerizing the above functional monomer in the presence of a variety of carboxamidrazones, were extensively tested to determine their selectivity. This was accomplished by incubating the MIP with analytes, washing and drying the particles, and resuspending them in methanol in a 96-well microplate to measure fluorescence. In most cases the target caused a decrease in the fluorescence of the imprinted polymer, but there were considerable differences in selectivity, depending on the nature of the template molecule. Templates with rigid aromatic units gave MIPs that, in general, exhibited the greatest fluorescence quenching in the presence of the target, and displayed good selectivity. On the other hand, MIPs prepared from templates with flexible substituents that had more freedom to rotate were less selective and often exhibited greater fluorescence quenching with analytes other than the target. Similarly, analytes with a greater degree of flexibility tended to cross-react to a greater extent than large and relatively inflexible compounds with fewer degrees of freedom. Analyte binding could be measured using a ratiometric measurement method employing two different sets of excitation and emission wavelengths, only one of which was affected by binding of the target molecule. This obviates the need to accurately weigh the imprinted polymer for a binding study, thereby decreasing the analysis time. Results using the mass-independent ratio method were similar to those obtained by carefully weighing the polymer.

L-Tryptophan. A somewhat different approach to determine target binding to a fluorescent MIP was used by Wang's group.[114] Imprinted binding sites for L-tryptophan were created using a dansyl derivative of dimethylacrylic acid as the fluorescent functional monomer (Table 6.3). Because L-tryptophan did not directly affect the fluorescence of the MIP, binding studies were performed in the presence of a small quencher molecule, *p*-nitrobenzaldehyde. The small size of the quencher allowed it to easily diffuse in and out of the imprinted sites, and quench the fluorescence of the dansyl moiety in the binding sites. Incubation with L-tryptophan resulted in displacement of *p*-nitrobenzaldehyde from the imprinted sites, increasing the fluorescence of the imprinted polymer. In contrast, L-tryptophan had no effect on the fluorescence of the nonimprinted polymer in the presence of *p*-nitrobenzaldehyde. The highest L-tryptophan concentration (10 mM) increased the fluorescence intensity of the imprinted polymer in the presence of the quencher by about 40%. The imprinted polymer discriminated between the D- and L-enantiomers of the target, as well as the side chains of the amino acids, since smaller increases in fluorescence intensity were observed with D-tryptophan, L-phenylalanine, L-alanine, indole and indole-3-propionic acid. Although unique, this approach has certain drawbacks, including low sensitivity (detection limit ~ 0.5 mM), the use of a two-phase system, and a long incubation time (4.0 h).

D-Fructose. The above group subsequently prepared a fluorescent imprinted polymer for detecting D-fructose using the covalent imprinting approach. The functional monomer was constructed by linking a boronate group to a polymerizable anthracene molecule with a nitrogen-containing spacer (Table 6.3, see also Fig. 6.3).[29, 191] D-fructose was covalently coupled to the functional monomer, and the complex used to prepare an MIP, from which D-fructose was subsequently removed by hydrolysis. Increasing concentrations of D-fructose increased the fluorescence emission intensity of imprinted polymer suspensions, because the binding of the target to the boronic acid group ties down the lone pair of electrons on the nitrogen of the spacer, eliminating their quenching effect on the anthracene moiety. Although the sensitivity of the MIP to D-fructose was rather low (~ 1 to 100 mM), it was selective for D-fructose over D-glucose.

2,4-D. A fluorescent anthracene derivative (3-[*N,N*-bis(9-anthrylmethyl)amino]propyl triethoxysilane) was again chosen as a functional monomer to prepare imprinted sites for 2,4-D in a silica polymer matrix by Leung et al. (Table 6.3, see also Fig. 6.6A).[168] The fluorescence emission of imprinted polymer suspensions increased in proportion to the concentration of 2,4-D, with the highest concentration (754 µM) producing a fluorescence enhancement of about 14%. This is similar in magnitude to the fluorescence change we observed with the cAMP-imprinted polymer, although they observed an enhancement, rather than a reduction, in fluorescence. The average affinity of their MIP for its target (K_a = 2.2 x 10^6 M^{-1}) was similar to that which we reported. They also observed a rather high sample-to-sample variability, analogous to our results.

6.3.2.1.2 MIPs That Utilize Fluorescent Metal Complexes. The incorporation of certain small organic molecules into the coordination sphere of some metals can produce luminescent/fluorescent complexes, or change the intensity and/or position of the luminescent/fluorescent spectrum of the metal or small organic molecule. Such interactions (e.g. Zn-phosphonate) can be highly selective,[192] and therefore are suitable for designing imprinted polymers selective for small molecules, or for the metal ions themselves. To prepare fluorescent or luminescent MIPs selective for small molecules, the metal ion is typically complexed to two molecules, the template and another molecule that has a polymerizable group to anchor it to the imprinted polymer matrix. These metal complexes are known as mixed ligand metal complexes. Lanthanide ions such as Eu^{3+} are particularly suited for this purpose, as they exhibit narrow spectral peaks, long luminescence lifetimes, and form intensely luminescent complexes when they coordinate with certain molecules.[193] On the other hand, fluorescent MIPs that specifically recognize the metal ion itself are generally produced with fluorescent metal complexes that consist of the metal template coordinated to only one molecule, which typically contains an anchoring polymerizable group.

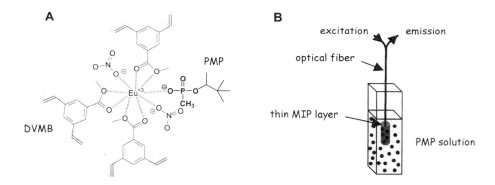

Figure 6.11. **A**. Potential structure of europium complex with PMP (pinacolyl methylphosphonate, depicted using thick bond lines) and DVMB (methyl-3,5-divinylbenzoate). Shaded double bonds indicate portion of monomer incorporated into polymer matrix. **B**. Fiber optic setup used for luminescence measurements.[42]

Pinacolyl methylphosphonate (PMP). A luminescent europium metal complex was chosen by Jenkins et al.[42, 43] to prepare a very sensitive and selective MIP-based fiber optic sensor for PMP, the hydrolysis product of the organofluorophosphorus nerve agent Soman (Fig. 6.11, Table 6.4). It was necessary to target the hydrolysis product, because of the extreme toxicity of the parent compound, and the fact that it is forbidden to perform tests using the actual chemical warfare agent. The europium complex employed for imprinting consisted of methyl-3,5-divinylbenzoate (DVMB), to hold the Eu^{3+} ion in the polymer matrix after polymerization, and the template PMP (Fig. 6.11A). The imprinted polymer was synthesized by dissolving the PMP-europium-DVMB complex directly in

Table 6.4. Association constants (K_a) and useful concentration ranges for the binding of various target molecules to their fluorescent MIPs. Fluorescent metal complex monomers used to prepare fluorescent MIPs are also shown.

Target	Fluorescent metal complex monomer	K_a (M^{-1})	Useful range (μM)	Ref
PMP (**27**)	Eu^{3+}–DVMB complex (**33**)	NA	0.000004 - 56[a]	42
Glyphosate (**28**)	Eu^{3+}–DVMB complex (**33**)	NA	0.000053 - 591	43
Chloropyrifos-methyl (**29**)	Eu^{3+}–DVMB complex (**33**)	NA	0.000016 - 310	43
Diazinon (**30**)	Eu^{3+}–DVMB complex (**33**)	NA	0.000023 - 329	43
9-Ethyladenine (**31**)	Zn^{2+}–porphyrin[b] complex (**34**)	7.5 x 10^5 [c]	NA	106
(-)-Cinchonidine (**3**)[d]	Zn^{2+}–porphyrin[b] complex (**34**)	1.1 x 10^7 [e]	NA	130
Histamine (**32**)	Zn^{2+}–porphyrin[f] complex (**35**)	4.5 x 10^3 [g]	NA	131
Pb^{2+}	Pb^{2+}–DVMB complex	NA	0.24 - 4.8	194
Al^{3+}	Al^{3+}–morin complex	NA	0.37 - 37	195

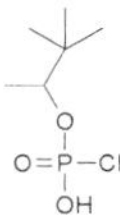

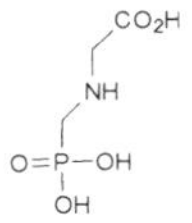

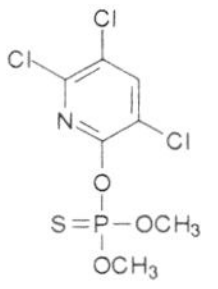

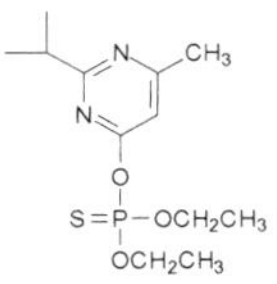

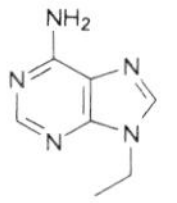

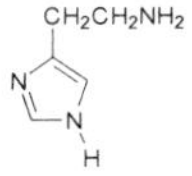

27   28   29   30   31   32

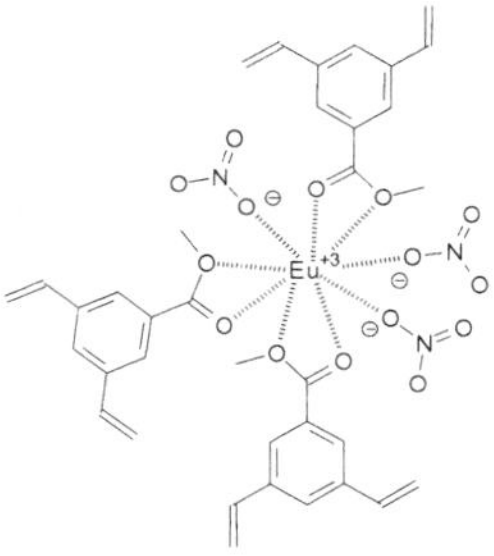

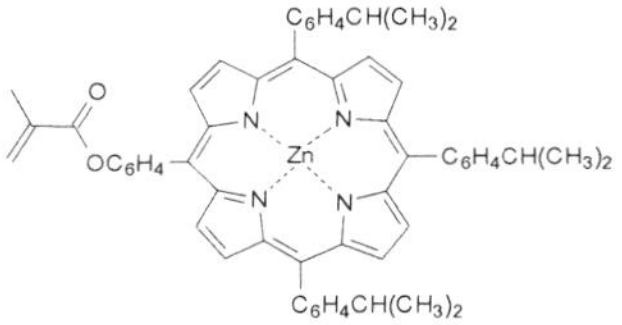

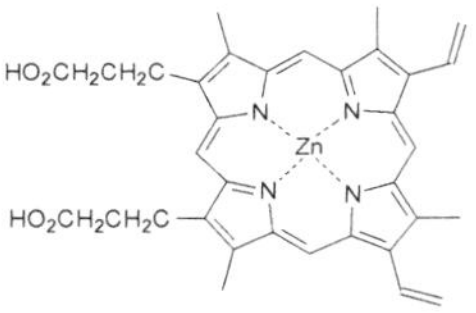

33   34   35

[a] a portable version was also constructed which displayed a useful range of 56 pM to 56 μM
[b] 5,10,15-tris(4-isopropylphenyl)-20-(4-methacryloyloxyphenyl)porphyrin Zn^{2+} complex
[c] site population density = 6.2 μmol/g polymer
[d] see Table 1 for structure
[e] site population density = 5.3 μmol/g polymer
[f] 8,13-divinyl-3,7,12,17-tetramethyl-21H,23H-porphine-2,18-dipropionic acid Zn^{2+} complex
[g] site population density = 180 μmol/g polymer; low affinity sites with K_a = 2.7 x 10^2 M^{-1} and site population density = 1100 μmol/g polymer were also reported

the styrene monomer without the use of a porogen.

To prepare the sensor, a thin layer (50-75 μm) of MIP was coated onto the tip of an optical fiber (400 μm) by dipping it into a viscous solution of partially polymerized polymer, and curing it overnight under a UV lamp. The 610 nm luminescence peak due to the PMP template in the initially formed imprinted polymer was dramatically reduced after washing out the PMP template, but reappeared once again upon placing the MIP-coated optical fiber in a solution of PMP at pH 13. The luminescence response was measured using an argon laser (465.8 nm) by placing the active end of the sensor into a cuvette containing a PMP sample and allowing it to equilibrate for 10 min (Fig. 6.11B). The sensor was regenerated between samples by rinsing with deionized water. The area of the luminescence band at 610 nm increased linearly with increasing concentrations of PMP between 4 pM (750 ppq) and 56 μM (10 ppm), giving a dynamic range of over 7 orders of magnitude. The cross-reactivities of several structurally similar organophosphorus pesticides and herbicides were tested, and, even at concentrations much higher than those found in the environment (i.e. 1000 ppm), none displayed a luminescence peak at 610 nm. Furthermore, poisoning of the sensor by irreversible binding of various potential interferents did not occur. However, results for a corresponding nonimprinted polymer control were not provided. A potential portable field MIP sensing device based on a miniature fiber optic spectrometer was also constructed and had a linear range between 56 pM (10 ppt) and 56 μM (10 ppm).

The effect of polymer thickness and pH on the response time (the time required to reach 80% of the maximal response) of the sensor was evaluated. The response time was reduced from 14 min to 8 min when the thickness of the polymer coat was reduced from 200 to 100 μm, presumably because the target binding sites were more accessible with the thinner layer. Increasing the pH from 6 to 13 also shortened the response time. These basic conditions not only enhance the performance of the assay, but also ensure complete hydrolysis of the highly toxic Soman nerve agents prior to measurement in the field.

Similar results were obtained with MIP-based fiber optic sensors selective for the nonhydrolyzed organophosphates and organothiophosphates, glyphosate, diazinon, and chloropyrifos-methyl, constructed by imprinting europium complexes with the target molecules as described above for the PMP sensor (Table 6.4).[43] These MIPs were also highly sensitive, having dynamic ranges between approximately 16 pM (~ 5 ppt) and 591 μM (~ 100 ppm), and 80% response times of ~ 15 min. Target molecules could induce a response in the corresponding nonimprinted polymer only at concentrations above about 0.8 μM (~ 250 ppb) with response times between 20 and 30 min.

9-Ethyladenine, (-)-cinchonidine, and histamine. The zinc (II) porphyrin complex has been quite popular as the fluorescence recognition and signal-generating component of imprinted binding sites for small organic molecules. A polymerizable analogue of this metal complex was used as a fluorescent functional monomer, together with MAA as an addition functional monomer, to imprint 9-ethyladenine,[106] (-)-cinchonidine,[130] and histamine[131] (note that (-)-cinchonidine was also imprinted without a fluorescent functional monomer, see section 13.3.1.1 above) (Table 6.4, see also Fig. 6.3). Suspensions of these MIPs all underwent fluorescence quenching of ~ 20% in the presence of relatively high concentrations (0.5 mM) of their respective target molecules, and were selective for the target. The histamine-imprinted polymer displayed the greatest response when it was prepared with DMSO, the porogen most similar to water, which was used in the fluorescence binding assay. This is consistent with other reports demonstrating that the best binding of a target occurs in the solvent used for imprinting.[50, 84, 100]

Reducing the polarity of the porogen by adding chloroform (1/1, v/v), produced an imprinted polymer that displayed poor quenching in the presence of histamine in aqueous solution. However, this could be reversed by increasing the polarity of the porogen by adding tetrabutylammonium hydroxide (DMSO/chloroform/TBA, 4.5/4.5/1, v/v/v).

Pb^{2+} and Al^{3+}. In addition to being used to develop imprinted polymers capable of detecting small molecules by fluorescence, metal-complexing functional monomers can also be employed to create fluorescent MIPs that are selective for the metal ion itself. Murray et al.[194] prepared one such MIP for the detection of Pb^{2+} in which a complex of polymerizable DVMB (methyl-3,5-divinylbenzoate) with Pb^{2+} was imprinted (Table 6.4). This complex has the advantage that the fluorescence of DVMB shifts to longer wavelengths (from blue to yellow-green) upon coordinating to lead. They fabricated a fiber optic sensor by dipping the end of an optical fiber into a partially polymerized imprinted polymer solution consisting of the Pb^{2+}-DVMB complex and divinylbenzene dissolved in styrene. Rebinding of Pb^{2+} increased the fluorescence intensity output of the sensor in a concentration-dependent manner. The response was linear between approximately 0.24 and 4.8 μM (50 – 1000 ppb). The selectivity of this sensor was not reported.

A flow-through sensing system for the detection of Al^{3+}, using a molecularly imprinted polymer prepared with a fluorescent Al^{3+}-morin complex was reported by Al-Kindy et al. (Table 6.4).[195] The MIP was packed into a flow cell for evaluation, a setup identical to the one they used for the detection of flavonol (Section 6.3.1.1). The fluorescence intensity of the imprinted polymer increased linearly between Al^{3+} concentrations of 0.37 to 37 μM ($\sim$ 10 to 1000 ppb). Other ions, such as Mg^{2+}, Ca^{2+}, and Be^{2+} had smaller effects on fluorescence intensity. However, since morin did not possess polymerizable groups to covalently anchor it to the polymer matrix, it could leach from the polymer over time, diminishing the response of the MIP, and thus compromising it's reusability.

6.3.2.2 Nonselective Fluorescent Monomer

If the use of a specialized fluorescent functional monomer is not possible or impractical, a nonselective fluorescent monomer that does not directly interact with the target molecule can still be used to prepare the MIP, as long as it undergoes a change in fluorescence as a result of target binding to the imprinted sites. This approach would, therefore, be applicable to a wider variety of target molecules.

(S)-Propranolol. Molecularly imprinted microspheres for (S)-propranolol were cleverly combined with proximity scintillation as a means to detect the binding of the ^{3}H-labeled target molecule by fluorescence (Table 6.5).[196, 197] The imprinted microspheres (0.6 to 1 μm) were prepared using a functional monomer (MAA) to interact with the target, as well as a scintillation monomer, 4-(hydroxymethyl)-2,5-diphenyloxazole acrylate, which is capable of emitting fluorescent light (i.e. a proximity scintillation signal) when excited by the β-emission of a radioactive molecule in its vicinity. To further enhance the transfer of energy from the bound radioactive target to the scintillator, an additional aromatic "antenna" unit, namely DVB, was also incorporated into the polymer matrix.

Table 6.5. Association constants (K_a) and useful concentration ranges for the binding of various target molecules to their fluorescent MIPs. The structures of some nonselective fluorescent monomers used to prepare fluorescent MIPs are also shown.

Target	Nonselective fluorescent monomer	K_a (M^{-1})	Useful range (μM)	Ref
(S)-Propranolol (36)	4-(hydroxymethyl)-2,5-diphenyloxazole acrylate (38)	1.8×10^5 (buffer)[a][b] 1.3×10^6 (ACN)[c][d]	~ 0.34 – 1690 (buffer)[a] ~ 0.34 – 3380 (ACN)[c]	196
DDT (37)[e]	4-chloro-7-nitrobenzofurazan (NBD) (39)	NA	0.00014 – 0.0028	138

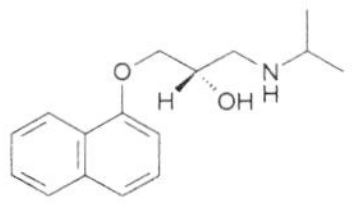

36	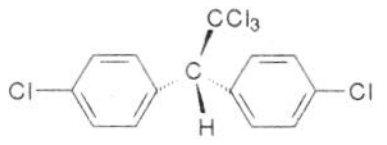 37	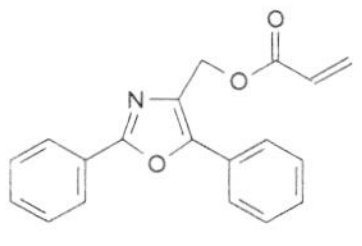 38	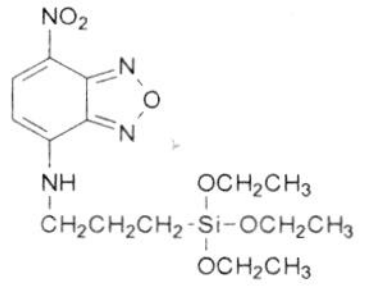 39

[a] citrate buffer (25 mM, pH 6.0)/acetonitrile (50/50, v/v)
[b] site population density = 38 μmol/g polymer
[c] acetonitrile (0.5% acetic acid)
[d] site population density = 4.4 μmol/g polymer
[e] MIP prepared using 4,4'-ethylenedianiline (EDA) as a surrogate template

Thus, when (S)-[³H]propranolol binds to the imprinted microspheres, radiation energy is transferred to the scintillator via the aromatic "antenna" component, and a fluorescence signal ismeasured using a scintillation counter. Unlabeled (S)-propranolol displaced (S)-[³H]propranolol from the binding sites, thus reducing the fluorescence emitted by the MIP, permitting measurement of the concentration of (S)-propranolol without the need to separate the MIP form the liquid phase. Average association constants of 1.8×10^5 M^{-1} and 1.3×10^6 M^{-1} were calculated for the binding of (S)-propranolol to the imprinted sites in buffer and organic solvent, respectively. The cross-reactivity of the R isomer was ~ 2%, indicating that the imprinted microspheres displayed a chiral selectivity for the target. The results were validated by centrifuging the MIP suspension and counting the free radioactive (S)-propranolol in the supernatant by liquid scintillation counting. Even though this approach is reasonably sensitive in aqueous and organic media, and does not require a separation step prior to analysis, the use of radioactivity may be of concern.

1,1-bis(4-chlorophenyl)-2,2,2-trichloroethane (DDT). Graham et al.[138] prepared a silane-based MIP containing the polarity-sensitive fluorescent molecule NBD (4-chloro-7-nitrobenzofurazan) to detect the polychlorinated pesticide, DDT, by fluorescence (Table 6.5). Since DDT lacks polar functional groups that can interact either covalently or noncovalently with functional monomers, the sacrificial spacer molecular imprinting approach using 4,4'-ethylenedianiline (EDA) as the surrogate template was employed. Both the EDA and NBD molecules were first covalently linked to silane molecules so that they could be incorporated into a silica polymer matrix. The EDA-silane and NBD-silane conjugates were then covalently attached together to locate the NBD fluorophore adjacent to the binding site for DDT in the imprinted polymer. Next, the main sol-gel component, 1,4-bis(trimethoxysilylethyl)benzene (BTEB) was added, and the imprinting solution was dip-coated onto glass microscope slides to create thin MIP films. Removal of the EDA molecule by reducing the two urea bonds linking it to the silane matrix exposed the imprinted sites. Measurement of the fluorescence of the MIP film in a cuvette, revealed linear fluorescence intensity increases for DDT concentrations in water between 28 pM (10 ppt) and about 2.8 nM (1 ppb), with a detection limit of approximately 140 pM (50 ppt). The response time was very rapid (60 s), and binding was easily reversed by rinsing with acetone and water. However, the maximal increase in fluorescence intensity (~ 2.8%) observed at 28 nM (10 ppb) DDT was very small, and there was a considerable degree of variability. Moreover, the imprinted film exhibited only a moderate degree of selectivity for DDT over other related molecules. The small fluorescence change observed upon DDT binding may be due to the lack of a change in the polarity of the microenvironment surrounding the fluorophore upon target binding. Another possibility is that relatively few fluorescent NBD-silane molecules may have reacted with the sacrificial spacer silane (i.e. EDA-silane), resulting in only a small number of fluorescent molecules being located near enough to the binding site to transduce the binding event.

Limitations of fluorescent MIP approach. Even though the use of fluorescent MIPs is more versatile than other approaches, because it may be applied to a larger population of target molecules, it is not without disadvantages. For fluorescent MIPs prepared using target-selective fluorescent functional monomers, each target requires the synthesis of a specific fluorescent functional monomer. This can be time-consuming, arduous, and requires some skill. Producing fluorescent MIPs with nonselective fluorescent monomers may avoid this problem, as they may be used for many different target molecules. Another drawback of using fluorescent MIPs is that the magnitude of the change in fluorescence intensity is often rather modest. This is true not only for MIPs prepared with nonselective fluorescent monomers, in which the fluorescent groups do not directly interact with the target molecule, but also for MIPs prepared with target-selective functional monomers as discussed above.

There are several possible explanations for the low magnitude of the fluorescence changes observed upon target binding. Photobleaching could be a potential problem, but is unlikely to be the major contributing factor. Another possibility is that the binding of the target molecule induces only relatively small changes in the environment (e.g. polarity) surrounding the fluorophore.[138] Finally, it is possible that only a small proportion of binding sites have the fluorophore perfectly oriented to permit signal transduction of the binding event, due to the existence of heterogeneous populations of binding sites. This could be a problem when the noncovalent imprinting approach is used, especially when the affinity of the functional monomer for the template is relatively low. Furthermore, many of the examples described above used polar solvents for imprinting (MIPs for cAMP, cGMP, L-tryptophan, 2,4-D), which would reduce the contribution of hydrogen bonding and electrostatic interactions to the binding of the templates to their functional monomers, further increasing the heterogeneity of the binding sites. However, even MIPs prepared with fluorescent metal complexes (MIPs for 9-ethyladenine, (-)-cinchonidine, and histamine), whose interactions are expected to be quite strong in polar solvents, displayed low responses to target binding.

Another problem is that in most cases, binding of the target molecule to the MIP results in fluorescence quenching, necessitating the measurement of a modest reduction in fluorescence against a bright background. A binding-induced fluorescence enhancement would be more desirable, as, in this case, a light signal is effectively turned on against a dark background signal, potentially increasing sensitivity.[100]

Improvements in the fluorescence response of an MIP to the binding event may be obtained by strengthening the interactions between the functional monomer and the target molecule, possibly through the use of covalent or stoichiometric noncovalent interactions, or by using a combination of different interactions (i.e. hydrogen bonding, metal-complexing, electrostatic, hydrophobic). This would create a more well-defined homogeneous population of binding sites, which could have a higher affinity for the target. An example of this would be the large fluorescent enhancement (> 10-fold) observed by Jenkins et al.[43] for the binding of glyphosate to a fluorescent MIP designed using a europium metal complex as the functional monomer. It may also be possible to enhance the fluorescence response by optimizing the fluorophore. Although this may be difficult, because it has to be done largely by trial and error, the selection of a fluorophore may be aided by molecular modeling and combinatorial chemistry approaches, together with the development of rapid high-throughput analytical methods. Ultimately, these improvements could lead to MIPs with enhanced fluorescence responses to target binding, as well as increased sensitivity, selectivity, and dynamic concentration range of the MIPs.

6.4 FLUORESCENT MIP-BASED BIOMIMETIC SENSORS

Sensors are defined as devices designed to provide "specific quantitative or semi-quantitative analytical information" about the complex environment in and around us in real time.[198] They contain a recognition element that binds the target molecule with a high degree of selectivity, producing a chemical or physical signal, which is converted by a transducer (i.e. detector) into a quantifiable output signal, usually electrical.[93, 151] The components are integrated into a self-contained device, usually of small size, such that the recognition and transducing elements are in direct contact. Sensors using biomolecules, such as antibodies, enzymes or receptors as recognition elements are referred to as biosensors. Common transducers use electrochemical, optical (e.g. absorbance, fluorescence), and mass-sensitive signal transduction mechanisms. The main advantages of sensors are simplicity and speed of measurement.

Important criteria for assessing the performance of a sensor include sensitivity, linear concentration range, selectivity, and response time.[198] Sensitivity depends on the affinity of the target molecule for the recognition element, and on the nature of its coupling to the transducer. Selectivity is determined mainly by the recognition element of the sensor. The response time is defined as the time necessary to reach 90% of the steady-state response, and is dependent on diffusion rates to and through the matrix containing the recognition sites. Other parameters that should be considered in evaluating a sensor's performance are: sample throughput, which depends not only on the response time, but also on the recovery time (time needed for signal to return to baseline); reproducibility; stability to variations in pH and temperature, resistance to organic solvents; and lifetime of operation and storage.

There are few examples of actual sensor devices using MIPs as recognition elements described in this chapter that strictly qualify as biomimetic sensors given the definition above. One example is the luminescence fiber optic sensor for PMP described by Jenkins et al. (Section 6.3.2.1.2).[42, 43] Their practical device would allow dipstick-type sampling, as the coated end of the optical fiber is dipped into a solution to take a measurement. However, it must be regenerated after each use, and therefore is suitable only for measuring separate samples. The MIP-coated planar waveguide integrated into a flow system for the fluorescence detection of pyrene and other PAHs by Dickert is another example of an MIP-based sensor (Section 6.3.1.1).[79, 143]

6.5 CONCLUSIONS AND FUTURE DIRECTIONS

The implementation of MIPs as biomimetic recognition elements in sensors is hampered by the fact that they still do not perform as well as their natural recognition counterparts (i.e. antibodies, receptors, enzymes). MIPs, for the most part, have not reached the levels of sensitivity and selectivity of biological recognition molecules, especially in aqueous systems. MIPs with a high density of homogeneous binding sites that display high affinities and selectivities for their targets are prerequisites for the construction of sensitive sensors. Although considerable progress has been made toward this end, more work is necessary, especially in developing imprinted polymers that perform well in aqueous media, such as biological fluids and environmental samples. While initial studies show some promise for the imprinting of a limited number of proteins, not much attention has been devoted to the development of MIPs that selectively recognize other important biological macromolecules, such as polysaccharides and nucleic acids. Given the tremendous advances in proteomics and genomics, as well as the surge of interest in biomarkers of disease,[199, 200] it is likely that MIPs toward important biological molecules (e.g. lipids) and macromolecules will be developed. In the majority of studies MIPs have not been tested under the complex conditions in which they would ultimately have to be used (e.g. biological fluids, cell and tissue extracts, ground water) to determine their compatibility with such environments, and, most importantly, their selectivity in the presence of many potential interfering substances and matrix components. Before MIPs can truly be considered as practical alternatives or complements to biological recognition molecules in sensors, they will have to prove their utility for the analysis of real samples.

In the construction of sensors, biological recognition molecules are often immobilized in a thin layer at the transducer surface, for example by entrapping them behind a permeable or semi-permeable membrane or within a polymeric matrix, or by covalent attachment.[3, 198] This brings the recognition and transducing components in close contact, allowing efficient transfer of information, which can help to increase the sensitivity of the sensor. For optical sensors based on MIPs, close contact between the imprinted polymer and the transducer (e.g. optical fiber or planar optical waveguide), has been achieved by creating a thin imprinted polymer film that coats the transducer surface.

Even though some procedures have been described to accomplish this, more methods are required that rapidly and cheaply produce thin MIP films. Ideally, these imprinted films should be robust, transparent, and compatible with the transducer, which are problems with the methacrylate, acrylate and vinyl monomer-based matrices that are currently used.

An approach that offers great potential for molecular imprinting is the soft lithography technique.[201] This is a microfabrication method that is similar to photolithography, which is used to produce microelectronic devices for information technology, in that it allows a micropattern to be generated on the surface of a substrate. An elastomeric stamp or mold of poly(dimethylsiloxane) (PDMS) is created that has a surface with a relief structure of a particular pattern. For example, the pattern could be a series of parallel microchannels, so that when the stamp is placed on a substrate (e.g. silicone, glass) a network of empty microchannels is created in the areas where the stamp does not contact the surface of the substrate. These empty microchannels may then be filled by capillary action by placing a small drop of a liquid prepolymerization solution at the channel openings. After allowing the solution to completely polymerize, the PDMS stamp is removed to reveal a pattern of polymeric microcapillaries on the surface of the substrate. In fact, this process, called micromolding in capillaries (MIMIC), was used to create polyurethane-based imprinted polymer microcapillaries or microfilaments for anthracene on silicon wafers.[202] Such microfilaments could be coupled to a laser light source to create optical waveguides, and since anthracene is fluorescent, target binding to the imprinted sites may then be determined. Although binding studies were not performed, coupling of the laser light into the imprinted waveguides was shown to be feasible. Since the possibility exists of filling each microchannel with an imprinted polymer against different target molecules, this microfabrication technique has the potential to create an array of imprinted polymer microstructures on a single chip that could be used to simultaneously analyze multiple analytes. One could ultimately envision the construction of microsensor arrays for multianalyte analysis. Even though further optimization of this technique with imprinted polymers is necessary, as the PDMS stamps are not always compatible with the imprinted materials in use, it offers great promise as it allows components to be miniaturized, a current trend in sensor technology.[90]

Several different ways have been demonstrated in which fluorescence can been used to detect the binding of a target molecule to an MIP. In almost all cases, fluorescence detection was based on measuring the fluorescence intensity of a fluorescent target molecule, a competitive fluorescent reporter, or a fluorescent imprinted polymer as a function of the target concentration. However, fluorescence intensity is sensitive to many factors, and can vary in complex samples for reasons other than changes in target concentration.[203] For instance, changes in excitation intensity, pH, and light scattering, as well as fluorophore photobleaching, can all affect fluorescence intensity. Using fluorescence lifetime,[187] or intensity-ratio measurements[105] is one way to circumvent these shortcomings, as such measurements are usually not affected by fluorophore concentration, photobleaching and instrumental factors.[7] Intensity-ratio measurements are possible if there is a shift of wavelength in the emission or excitation spectrum in the presence of the target, or if two bands exist in either the emission or excitation spectrum whose intensity responds differently to the presence of the target.

The possibility also exists for the measurement of fluorescence resonance energy transfer (FRET), and fluorescence anisotropy.[101] Interestingly, some time ago

Sherrington's group[204] reported the preparation of a molecularly imprinted anisotropic polymer whose sites were oriented perpendicular to the polarization plane of UV light. The binding of target molecules to such imprinted sites may then be detected by measuring the difference in the UV absorption of the polymer in the parallel and perpendicular orientation (i.e. $\Delta A = A_\perp - A_\parallel$). Another promising approach is the use of quantum dots, which are luminescent inorganic nanoparticles that are currently generating much excitement.[205] The measurement of phosphorescence to determine target binding has also been presented.[206] Given the high sensitivity of fluorescence measurements, the multiple parameters that can be measured, and the ease with which it can be combined with optical fibers or planar optical waveguides (e.g. quartz plate), fluorescence will undoubtedly play a large role in the development of MIP-based optical sensors. Since fluorescence is also very rapid and nondestructive, and is nontoxic to living organisms, it could conceivably be used in the creation of micro- and nano-sized sensors suitable for continuous *in vivo* measurements, especially for measurements inside living cells as they are undergoing physiological responses.

Enzymes are commonly used as recognition elements in sensors.[3] Such biosensors may display enhanced target molecule dissociation rates compared to recognition elements that do not possess catalytic properties (e.g. antibodies), because the enzyme not only binds the target molecule, but also transforms it into a product which could have a lower affinity for the binding site. Enzyme-based biosensors may therefore be regenerated relatively quickly, and thus capable of high sample throughput and real-time continuous monitoring. In addition, increased selectivity and lower interference from nonspecific binding may also be observed. None of the reports discussed in this chapter present MIPs able to detect target binding based on a catalytic reaction. However, MIPs have been prepared that catalyze chemical reactions, which could be used in such an approach.[91] Further work is required before catalytic MIPs may be applied in sensor technology as possible alternatives to enzymes, because the catalytic activity of these MIPs has not yet reached the efficiency of enzymes.

A considerable amount of work has been done in developing MIPs in which target binding may be detected using fluorescence. However, only very few actual biomimetic sensors based on these approaches have been reported, largely because the affinity and selectivity of MIPs still do not match those of natural recognition molecules, especially in an aqueous environment. The impetus for further development of MIPs as biomimetic recognition elements for use in sensors are their stability compared to biological molecules, and their ability to be tailor-made for a wide variety of target molecules, especially small molecules for which it is difficult to obtain biological recognition entities. The simple manner in which MIPs can be prepared permits their mass production at low cost, an important aspect in sensor technology given the current interest in low-cost disposable transducers. Although further work is required before practical biomimetic sensors can be constructed using fluorescence-based MIPs, this approach clearly holds enormous promise for the future.

6.6 ACKNOWLEDGEMENTS

We wish to thank Dr. Irene Idziak and Amina Benrebouh for their advice and stimulating discussions.

6.7 REFERENCES

1. C. P. Price and D. J. Newman, *Principles and Practice of Immunoassay, 2nd Ed.*, (Stockton Press, New York, 1997).
2. M. L. Yarmush, A. M. Weiss, K. P. Antonsen, D. J. Odde, and D. M. Yarmush, Immunoaffinity purification: Basic principles and operational considerations, *Biotechnol. Adv.* **10**, 413-446 (1992).
3. M. D. Marazuela and M. C. Moreno-Bondi, Fiber-optic biosensors - An overview, *Anal. Bioanal. Chem.* **372**, 664-682 (2002).
4. U. E. Spichiger-Keller, *Chemical Sensors and Biosensors for Medical and Biological Applications*, (Wiley-VCH, Toronto, 1998).
5. B. Sellergren (Editor) *Molecularly Imprinted Polymers. Man-made Mimics of Antibodies and their Applications in Analytical Chemistry*, (Elsevier, New York, 2001).
6. J.-M. Lehn, *Supramolecular Chemistry*, (VCH, New York, 1995).
7. A. W. Czarnik (Editor) *Fluorescent Chemosensors for Ion and Molecule Recognition*, (American Chemical Society, Washington, DC, 1992).
8. G. Grynkiewcz, M. Poenie, and R. Y. Tsien, A new generation of Ca^{2+} indicators with greatly improved fluorescence properties, *J. Biol. Chem.* **260**, 3440-3450 (1985).
9. R. P. Haugland, *Handbook of Fluorescent Probes and Research Products*, 9th ed., (Molecular Probes, Eugene, OR, 2002).
10. G. Wulff, Molecular imprinting in cross-linked materials with the aid of molecular templates - A way towards artificial antibodies, *Angew. Chem. -Int. Edit.* **34**, 1812-1832 (1995).
11. E. Fischer, *Ber. Dtsh. Chem. Ges.* **27**, 2985-2993 (1894).
12. H. S. Andersson and I. A. Nicholls, A historical perspective of the development of molecular imprinting, in: *Molecularly Imprinted Polymers. Man-made Mimics of Antibodies and their Applications in Analytical Chemistry*, edited by B. Sellergren (Elsevier, New York, 2001), Chapter 1, pp. 1-19.
13. F. H. Dickey, The preparation of specific adsorbents, *Proc. Natl. Acad. Sci. U. S. A.* **35**, 227-229 (1949).
14. L. Pauling, A theory of the structure and process of formation of antibodies, *J. Am. Chem. Soc.* **62**, 2643-2657 (1940).
15. G. Wulff, A. Sarhan, and D. Zabrocki, Enzyme-analogue built polymers and their use for the resolution of racemates, *Tetrahedron Lett.* **44**, 4329-4332 (1973).
16. G. Wulff and A. Sarhan, The use of polymers with enzyme-analogous structures for the resolution of racemates, *Angew. Chem. -Int. Edit.* **11**, 341 (1972).
17. T. Takeuchi and J. Haginaka, Separation and sensing based on molecular recognition using molecularly imprinted polymers, *J. Chromatogr. B* **728**, 1-20 (1999).
18. K. Mosbach and O. Ramstrom, The emerging technique of molecular imprinting and its future impact on biotechnology, *Biotechnology* **14**, 163-170 (1996).
19. L. I. Andersson, B. Sellergren, and K. Mosbach, Imprinting of amino acid derivatives in macroporous polymers, *Tetrahedron Lett.* **25**, 5211-5214 (1984).
20. R. Arshady and K. Mosbach, Synthesis of substrate-selective polymers by host-guest polymerization, *Makromol. Chem.* **182**, 687-692 (1981).
21. B. Ekberg and K. Mosbach, Molecular imprinting - A technique for producing specific separation materials, *Trends Biotechnol.* **7**, 92-96 (1989).
22. B. Sellergren, M. Lepisto, and K. Mosbach, Highly enantioselective and substrate-selective polymers obtained by molecular imprinting utilizing noncovalent interactions - NMR and chromatographic studies on the nature of recognition, *J. Am. Chem. Soc.* **110**, 5853-5860 (1988).
23. B. Sellergren, Molecular imprinting by noncovalent interactions - Enantioselectivity and binding-capacity of polymers prepared under conditions favoring the formation of template complexes, *Makromol. Chem.* **190**, 2703-2711 (1989).
24. K. J. Shea, D. A. Spivak, and B. Sellergren, Polymer complements to nucleotide bases. Selective binding of adenine derivatives to imprinted polymers, *J. Am. Chem. Soc.* **115**, 3368-3369 (1993).
25. L. I. Andersson and K. Mosbach, Enantiomeric resolution on molecularly imprinted polymers prepared with only noncovalent and nonionic interactions, *J. Chromatogr.* **516**, 313-322 (1990).
26. K. Haupt, Molecularly imprinted polymers in analytical chemistry, *Analyst* **126**, 747-756 (2001).
27. M. Kempe and K. Mosbach, Binding studies on substrate- and enantio-selective molecularly

imprinted polymers, *Anal. Lett.* **24**, 1137-1145 (1991).

28.   B. Deore, Z. Chen, and T. Nagaoka, Potential-induced enantioselective uptake of amino acid into molecularly imprinted overoxidized polypyrrole, *Anal. Chem.* **72**, 3989-3994 (2000).

29.   S. H. Gao, W. Wang, and B. H. Wang, Building fluorescent sensors for carbohydrates using template-directed polymerizations, *Bioorg. Chem.* **29**, 308-320 (2001).

30.   A. G. Mayes, L. I. Andersson, and K. Mosbach, Sugar binding polymers showing high anomeric and epimeric discrimination obtained by noncovalent molecular imprinting, *Anal. Biochem.* **222**, 483-488 (1994).

31.   G. H. Chen, Z. B. Guan, C. T. Chen, L. T. Fu, V. Sundaresan, and F. H. Arnold, A glucose-sensing polymer, *Nat. Biotechnol.* **15**, 354-357 (1997).

32.   S. A. Piletsky, K. Piletskaya, E. V. Piletskaya, K. Yano, A. Kugimiya, A. V. Elgersma, R. Levi, U. Kahlow, T. Takeuchi, I. Karube, T. I. Panasyuk, and A. V. Elskaya, A biomimetic receptor system for sialic acid based on molecular imprinting, *Anal. Lett.* **29**, 157-170 (1996).

33.   N. Sallacan, M. Zayats, T. Bourenko, A. B. Kharitonov, and I. Willner, Imprinting of nucleotide and monosaccharide recognition sites in acrylamidephenylboronic acid-acrylamide copolymer membranes associated with electronic transducers, *Anal. Chem.* **74**, 702-712 (2002).

34.   L. I. Andersson, R. Muller, G. Vlatakis, and K. Mosbach, Mimics of the binding sites of opioid receptors obtained by molecular imprinting of enkephalin and morphine, *Proc. Natl. Acad. Sci. U. S. A.* **92**, 4788-4792 (1995).

35.   L. I. Andersson, Application of molecular imprinting to the development of aqueous buffer and organic solvent based radioligand binding assays for (S)-propranolol, *Anal. Chem.* **68**, 111-117 (1996).

36.   O. Ramstrom, L. Ye, and K. Mosbach, Artificial antibodies to corticosteroids prepared by molecular imprinting, *Chem. Biol.* **3**, 471-477 (1996).

37.   G. Vlatakis, L. I. Andersson, R. Muller, and K. Mosbach, Drug assay using antibody mimics made by molecular imprinting, *Nature* **361**, 645-647 (1993).

38.   R. Levi, S. McNiven, S. A. Piletsky, S. H. Cheong, K. Yano, and I. Karube, Optical detection of chloramphenicol using molecularly imprinted polymers, *Anal. Chem.* **69**, 2017-2021 (1997).

39.   C. J. Allender, K. R. Brain, C. Ballatore, D. Cahard, A. Siddiqui, and C. McGuigan, Separation of individual antiviral nucleotide prodrugs from synthetic mixtures using cross-reactivity of a molecularly imprinted stationary phase, *Anal. Chim. Acta* **435**, 107-113 (2001).

40.   J. G. Karlsson, L. I. Andersson, and I. A. Nicholls, Probing the molecular basis for ligand-selective recognition in molecularly imprinted polymers selective for the local anaesthetic bupivacaine, *Anal. Chim. Acta* **435**, 57-64 (2001).

41.   J. Matsui, I. A. Nicholls, and T. Takeuchi, Molecular recognition in cinchona alkaloid molecular imprinted polymer rods, *Anal. Chim. Acta* **365**, 89-93 (1998).

42.   A. L. Jenkins, O. M. Uy, and G. M. Murray, Polymer-based lanthanide luminescent sensor for detection of the hydrolysis product of the nerve agent Soman in water, *Anal. Chem.* **71**, 373-378 (1999).

43.   A. L. Jenkins, R. Yin, and J. L. Jensen, Molecularly imprinted polymer sensors for pesticide and insecticide detection in water, *Analyst* **126**, 798-802 (2001).

44.   J. Matsui, Y. Miyoshi, O. DoblhoffDier, and T. Takeuchi, A molecularly imprinted synthetic polymer receptor selective for atrazine, *Anal. Chem.* **67**, 4404-4408 (1995).

45.   J. Matsui, M. Okada, M. Tsuruoka, and T. Takeuchi, Solid-phase extraction of a triazine herbicide using a molecularly imprinted synthetic receptor, *Anal. Comm.* **34**, 85-87 (1997).

46.   M. T. Muldoon and L. H. Stanker, Molecularly imprinted solid-phase extraction of atrazine from beef liver extracts, *Anal. Chem.* **69**, 803-808 (1997).

47.   M. T. Muldoon and L. H. Stanker, Development and application of molecular imprinting technology for residue analysis, in: *Immunochemical Technology for Environmental Applications*, 657 ed., edited by D. S. Aga and E. M. Thurman (American Chemical Society, Washington, DC, 1997), pp. 314-330.

48.   T. Panasyuk-Delaney, V. M. Mirsky, M. Ulbricht, and O. S. Wolfbeis, Impedometric herbicide chemosensors based on molecularly imprinted polymers, *Anal. Chim. Acta* **435**, 157-162 (2001).

49.   F. L. Dickert, P. Achatz, and K. Halikias, Double molecular imprinting - A new sensor concept for improving selectivity in the of polycyclic aromatic hydrocarbons (PAHs) in water, *Fresenius J. Anal. Chem.* **371**, 11-15 (2001).

50.   D. Spivak, M. A. Gilmore, and K. J. Shea, Evaluation of binding and origins of specificity of 9-

ethyladenine imprinted polymers, *J. Am. Chem. Soc.* **119**, 4388-4393 (1997).

51. D. A. Spivak and K. J. Shea, Binding of nucleotide bases by imprinted polymers, *Macromolecules* **31**, 2160-2165 (1998).

52. H. Tsunemori, K. Araki, K. Uezu, M. Goto, and S. Furusaki, Surface imprinting polymers for the recognition of nucleotides, *Bioseparation* **10**, 315-321 (2001).

53. P. Turkewitsch, B. Wandelt, G. D. Darling, and W. S. Powell, Fluorescent functional recognition sites through molecular imprinting. A polymer-based fluorescent chemosensor for aqueous cAMP, *Anal. Chem.* **70**, 2025-2030 (1998).

54. K. Yano, K. Tanabe, T. Takeuchi, J. Matsui, K. Ikebukuro, and I. Karube, Molecularly imprinted polymers which mimic multiple hydrogen bonds between nucleotide bases, *Anal. Chim. Acta* **363**, 111-117 (1998).

55. J. Mathew and O. Buchardt, Molecular imprinting approach for the recognition of adenine in aqueous medium and hydrolysis of adenosine 5'-triphosphate, *Bioconjug. Chem.* **6**, 524-528 (1995).

56. H. Dong, A. J. Tong, and L. D. Li, Syntheses of steroid-based molecularly imprinted polymers and their molecular recognition study with spectrometric detection, *Spectroc. Acta Pt. A-Molec. Biomolec. Spectr.* **59**, 279-284 (2003).

57. I. Idziak and A. Benrebouh, A molecularly imprinted polymer for 17 α-ethynylestradiol evaluated by immunoassay, *Analyst* **125**, 1415-1417 (2000).

58. A. Rachkov, S. McNiven, S. H. Cheong, A. El'skaya, K. Yano, and I. Karube, Molecularly imprinted polymers selective for β-estradiol, *Supramol. Chem.* **9**, 317-323 (1998).

59. O. Ramstrom, L. Ye, M. Krook, and K. Mosbach, Screening of a combinatorial steroid library using molecularly imprinted polymers, *Anal. Comm.* **35**, 9-11 (1998).

60. L. Ye, Y. Yu, and K. Mosbach, Towards the development of molecularly imprinted artificial receptors for the screening of estrogenic chemicals, *Analyst* **126**, 760-765 (2001).

61. N. Perez, M. J. Whitcombe, and E. N. Vulfson, Surface imprinting of cholesterol on submicrometer core-shell emulsion particles, *Macromolecules* **34**, 830-836 (2001).

62. M. J. Whitcombe, M. E. Rodriguez, P. Villar, and E. N. Vulfson, A new method for the introduction of recognition site functionality into polymers prepared by molecular imprinting - Synthesis and characterization of polymeric receptors for cholesterol, *J. Am. Chem. Soc.* **117**, 7105-7111 (1995).

63. S. Dai, Y. S. Shin, L. M. Toth, and C. E. Barnes, Spectroscopic probing of adsorption of uranyl to uranyl-imprinted silica sol-gel glass via steady-state and time-resolved fluorescence measurement, *J. Phys. Chem. B* **101**, 5521-5524 (1997).

64. S. Dai, M. C. Burleigh, Y. H. Ju, H. J. Gao, J. S. Lin, S. J. Pennycook, C. E. Barnes, and Z. L. Xue, Hierarchically imprinted sorbents for the separation of metal ions, *J. Am. Chem. Soc.* **122**, 992-993 (2000).

65. A. L. Jenkins and G. M. Murray, Ultratrace determination of selected lanthanides by luminescence enhancement, *Anal. Chem.* **68**, 2974-2980 (1996).

66. T. Rosatzin, L. I. Andersson, W. Simon, and K. Mosbach, Preparation of Ca^{2+} Selective Sorbents by Molecular Imprinting Using Polymerizable Ionophores, *J. Chem. Soc. -Perkin Trans. 2* 1261-1265 (1991).

67. K. Tsukagoshi, M. Murata, and M. Maeda, Imprinting polymerisation for recognition and separation of metal ions, in: *Molecularly Imprinted Polymers. Man-made Mimics of Antibodies and their Applications in Analytical Chemistry*, edited by B. Sellergren (Elsevier, New York, 2001), Chapter 9, pp. 245-269.

68. K. Uezu, H. Nakamura, J. Kanno, T. Sugo, M. Goto, and F. Nakashio, Metal ion-imprinted polymer prepared by the combination of surface template polymerization with postirradiation by gamma-rays, *Macromolecules* **30**, 3888-3891 (1997).

69. H. Asanuma, T. Akiyama, K. Kajiya, T. Hishiya, and M. Komiyama, Molecular imprinting of cyclodextrin in water for the recognition of nanometer-scaled guests, *Anal. Chim. Acta* **435**, 25-33 (2001).

70. B. R. Hart and K. J. Shea, Molecular imprinting for the recognition of N-terminal histidine peptides in aqueous solution, *Macromolecules* **35**, 6192-6201 (2002).

71. M. Kempe, Antibody-mimicking polymers as chiral stationary phases in HPLC, *Anal. Chem.* **68**, 1948-1953 (1996).

72. J. U. Klein, M. J. Whitcombe, F. Mulholland, and E. N. Vulfson, Template-mediated synthesis of a

polymeric receptor specific to amino acid sequences, *Angew. Chem. -Int. Edit.* **38**, 2057-2060 (1999).

73. A. Rachkov and N. Minoura, Towards molecularly imprinted polymers selective to peptides and proteins. The epitope approach, *Biochim. Biophys. Acta-Protein Struct. Molec. Enzym.* **1544**, 255-266 (2001).

74. I. Chianella, M. Lotierzo, S. A. Piletsky, I. E. Tothill, B. N. Chen, K. Karim, and A. P. F. Turner, Rational design of a polymer specific for microcystin-LR using a computational approach, *Anal. Chem.* **74**, 1288-1293 (2002).

75. P. K. Dhal, M. G. Kulkarni, and R. A. Mashelkar, Bio-imprinting: Polymeric receptors with and for biological macromolecules, in: *Molecularly Imprinted Polymers. Man-made Mimics of Antibodies and their Applications in Analytical Chemistry*, edited by B. Sellergren (Elsevier, New York, 2001), Chapter 10, pp. 271-294.

76. S. Mallik, S. D. Plunkett, P. K. Dhal, R. D. Johnson, D. Pack, D. Shnek, and F. H. Arnold, Towards materials for the specific recognition and separation of proteins, *New J. Chem.* **18**, 299-304 (1994).

77. B. Sellergren, Imprinted polymers with memory for small molecules, proteins, or crystals, *Angew. Chem. -Int. Edit.* **39**, 1031-1037 (2000).

78. A. Aherne, C. Alexander, M. J. Payne, N. Perez, and E. N. Vulfson, Bacteria-mediated lithography of polymer surfaces, *J. Am. Chem. Soc.* **118**, 8771-8772 (1996).

79. F. L. Dickert, O. Hayden, and K. P. Halikias, Synthetic receptors as sensor coatings for molecules and living cells, *Analyst* **126**, 766-771 (2001).

80. F. L. Dickert, P. A. Lieberzeit, O. Hayden, S. Gazda-Miarecka, K. Halikias, K. J. Mann, and C. Palfinger, Chemical sensors - from molecules, complex mixtures to cells - supramolecular imprinting strategies, *Sensors* **3**, 381-392 (2003).

81. N. Perez, C. Alexander, and E. N. Vulfson, Surface imprinting of microorganisms, in: *Molecularly Imprinted Polymers. Man-made Mimics of Antibodies and their Applications in Analytical Chemistry*, edited by B. Sellergren (Elsevier, New York, 2001), Chapter 11, pp. 295-304.

82. S. A. Piletsky, E. V. Piletska, A. Bossi, K. Karim, P. Lowe, and A. P. F. Turner, Substitution of antibodies and receptors with molecularly imprinted polymers in enzyme-linked and fluorescent assays, *Biosens. Bioelectron.* **16**, 701-707 (2001).

83. H. Q. Shi, W. B. Tsai, M. D. Garrison, S. Ferrari, and B. D. Ratner, Template-imprinted nanostructured surfaces for protein recognition, *Nature* **398**, 593-597 (1999).

84. L. I. Andersson, Molecular imprinting for drug bioanalysis - A review on the application of imprinted polymers to solid-phase extraction and binding assay, *J. Chromatogr. B* **739**, 163-173 (2000).

85. L. I. Andersson, Selective solid-phase extraction of bio- and environmental samples using molecularly imprinted polymers, *Bioseparation* **10**, 353-364 (2001).

86. V. T. Remcho and Z. J. Tan, MIPs as chromatographic stationary phases for molecular recognition, *Anal. Chem.* 248A-255A (1999).

87. C. Chassaing, J. Stokes, R. F. Venn, F. Lanza, B. Sellergren, A. Holmberg, and C. Berggren, Molecularly imprinted polymers for the determination of a pharmaceutical development compound in plasma using 96-well MISPE technology, *J. Chromatogr. B* **804**, 71-81 (2004).

88. R. J. Ansell, MIP-ligand binding assays (pseudo-immunoassays), *Bioseparation* **10**, 365-377 (2001).

89. K. Haupt and K. Mosbach, Plastic antibodies: Developments and applications, *Trends Biotechnol.* **16**, 468-475 (1998).

90. K. Haupt, Imprinted polymers: The next generation, *Anal. Chem.* 377A-383A (2003).

91. G. Wulff, Enzyme-like catalysis by molecularly imprinted polymers, *Chem. Rev.* **102**, 1-27 (2002).

92. R. J. Ansell, Molecularly imprinted polymers in pseudoimmunoassay, *J. Chromatogr. B* **804**, 151-165 (2004).

93. K. Haupt and K. Mosbach, Molecularly imprinted polymers and their use in biomimetic sensors, *Chem. Rev.* **100**, 2495-2504 (2000).

94. D. Kriz, O. Ramstrom, and K. Mosbach, Molecular imprinting - New possibilities for sensor technology, *Anal. Chem.* **69**, A345-A349 (1997).

95. K. Yano and I. Karube, Molecularly imprinted polymers for biosensor applications, *Trends Anal. Chem.* **18**, 199-204 (1999).

96. S. McNiven and I. Karube, Toward optical sensors for biologically active molecules, in: *Molecularly Imprinted Polymers. Man-made Mimics of Antibodies and their Applications in*

Analytical Chemistry, edited by B. Sellergren (Elsevier, New York, 2001), Chapter 20, pp. 467-501.

97.   K. Ensing and T. de Boer, Tailor-made materials for tailor-made applications: Application of molecular imprints in chemical analysis, *Trends Anal. Chem.* **18**, 138-145 (1999).

98.   P. K. Owens, L. Karlsson, E. S. M. Lutz, and L. I. Andersson, Molecular imprinting for bio-and pharmaceutical analysis, *Trends Anal. Chem.* **18**, 146-154 (1999).

99.   J. R. Lakowicz (Editor) *Topics in Fluorescence Spectroscopy. Probe Design and Chemical Sensing*, (Plenum Press, New York, 1994).

100. S. Al Kindy, R. Badia, J. L. Suarez-Rodriguez, and M. E. Diaz-Garcia, Molecularly imprinted polymers and optical sensing applications, *Crit. Rev. Anal. Chem.* **30**, 291-309 (2000).

101. O. S. Wolfbeis, E. Terpetschnig, S. A. Piletsky, and E. Pringsheim, Fluorescence techniques for probing molecular interactions in imprinted polymers, in: *Applied Fluorescence in Chemistry, Biology and Medicine*, edited by W. Rettig (Springer, Berlin, 1999), Chapter 12, pp. 277-295.

102. B. Sellergren, The non-covalent approach to molecular imprinting, in: *Molecularly Imprinted Polymers. Man-made Mimics of Antibodies and their Applications in Analytical Chemistry*, edited by B. Sellergren (Elsevier, New York, 2001), Chapter 5, pp. 113-184.

103. G. Wulff and A. Biffis, Molecular imprinting with covalent or stoichiometric non-covalent interactions, in: *Molecularly Imprinted Polymers. Man-made Mimics of Antibodies and their Applications in Analytical Chemistry*, edited by B. Sellergren (Elsevier, New York, 2001), Chapter 4, pp. 71-111.

104. J. L. Suarez-Rodriguez and M. E. Diaz-Garcia, Flavonol fluorescent flow-through sensing based on a molecular imprinted polymer, *Anal. Chim. Acta* **405**, 67-76 (2000).

105. D. L. Rathbone and Y. Ge, Selectivity of response in fluorescent polymers imprinted with N^1-benzylidene pyridine-2-carboxamidrazones, *Anal. Chim. Acta* **435**, 129-136 (2001).

106. J. Matsui, M. Higashi, and T. Takeuchi, Molecularly imprinted polymer as 9-ethyladenine receptor having a porphyrin-based recognition center, *J. Am. Chem. Soc.* **122**, 5218-5219 (2000).

107. P. Turkewitsch, B. Wandelt, G. D. Darling, and W. S. Powell, Nucleotides enhance the fluorescence of trans-4-(p-N,N-dimethylaminostyryl)-N-vinylbenzylpyridinium chloride, *J. Photochem. Photobiol. A-Chem.* **117**, 199-207 (1998).

108. L. Q. Wu, B. W. Sun, Y. Z. Li, and W. B. Chang, Study properties of molecular imprinting polymer using a computational approach, *Analyst* **128**, 944-949 (2003).

109. I. Idziak, A. Benrebouh, and F. Deschamps, Simple NMR experiments as a means to predict the performance of an anti-17 alpha-ethynylestradiol molecularly imprinted polymer, *Anal. Chim. Acta* **435**, 137-140 (2001).

110. F. Lanza, A. J. Hall, B. Sellergren, A. Bereczki, G. Horvai, S. Bayoudh, P. A. G. Cormack, and D. Sherrington, Development of a semiautomated procedure for the synthesis and evaluation of molecularly imprinted polymers applied to the search for functional monomers for phenytoin and nifedipine, *Anal. Chim. Acta* **435**, 91-106 (2001).

111. T. Takeuchi, D. Fukuma, and J. Matsui, Combinatorial molecular imprinting: An approach to synthetic polymer receptors, *Anal. Chem.* **71**, 285-290 (1999).

112. T. Takeuchi, A. Seko, J. Matsui, and T. Mukawa, Molecularly imprinted polymer library on a microtiter plate. High-throughput synthesis and assessment of cinchona alkaloid-imprinted polymers, *Instrum. Sci. Technol.* **29**, 1-9 (2001).

113. B. Dirion, Z. Cobb, E. Schillinger, L. I. Andersson, and B. Sellergren, Water-compatible molecularly imprinted polymers obtained via high-throughput synthesis and experimental design, *J. Am. Chem. Soc.* **125**, 15101-15109 (2003).

114. Y. Liao, W. Wang, and B. H. Wang, Building fluorescent sensors by template polymerization: The preparation of a fluorescent sensor for L-tryptophan, *Bioorg. Chem.* **27**, 463-476 (1999).

115. G. Wulff, B. Heide, and G. Helfmeier, Molecular recognition through the exact placement of functional groups on rigid matrices via a template approach, *J. Am. Chem. Soc.* **108**, 1089-1091 (1986).

116. K. J. Shea and D. Y. Sasaki, On the control of microenvironment shape of functionalized network polymers prepared by template polymerization, *J. Am. Chem. Soc.* **111**, 3442-3444 (1989).

117. B. Sellergren, Noncovalent molecular imprinting: Antibody-like molecular recognition in polymeric network materials, *Trends Anal. Chem.* **16**, 310-320 (1997).

118. G. Wulff and K. Knorr, Stoichiometric noncovalent interaction in molecular imprinting, *Bioseparation* **10**, 257-276 (2001).

119. P. K. Dhal, Metal-ion coordination in designing molecularly imprinted polymer receptors, in:

Molecularly Imprinted Polymers. Man-made Mimics of Antibodies and their Applications in Analytical Chemistry, edited by B. Sellergren (Elsevier, New York, 2001), Chapter 6, pp. 185-201.

120. S. Striegler, Designing selective sites in templated polymers utilizing coordinative bonds, *J. Chromatogr. B* **804**, 183-195 (2004).

121. H. Kubo, H. Nariai, and T. Takeuchi, Multiple hydrogen bonding-based fluorescent imprinted polymers for cyclobarbital prepared with 2,6-bis(acrylamido)pyridine, *Chem. Commun.* 2792-2793 (2003).

122. C. Lubke, M. Lubke, M. J. Whitcombe, and E. N. Vulfson, Imprinted polymers prepared with stoichiometric template-monomer complexes: Efficient binding of ampicillin from aqueous solutions, *Macromolecules* **33**, 5098-5105 (2000).

123. K. Tanabe, T. Takeuchi, J. Matsui, K. Ikebukuro, K. Yano, and I. Karube, Recognition of barbiturates in molecularly imprinted copolymers using multiple hydrogen bonding, *J. Chem. Soc. , Chem. Commun.* 2303-2304 (1995).

124. N. T. K. Thanh, D. L. Rathbone, D. C. Billington, and N. A. Hartell, Selective recognition of cyclic GMP using a fluorescence-based molecularly imprinted polymer, *Anal. Lett.* **35**, 2499-2509 (2002).

125. G. Wulff and R. Schonfeld, Polymerizable amidines - Adhesion mediators and binding sites for molecular imprinting, *Adv. Mater.* **10**, 957-+ (1998).

126. H. Asanuma, T. Hishiya, and M. Komiyama, Tailor-made receptors by molecular imprinting, *Adv. Mater.* **12**, 1019-1030 (2000).

127. T. Hishiya, H. Asanuma, and M. Komiyama, Spectroscopic anatomy of molecular-imprinting of cyclodextrin. Evidence for preferential formation of ordered cyclodextrin assemblies, *J. Am. Chem. Soc.* **124**, 570-575 (2002).

128. P. K. Dhal and F. H. Arnold, Metal-coordination interactions in the template-mediated synthesis of substrate-selective polymers: Recognition of bis(imidazole) substrates by copper(II) iminodiacetate containing polymers, *Macromolecules* **25**, 7051-7059 (1992).

129. J. Matsui, Y. Tachibana, and T. Takeuchi, Molecularly imprinted receptor having metalloporphyrin-based signaling binding site, *Anal. Comm.* **35**, 225-227 (1998).

130. T. Takeuchi, T. Mukawa, J. Matsui, M. Higashi, and K. D. Shimizu, Molecularly imprinted polymers with metalloporphyrin-based molecular recognition sites coassembled with methacrylic acid, *Anal. Chem.* **73**, 3869-3874 (2001).

131. A. J. Tong, H. Dong, and L. D. Li, Molecular imprinting-based fluorescent chemosensor for histamine using zinc (II)-protoporphyrin as a functional monomer, *Anal. Chim. Acta* **466**, 31-37 (2002).

132. M. J. Whitcombe, L. Martin, and E. N. Vulfson, Predicting the selectivity of imprinted polymers, *Chromatographia* **47**, 457-464 (1998).

133. R. J. Ansell, O. Ramstrom, and K. Mosbach, Towards artificial antibodies prepared by molecular imprinting, *Clin. Chem.* **42**, 1506-1512 (1996).

134. J. Matsui, T. Kato, T. Takeuchi, M. Suzuki, K. Yokoyama, E. Tamiya, and I. Karube, Molecular recognition in continuous polymer rods prepared by a molecular imprinting technique, *Anal. Chem.* **65**, 2223-2224 (1993).

135. T. Takeuchi and J. Matsui, Miniaturized molecularly imprinted continuous polymer rods, *HRC-J. High Resolut. Chromatogr.* **23**, 44-46 (2000).

136. S. McNiven, M. Kato, R. Levi, K. Yano, and I. Karube, Chloramphenicol sensor based on an in situ imprinted polymer, *Anal. Chim. Acta* **365**, 69-74 (1998).

137. L. Schweitz, L. I. Andersson, and S. Nilsson, Rapid electrochromatographic enantiomer separations on short molecularly imprinted polymer monoliths, *Anal. Chim. Acta* **435**, 43-47 (2001).

138. A. L. Graham, C. A. Carlson, and P. L. Edmiston, Development and characterization of molecularly imprinted sol-gel materials for the selective detection of DDT, *Anal. Chem.* **74**, 458-467 (2002).

139. E. Hedborg, F. Winquist, I. Lundstrom, L. I. Andersson, and K. Mosbach, Some studies of molecularly imprinted polymer membranes in combination with field-effect devices, *Sens. Actuator A-Phys.* **37-8**, 796-799 (1993).

140. M. Jakusch, M. Janotta, B. Mizaikoff, K. Mosbach, and K. Haupt, Molecularly imprinted polymers and infrared evanescent wave spectroscopy. A chemical sensors approach, *Anal. Chem.* **71**, 4786-4791 (1999).

141. S. A. Piletsky, H. Matuschewski, U. Schedler, A. Wilpert, E. V. Piletska, T. A. Thiele, and M. Ulbricht, Surface functionalization of porous polypropylene membranes with molecularly

imprinted polymers by photograft copolymerization in water, *Macromolecules* **33**, 3092-3098 (2000).

142. S. A. Piletsky, E. V. Piletskaya, T. L. Panasyuk, A. V. El'skaya, R. Levi, I. Karube, and G. Wulff, Imprinted membranes for sensor technology: Opposite behavior of covalently and noncovalently imprinted membranes, *Macromolecules* **31**, 2137-2140 (1998).

143. F. L. Dickert, M. Tortschanoff, W. E. Bulst, and G. Fischerauer, Molecularly imprinted sensor layers for the detection of polycyclic aromatic hydrocarbons in water, *Anal. Chem.* **71**, 4559-4563 (1999).

144. K. Haupt, K. Noworyta, and W. Kutner, Imprinted polymer-based enantioselective acoustic sensor using a quartz crystal microbalance, *Anal. Comm.* **36**, 391-393 (1999).

145. K. Das, J. Penelle, and V. M. Rotello, Selective picomolar detection of hexachlorobenzene in water using a quartz crystal microbalance coated with a molecularly imprinted polymer thin film, *Langmuir* **19**, 3921-3925 (2003).

146. C. J. Percival, S. Stanley, M. Galle, A. Braithwaite, M. I. Newton, G. McHale, and W. Hayes, Molecular-imprinted, polymer-coated quartz crystal microbalances for the detection of terpenes, *Anal. Chem.* **73**, 4225-4228 (2001).

147. C. Malitesta, I. Losito, and P. G. Zambonin, Molecularly imprinted electrosynthesized polymers: New materials for biomimetic sensors, *Anal. Chem.* **71**, 1366-1370 (1999).

148. T. L. Panasyuk, V. M. Mirsky, S. A. Piletsky, and O. S. Wolfbeis, Electropolymerized molecularly imprinted polymers as receptor layers in capacitive chemical sensors, *Anal. Chem.* **71**, 4609-4613 (1999).

149. S. A. Piletsky, E. V. Piletska, B. N. Chen, K. Karim, D. Weston, G. Barrett, P. Lowe, and A. P. F. Turner, Chemical grafting of molecularly imprinted homopolymers to the surface of microplates. Application of artificial adrenergic receptor in enzyme-linked assay for beta-agonists determination, *Anal. Chem.* **72**, 4381-4385 (2000).

150. F. L. Dickert and O. Hayden, Noncovalent molecularly imprinted sensors for vapours, polyaromatic hydrocarbons and complex mixtures, in: *Molecularly Imprinted Polymers. Man-made Mimics of Antibodies and their Applications in Analytical Chemistry*, edited by B. Sellergren (Elsevier, New York, 2001), Chapter 21, pp. 503-525.

151. D. Kriz and R. J. Ansell, Biomimetic electrochemical sensors based on molecular imprinting, in: *Molecularly Imprinted Polymers. Man-made Mimics of Antibodies and their Applications in Analytical Chemistry*, edited by B. Sellergren (Elsevier, New York, 2001), Chapter 18, pp. 417-440.

152. A. G. Mayes and K. Mosbach, Molecularly imprinted polymer beads: Suspension polymerization using a liquid perfluorocarbon as the dispersing phase, *Anal. Chem.* **68**, 3769-3774 (1996).

153. L. Ye, R. Weiss, and K. Mosbach, Synthesis and characterization of molecularly imprinted microspheres, *Macromolecules* **33**, 8239-8245 (2000).

154. A. G. Mayes and K. Mosbach, Molecularly imprinted polymers: Useful materials for analytical chemistry?, *Trends Anal. Chem.* **16**, 321-332 (1997).

155. L. Ye, P. A. G. Cormack, and K. Mosbach, Molecular imprinting on microgel spheres, *Anal. Chim. Acta* **435**, 187-196 (2001).

156. A. G. Mayes, Polymerisation techniques for the formation of imprinted beads, in: *Molecularly Imprinted Polymers. Man-made Mimics of Antibodies and their Applications in Analytical Chemistry*, edited by B. Sellergren (Elsevier, New York, 2001), Chapter 12, pp. 305-324.

157. M. Glad, P. Reinholdsson, and K. Mosbach, Molecularly imprinted composite polymers based on trimethylolpropane trimethacrylate (TRIM) particles for efficient enantiomeric separations, *Reactive Polymers* **25**, 47-54 (1995).

158. C. Sulitzky, B. Ruckert, A. J. Hall, F. Lanza, K. Unger, and B. Sellergren, Grafting of molecularly imprinted polymer films on silica supports containing surface-bound free radical initiators, *Macromolecules* **35**, 79-91 (2002).

159. L. Ye, P. A. G. Cormack, and K. Mosbach, Molecularly imprinted monodisperse microspheres for competitive radioassay, *Anal. Comm.* **36**, 35-38 (1999).

160. I. Surugiu, L. Ye, E. Yilmaz, A. Dzgoev, B. Danielsson, K. Mosbach, and K. Haupt, An enzyme-linked molecularly imprinted sorbent assay, *Analyst* **125**, 13-16 (2000).

161. J. Matsui, K. Fujiwara, S. Ugata, and T. Takeuchi, Solid-phase extraction with a dibutylmelamine-imprinted polymer as triazine herbicide-selective sorbent, *J. Chromatogr. A* **889**, 25-31 (2000).

162. F. G. Tamayo, J. L. Casillas, and A. Martin-Esteban, Highly selective fenuron-imprinted polymer with a homogeneous binding site distribution prepared by precipitation polymerisation and its

application to the clean-up of fenuron in plant samples, *Anal. Chim. Acta* **482**, 165-173 (2003).

163. R. J. Ansell and K. Mosbach, Magnetic molecularly imprinted polymer beads for drug radioligand binding assay, *Analyst* **123**, 1611-1616 (1998).

164. E. Yilmaz, K. Haupt, and K. Mosbach, The use of immobilized templates - A new approach in molecular imprinting, *Angew. Chem. -Int. Edit.* **39**, 2115-+ (2000).

165. M. Yoshida, K. Uezu, M. Goto, S. Furusaki, and M. Takagi, An enantioselective polymer prepared by the surface molecular-imprinting technique, *Chem. Lett.* 925-926 (1998).

166. H. Q. Shi and B. D. Ratner, Template recognition of protein-imprinted polymer surfaces, *J. Biomed. Mater. Res.* **49**, 1-11 (2000).

167. M. F. Lulka, J. P. Chambers, E. R. Valdes, R. G. Thompson, and J. J. Valdes, Molecular imprinting of small molecules with organic silanes: Fluorescence detection, *Anal. Lett.* **30**, 2301-2313 (1997).

168. M. K. P. Leung, C. F. Chow, and M. H. W. Lam, A sol-gel derived molecular imprinted luminescent PET sensing material for 2,4-dichlorophenoxyacetic acid, *J. Mater. Chem.* **11**, 2985-2991 (2001).

169. M. F. Lulka, S. S. Iqbal, J. P. Chambers, E. R. Valdes, R. G. Thompson, M. T. Goode, and J. J. Valdes, Molecular imprinting of ricin and its A and B chains to organic silanes: Fluorescence detection, *Mater. Sci. Eng. C-Biomimetic Supramol. Syst.* **11**, 101-105 (2000).

170. D. Y. Sasaki, Molecular imprinting approaches using inorganic matrices, in: *Molecularly Imprinted Polymers. Man-made Mimics of Antibodies and their Applications in Analytical Chemistry*, edited by B. Sellergren (Elsevier, New York, 2001), Chapter 8, pp. 213-244.

171. D. Kriz, O. Ramstrom, A. Svensson, and K. Mosbach, Introducing biomimetic sensors based on molecularly imprinted polymers as recognition elements, *Anal. Chem.* **67**, 2142-2144 (1995).

172. C. F. Chow, M. H. W. Lam, and M. K. P. Leung, Fluorescent sensing of homocysteine by molecular imprinting, *Anal. Chim. Acta* **466**, 17-30 (2002).

173. J. Matsui, H. Kubo, and T. Takeuchi, Molecularly imprinted fluorescent-shift receptors prepared with 2-(trifluoromethyl)acrylic acid, *Anal. Chem.* **72**, 3286-3290 (2000).

174. K. A. Connors, *Binding Constants. The Measurement of Molecular Complex Stability*, (John Wiley & Sons, Toronto, 1987).

175. A. Rachkov, S. McNiven, A. El'skaya, K. Yano, and I. Karube, Fluorescence detection of β-estradiol using a molecularly imprinted polymer, *Anal. Chim. Acta* **405**, 23-29 (2000).

176. S. Hart, Dual fiber optic capillary probe for fluorescence detection using molecularly imprinted polymers, *Proceedings of SPIE* **4201**, 112-117 (2001).

177. S. A. Piletsky, E. V. Piletskaya, A. V. Elskaya, R. Levi, K. Yano, and I. Karube, Optical detection system for triazine based on molecularly imprinted polymers, *Anal. Lett.* **30**, 445-455 (1997).

178. K. Haupt, A. G. Mayes, and K. Mosbach, Herbicide assay using an imprinted polymer based system analogous to competitive fluoroimmunoassays, *Anal. Chem.* **70**, 3936-3939 (1998).

179. S. A. Piletsky, E. Terpetschnig, H. S. Andersson, I. A. Nicholls, and O. S. Wolfbeis, Application of non-specific fluorescent dyes for monitoring enantio-selective ligand binding to molecularly imprinted polymers, *Fresenius J. Anal. Chem.* **364**, 512-516 (1999).

180. J. L. Suarez-Rodriguez and M. E. Diaz-Garcia, Fluorescent competitive flow-through assay for chloramphenicol using molecularly imprinted polymers, *Biosens. Bioelectron.* **16**, 955-961 (2001).

181. K. Haupt, Molecularly imprinted sorbent assays and the use of non-related probes, *React. Funct. Polym.* **41**, 125-131 (1999).

182. K. Haupt, A. Dzgoev, and K. Mosbach, Assay system for the herbicide 2,4-dichlorophenoxyacetic acid using a molecularly imprinted polymer as an artificial recognition element, *Anal. Chem.* **70**, 628-631 (1998).

183. M. Franek, V. Kolar, M. Granatova, and Z. Nevorankova, Monoclonal ELISA for 2,3-dichlorophenoxyacetic acid: Characterization of antibodies and assay optimization, *J. Agric. Food Chem.* **42**, 1369-1374 (1994).

184. S. Kroger, A. P. F. Turner, K. Mosbach, and K. Haupt, Imprinted polymer based sensor system for herbicides using differential-pulse voltammetry on screen printed electrodes, *Anal. Chem.* **71**, 3698-3702 (1999).

185. P. Turkewitsch, B. Wandelt, R. R. Ganju, G. D. Darling, and W. S. Powell, Effect of nucleotides on the charge-transfer fluorescence of trans-4-(p-N,N-dimethylaminostyryl)-N-vinylbenzylpyridinium chloride, *Chem. Phys. Lett.* **260**, 142-146 (1996).

186. P. Turkewitsch, G. D. Darling, and W. S. Powell, Enhanced fluorescence of 4-(p-dimethylaminostyryl)pyridinium salts in the presence of biological macromolecules, *Journal of the Chemical Society-Faraday Transactions* **94**, 2083-2087 (1998).

187. B. Wandelt, P. Turkewitsch, S. Wysocki, and G. D. Darling, Fluorescent molecularly imprinted polymer studied by time-resolved fluorescence spectroscopy, *Polymer* **43**, 2777-2785 (2002).

188. B. Wandelt, A. Mielniczak, P. Turkewitsch, and S. Wysocki, Steady-state and time-resolved fluorescence studies of fluorescent imprinted polymers, *J. Lumin.* **102**, 774-781 (2003).

189. J. H. G. Steinke, I. R. Dunkin, and D. C. Sherrington, A simple polymerisable carboxylic acid receptor: 2-acrylamido pyridine, *Trends Anal. Chem.* **18**, 159-164 (1999).

190. D. L. Rathbone, D. Su, Y. Wang, and D. C. Billington, Molecular recognition by fluorescent imprinted polymers, *Tetrahedron Lett.* **41**, 123-126 (2000).

191. W. Wang, S. H. Gao, and B. H. Wang, Building fluorescent sensors by template polymerization: The preparation of a fluorescent sensor for D-Fructose, *Org. Lett.* **1**, 1209-1212 (1999).

192. I. Idziak and A. Benrebouh, Extraction of phosphonates, 6616846 (2003).

193. G. M. Murray and O. M. Uy, Ionic sensors based on molecularly imprinted polymers, in: *Molecularly Imprinted Polymers. Man-made Mimics of Antibodies and their Application in Analytical Chemistry*, edited by B. Sellergren (Elsevier, New York, 2001), Chapter 19, pp. 441-465.

194. G. M. Murray, A. L. Jenkins, A. Bzhelyansky, and O. M. Uy, Molecularly imprinted polymers for the selective sequestering and sensing of ions, *Johns Hopkins Apl Technical Digest* **18**, 464-472 (1997).

195. S. Al Kindy, R. Badia, and M. E. Diaz-Garcia, Fluorimetric monitoring of molecular imprinted polymer recognition events for aluminium, *Anal. Lett.* **35**, 1763-1774 (2002).

196. L. Ye, I. Surugiu, and K. Haupt, Scintillation proximity assay using molecularly imprinted microspheres, *Anal. Chem.* **74**, 959-964 (2002).

197. L. Ye and K. Mosbach, Polymers recognizing biomolecules based on a combination of molecular imprinting and proximity scintillation: A new sensor concept, *J. Am. Chem. Soc.* **123**, 2901-2902 (2001).

198. D. R. Thevenot, K. Toth, R. A. Durst, and G. S. Wilson, Electrochemical biosensors: Recommended definitions and classification, *International Journal of Pure and Applied Chemistry* **71**, 2333-2348 (1999).

199. S. Bonassi, M. Neri, and R. Puntoni, Validation of biomarkers as early predictors of disease, *Mutat. Res.* **480-481**, 349-358 (2001).

200. R. Massé and B. F. Gibbs, Proteomics discovery of biomarkers in: *Industrial Proteomics Applications for Biotechnology and Pharmaceuticals*, edited by D. Figeys and M.M. Ross (John Wiley & Sons Inc., Hoboken, NJ), in press.

201. Y. Xia and G. M. Whitesides, Soft lithography, *Angew. Chem. -Int. Edit.* **37**, 550-575 (1998).

202. J. J. Brazier, M. Yan, S. Prahl, and Y.-C. Chen, Molecularly imprinted polymers used as optical waveguides for the detection of fluorescent analytes, *Mat. Res. Soc. Symp. Proc.* 115-120 (2002).

203. J. R. Lakowicz, *Principles of Fluorescence Spectroscopy*, (Plenum, New York, 1984).

204. J. H. G. Steinke, I. R. Dunkin, and D. C. Sherrington, Molecularly imprinted anisotropic polymer monoliths, *Macromolecules* **29**, 407-415 (1996).

205. C. I. Lin, A. K. Joseph, C. K. Chang, and Y. D. Lee, Molecularly imprinted polymeric film on semiconductor nanoparticles. Analyte detection by quantum dot photoluminescence, *J. Chromatogr. A* **1027**, 259-262 (2004).

206. A. Fernandez-Gonzalez, R. Badia Laino, M. E. Diaz-Garcia, L. Guardia, and A. Viale, Assessment of molecularly imprinted sol-gel materials for selective room temperature phosphorescence recognition of nafcillin, *J. Chromatogr. A* **804**, 247-254 (2004).

EXCIMER SENSING

Valentine I. Vullev, Hui Jiang, and Guilford Jones, II[*]

7.1. INTRODUCTION

The term, excimer, was defined in the 1960's in terms of a molecular complex formed by interaction of an excited molecule and its ground state counterpart.[1] The early assumption was that the species did not have a stable "dimer" ground state. Such "excited dimers" were expected, and in fact shown, to exhibit dual fluorescence, that associated with monomeric fluorophores and a broad, readily distinguished, red-shifted emission ascribed to the complexed excited molecular pair. The molecule first associated with this phenomenon is pyrene, whose fluorescence properties, first reported by Forster and Kasper,[2] ushered in an important chapter in molecular photophysics, the study of excited bimolecular complexes (more generally, exciplexes) in solution. Pyrene has in fact become one of the most studied of all organic molecules in terms of its photophysical properties. These studies have brought the pyrene structure to the status of gold standard as a molecular probe of microenvironments. It displays, probably better than any other organic molecule, the most attractive combination of properties for study of association phenomena, including long emission lifetimes for excited monomer and excimer states, and exceptional separation of fluorescence bands. In this treatment of excimer sensing, we will devote our attention only to systems that utilize the pyrene chromophore and its closely related derivatives, and focus especially on more recently reported examples from the "trace and track" history of this famous molecule.

In this effort, we will define the term, sensor, very broadly and attempt to strike two rather different themes: (1) one having to do with the role of pyrene as a structure probe that can sense the proximity or interaction of molecules, particularly biological or bio-inspired macromolecules; and (2) recent use of pyrene derivatives in myriad elegant architectures (cyclodextrin derivatives, coronands, calixarenes and others) that utilize the chromophore and its monomer/excimer emission ratio for sensitive monitoring of chemical analytes, temperature, pressure and other properties. For the first objective, examples will be chosen that show how pyrene intra- or inter-molecular stacking is both a blessing and a limitation. The ground state association of pyrene pairs has become more widely appreciated in terms of an exceptional range of applications including probe strategies and the construction of *de novo* biomimetic molecules. On the other hand, for

---

[*] Boston University, Chemistry Department and Photonics Center, Boston, MA 02215.

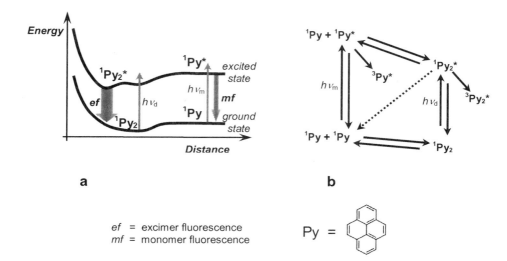

Scheme 7.1. Ground- and excited-state aggregation: (a) interaction energy dependence on the distance;[15] and (b) kinetic scheme.[3]

aqueous media the hydrophobic contacts of this simple condensed ring aromatic molecule provide energetic contributions to the molecular interactions that are targets for study (peptide aggregation, lipid diffusion, DNA recognition, protein folding), and, therefore, must be taken into account when binding phenomena or kinetic parameters are quantitatively assessed. For simplicity in terminology, we will not distinguish in this review between "excimers" formed from ground state vs. excited state association.

7.2. PYRENE EXCIMER AS SENSORY STRUCTURE PROBE OF THE ASSOCI-ATION OF BIOMOLECULES

7.2.1.   Excimers of Pyrene Derivatives on Macromolecular Templates

The tendency for pyrene and its derivatives to form excimers has been widely used for supramolecular design and for probing the structural properties of macromolecular systems. The excimer emission observed upon either aggregation between an excited- and a ground-state chromophore, or direct excitation of a ground-state aggregate of the same chromophore can be illustrated in an energy diagram (Scheme 7.1). For pyrene derivatives the latter process is more commonly observed in aqueous media where the hydrophobic and π-stacking interactions between the fluorophores are significant at relatively low concentrations[3] or for macromolecular scaffolds that retain the pyrene moieties in an aggregated state, even in organic media.[4] The complex of ground and excited state, associated and dissociated species is shown in a kinetic scheme (Scheme 7.1b).

Ground-state aggregation of simply substituted pyrenes can be readily detected by monitoring perturbations in the absorption spectrum: (1) a decrease in the apparent

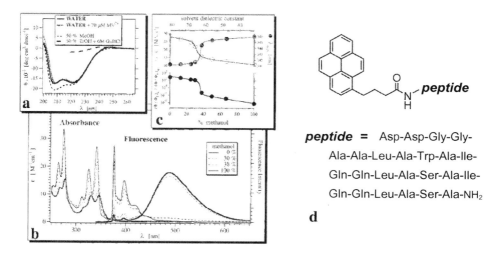

peptide = Asp-Asp-Gly-Gly-
Ala-Ala-Leu-Ala-Trp-Ala-Ile-
Gln-Gln-Leu-Ala-Ser-Ala-Ile-
Gln-Gln-Leu-Ala-Ser-Ala-NH$_2$

d

Figure 7.1. Absorption and emission of 10 μM TT1 including (**a**) circular dichroism (CD) spectra for various compositions, (**b**) recorded spectra for various water/methanol mixtures, and (**c**) titration data for pyrene absorption and monomer/dimer fluorescence ratios (λ_{exc} = 333 nm). (**d**) Sequence of TT1. (**a**, **b**, and **c** reproduced with permission from Reference 6.)

extinction coefficient, (2) a broadening of the absorption bands, and (3) a red shift of a few nanometers. Occasionally, absorption spectra cannot be collected because of the intense light-scattering properties of the sample or the impossibility to conduct transmission measurements in the UV region (e.g., modified surfaces, biocultures and cells). Under such conditions, comparison of the excitation spectra collected for the monomer vs. the excimer emission bands can shed light on whether ground-state aggregation occurs and to what extend.

The excimer emission of pyrene and pyrene derivatives appears as a broad featureless band and is shifted about 50 – 100 nm to the red of monomer emission. An example for a typical excimer emission from an alkylpyrene chromophore is shown in Figure 7.1. An amphipathic helical 24-residue polypeptide, TT1, that has alkylpyrene attached to its N-terminus exists as a monomer in organic solvents (e.g., methanol); however, in aqueous media it has a high propensity for aggregation.[5,6] For the polypeptide in methanol absorption and emission spectra are typical of monomeric alkylpyrene (Figure 7.1b). In buffered water solution, however, the intensities of the sharp emission bands at 380 and 400 nm are decreased and a broad featureless band of the alkylpyrene excimer appears at 485 nm. Furthermore, the changes in the absorption spectra (i.e., red shift and decrease in the molar absorptivity) are a clear indication of ground-state aggregation between the pyrene moieties. The distinct spectral separation of the fluorescence features of the alkylpyrene monomer and excimer and the relatively large emission quantum yield of the excimer, i.e., ~ 0.6 – 0.8,[7] make this particular chromophore a good choice for a photoprobe for excimer-sensing techniques.

The alkanoylpyrene (ketone) moiety is provided for illustration of a pyrene derivative whose somewhat altered excimer photophysics has been more recently examined.[8,9] It has been incorporated in synthetic polypeptides[9,10] including highly charged amphipathic helical structures, TT2, which like TT1 show a propensity for dimerization in aqueous media in such a way that the pyrene chromophores from the two helices are brought into proximity.[10] Hence, the observed alkanoylpyrene emission has two components, a monomer band at ~ 440 nm and an excimer band at ~ 530 nm (Figure 7.2a). The emission intensity of the alkanoylpyrene excimer proved to be extremely sensitive to the microenvironment of the chromophore pair. Upon addition of a negatively charged surfactant, SDS, which would bind predominantly to the positively charged N-terminus where the pyrenyl moiety is located, the intensity of the excimer emission increased more than two-fold (Figure 7.2b). Concurrently, similar addition of SDS did not lead to significant perturbations in the structure of the polypeptide (Figure 7.2c). When the concentration of the surfactant exceeded ~ 1 mM and the polypeptide was disaggregated, the excimer emission was completely depleted and a > 10-fold increase in the intensity of the monomer emission was observed (Figure 7.2). The emission quantum yield of the excimer of the alkanoylpyrene, however, is relatively low, i.e., ~ 0.02 – 0.05 [10], due to electron-transfer between the two pyrene moieties composing the excited dimer.[9] The monomer emission of the alkanoylpyrene, on the other hand, has a fluorescence quantum yield in the vicinity of ~ 0.3 – 0.6.[10] This significant difference between the emission efficiencies of the monomer and the excimer is illustrated with their spectral features as shown in Figure 7.2a.

An advantage provided by alkanoylpyrene derivatives is that they can be excited at ~ 400 nm,[8] which is a wavelength accessible for many imaging techniques. Also, the

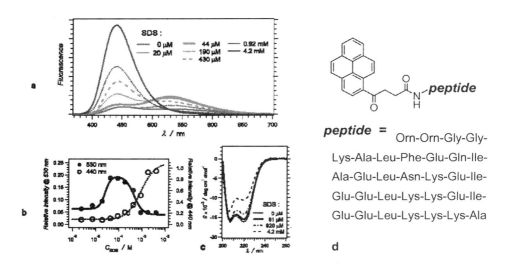

Figure 7.2. Emission and circular dichroism (CD) properties of TT2, 10 μM, in aqueous media (100 mM phosphate buffer, pH 7) in the presence of various amounts of SDS: **(a)** fluorescence spectra, λ_{ex}=355 nm; **(b)** the change of the intensities of the monomer and aggregate emission bands at 450 and 530 nm, respectively, with addition of SDS; and **(c)** CD spectra. **(d)** Sequence of TT2. (**a**, **b**, and **c** reproduced with permission from Reference 10.)

sensitivity of these chromophores toward the polarity and the hydrogen-bonding properties of the medium can be used to probe local microenvironment. The breadth of the monomer emission band of the pyrenyl chromophore and the relatively low fluorescence quantum yield of its excimer, however, can result in some ambiguities in excimer sensing experiments.

7.2.2. Considerations Regarding Pyrene Derivatives as Emission Probes for Macro-molecules

The two principal observables for fluorescence studies of pyrene-labeled molecules are: (1) relative emission intensity; and (2) the ratio between the emission intensities of the monomer and excimer.

Overall, using pyrene derivatives as emission probes have the following advantages: (1) the excimer and monomer fluorescence bands are well separated and easy to identify; (2) the emission quantum yields for both, the monomer and the excimer, are relatively large, e.g., between 0.1 and 0.9, depending on the pyrene substitutions and the surrounding environment; (3) the broad excimer fluorescence bands allow the use of a wide variety of chromophores as acceptors for energy-transfer experiments; (4) the relatively long lifetimes of the monomer and excimer excited states of the pyrene derivatives allows the measurement of dynamic processes with time constants as long as ~ 100 ns; (5) the derivatization of macromolecules with pyrenyl florescence probes is a relatively easy task. The first and second features have significant relevance for biological studies because of the capability to visualize qualitatively and image pyrene-containing systems without instrumentation. When photoexcited in the near UV region, an abundance of the pyrene monomer will result in a distinct blue fluorescence; if the excimer is the predominant fluorophore, however, the system will appear green.

The fluorescence quantum yield of the pyrene chromophore is sensitive to the polarity of the environment and the proximity to emission quenchers. Therefore, any conformational change or a process of aggregation in a macromolecular system that leads to alterations in the microenvironment around the pyrene probe can be followed by monitoring of its relative emission intensity. The reproducibility of the measurements of the absolute fluorescence intensities, however, are compromised by the sensitivity of many of the long-lived excited pyrene derivatives to quenching by oxygen.[11]

The appearance of an excimer is an indication that at least two pyrene chromophores have achieved a π-stacking geometry. In other words, if two distant sites of a macromolecule are labeled with pyrene probes, a conformational change that will bring these sites into proximity will result in excimer formation. Alternatively, excimers will appear when two or more macromolecules, each labeled with a single pyrene probe, aggregate in a manner that brings the pyrenes together.

Pyrene emission probes have, also, many shortcomings that ought to be considered when designing an experiment: (1) they are very hydrophobic, hence, they cannot be used unambiguously in aqueous media; (2) the pyrenyl group is very large (i.e., it can be approximated to a disk with a diameter of ~ 7 – 9 A), thus, placing the probe in the hydrophobic interior of a macromolecule to avoid contact with water, can result in steric structural perturbations; (3) pyrene requires UV excitation, which is not convenient for biological samples because of the presence of other species that absorb in the same region and also, fluorescent microscopes are not equipped with a UV light source and quartz lenses; (4) long excited-state lifetimes can also be a disadvantage because makes the chromophore more sensitive to the presence of small amounts of oxygen and other quenchers.

For aqueous media, which are most relevant for biochemical studies, pyrene emission probes should be used with caution because of the inherent driving forces for hydrophobic interaction. The *hydrophobic collapse* of pyrene moieties attached to water-soluble macromolecules can result in an alteration in the aggregation propensity of the macromolecules.[5,12,13] For instance, Garcia-Echeverria designed a simple experiment to test the geometry of a dimer of a polypeptide based on a sequence of a leucine-zipper-forming protein.[14] The author attached an alkylpyrene moiety to the N-terminus of the peptide and interpreted the appearance of a strong excimer emission band as an indication of the formation of parallel, rather than anti-parallel, helix dimers. Daughery and Gellman analyzed the aggregation of the same pyrene-labeled α-helical polypeptide and discovered that, in aqueous solution, it exists in the form of multimers considerably larger than a dimer.[12]

The aggregation of alkylpyrene in aqueous media is driven by hydrophobic and π-stacking interactions.[3] A recent study of the familiar labeling agent, 4-(1-pyrenyl)-butanoic acid (PBA), showed that the association constant for ground-state dimerization of PBA in water solution is 150 M^{-1}.[3] This value is about four orders of magnitude smaller than the value of the association constant for formation of PBA excimers that originate from an excited- and a ground-state species in aqueous medium. Although relatively small, the ground-state stacking propensity of pyrene chromophores added to an amphipathic peptide structures can result in an increase by two-to-three orders of magnitude in the values of the association constants for dimerization.[13] Apparently, the aggregation propensity of the pyrene probes can significantly modify the thermodynamics of a macromolecular system.

Complications can also be encountered during kinetic analysis of the photophysics of excimers when partial ground-state aggregation of the pyrene chromophores takes place (Scheme 7.1).[15] For example, lipid conjugates derivatized with alkylpyrenes are used for investigation of lateral diffusion in biologically related membranes. A rise in the excimer emission can be viewed as an indication that within the lifetime of the singlet-excited pyrene probe (a few hundreds of nanoseconds), the lipid to which it is attached, diffuses next to another lipid that contains a ground-state chromophore. A consideration of ground-state aggregation between pyrene moieties makes the interpretation of the excimer-emission results considerably more complicated.[16]

7.2.3. Application of Pyrene Fluorescence Probes to Studies of Proteins and Peptides

Pyrene probes have been extensively used in structural studies of proteins and peptides. Examples for ways to introduce these fluorescence probes to macromolecules are depicted in Scheme 7.2. Goedeweeck, et al., studied conformational dynamics of dipeptides labeled with two alkylpyrenes in series of organic solvents by monitoring the excimer emission with steady-state and time-resolved spectroscopy.[17] The authors reported a preference for extended conformations in hydrogen-bond-accepting solvents (e.g., ethyl acetate) and a preference for folded conformation in solvents that do not form hydrogen bonds (e.g., toluene).

Hammarstrom, et al., labeled the interior of human carbonic anhydrase II with two alkylpyrene chromophores and studied the folding and unfolding of the protein by monitoring the changes in the monomer and excimer fluorescence.[18] The pyrene moieties were introduced by chemical attachment of *N*-(1-pyrenemethyl)iodoacetamide to the thiol groups of a native cysteine (Cys, 206) and a Cys incorporated via a mutation (N67C). The investigators reported two stages of denaturation: (1) transition from folded to multi-globular conformation that resulted in increase in the excimer emission intensity (apparently, the increase in the flexibility of the peptide chains in the interior of the protein allows the two pyrene moieties to interact better with each other); and (2) transition from multi-globular to unfolded conformation that resulted in complete depletion of the excimer fluorescence and recovery of the monomer emission bands.

A similar approach was undertaken by Sahoo and co-workers who studied the change in the tertiary structure of apolipophorin III that occurs when this five-helix protein was transferred from aqueous media into phospholipid bi-layer membranes.[19,20] The five-helix bundle is formed from a single-chain polypeptide. One or two Cys residues were introduced in different helices in positions that are brought into proximity when the polypeptide folds into a helix bundle. The cysteine sites were selectively labeled with *N*-(1-pyrene)maleimide (Scheme 7.2b). In buffered water solutions, strong excimer emission was recorded only for the *bis*-labeled polypeptides, while all the *mono*-labeled derivatives exhibited only pyrene monomer fluorescence, indicating that no inter-peptide aggregation was taking place under the conditions of the experiment. When

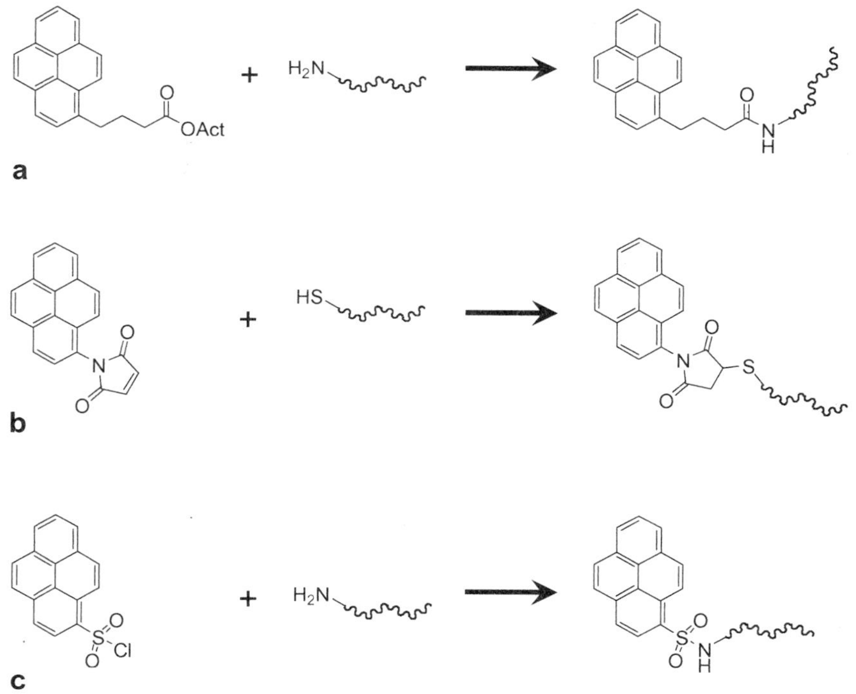

Scheme 7.2. Examples of derivatization of amine and thiol functional groups of macromolecules with pyrene photoprobes.

introduced into lipid bi-layers, however, the *bis*-labeled polypeptide still exhibited strong excimer emission that was quenched, in favor of monomer fluorescence, only when mono-labeled apolipophorin III was added. This finding suggests a strong propensity for aggregation of the polypeptide in lipophilic media, after its helix bundles are dissociated and converted into extended conformations.

Mihara et al. used two alkylpyrene probes in *de novo* designed polypeptide four-helix bundles as a proof for the correct tertiary structure.[21] An important issue, however, that the authors failed to address was what portion of the stabilization energy of the helix bundle was due to pyrene-pyrene interaction[3,13] and whether the same tertiary structure could be observed in the absence of these fluorescence probes.

For excimer sensing of properties of proteins the chromophores do not always need to be attached to the macromolecular structures. For instance, utilization of small pyrene-containing photoactive molecules that can selectively bind to certain proteins is a good approach to study molecular recognition and enzymatic activity.[22,23] To study the enzymatic activity of bovine trypsin, Ahn et al. prepared a series of short peptides that are substrates for the enzyme with two pyrene maleimide probes attached to their C- and N-termini.[23] The digestion of the substrates was followed by monitoring the disappearance of the excimer emission and the increase in the intensity of the monomer fluorescence as the peptide chains were hydrolyzed by the enzyme and the pyrene probes were separated from each other. In this type of experiment, to prevent the aggregation of the pyrene-labeled conjugates in aqueous media, the concentration of the probing molecules should be kept low ($< \sim 10^{-4}$ M^{-1}) and charged groups can be added for solubilization.

7.2.4.   Pyrene-labeled DNA Strands Used for Molecular Recognition

Fluorescence analysis has broad application in genomics, where high sensitivity is required for analysis of ultra-low sample concentrations. There are numerous examples where pyrene excimer assays have been successfully used for recognition of complimentary strands of DNA.

Paris and co-workers reported an elegant way for determination of DNA sequence with high sensitivity and specificity by using two pyrene-labeled strands complimentary to the sequence of the analyte.[24] The authors prepared a series of singly labeled oligodeoxynucleotide probes where alkylpyrene is introduced as a part of a nucleotide to the 5' or 3' end of the sequences. Maximum intensity of the excimer emission was observed when a complimentary strand with exactly matched spacing was added, i.e., N-2 spacing (Figure 7.3a). The specificity of this approach was tested by addition of a single point-mutation in the complementary strand. In the case of exact match in the sequences, strong green fluorescence resulting from the formation of excimer was observed (Figure 7.3b, Image B). Under the same conditions, in the presence or absence of a singly mismatched sequence, only pale bluish fluorescence from the pyrene monomer appeared (Figure 7.3b, Images A and C)

Some reports have indicated that an increase in the sensitivity of the DNA emission analysis can be achieved by placing two pyrene labels adjacent to each other.[25-27] For example, terminally labeled oligo-DNA strands with *bis*-pyrene probes exhibit about 20 – 30 fold increase in the ratios between the excimer and monomer emission intensities.[26]

A more complicated approach for emission assays of DNA binding has been described by Tong et al.[28] The researchers attached oligopeptides multiply-labeled with pyrenebutyramide (Scheme 7.2a) to short DNA strands. The specific interaction with

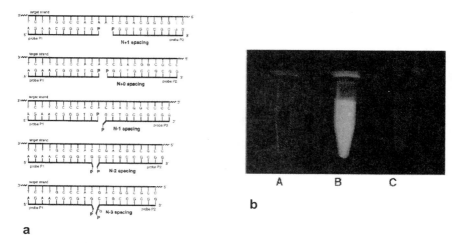

Figure 7.3. (a) Illustration of the relative spacing of dual pyrene probes P1 and P2 on the series of target DNAs designated N+1, N+0, N–1, N–2, N–3. Note that this relative spacing is formal (based on sequence only) and does not necessarily correspond to actual interchromophore distances. **(b)** Photograph of solutions of P3 and P4 probes in the absence of target (**A**), and in the presence of H-*ras* codon 12 mutant target MT (**B**) and unmutated target WT (**C**). Each probe and target is present at 10 mM in a pH 7.0 buffer (20 mM Na-PIPES) containing 20 mM Mg^{2+} and 200 mM Na$^+$ at 23 °C. Fluorescence is observed over a UV transilluminator (310 nm). (Reproduced with permission from Reference 24.)

another strand having a complementary sequence competed with the non-specific interaction between the pyrene moieties and the nucleotides, resulting in an increase in the excimer emission upon duplex formation. The reported changes of the fluorescence intensities, however, were only about two-fold or less.

7.2.5.   Pyrenyl-containing Lipid Membranes

Pyrene labeled lipids have been incorporated into bilayers to study lateral diffusion,[29,30] phase separation,[31,32] binding to the surface of the membrane,[30,33] conformational properties of the lipids,[34] and membrane fusion.[35-38] Because of their interfacial nature, fluorescently labeled lipid membranes have the potential for incorporation in analytical devices and sensors.[39]

Pyrene photoprobes are introduced to the lipid membranes by chemical attachment to the ω-position of one or both of the acyl chains of the phospholipids. The *mono*-labeled phospholipids are more suitable for investigation of intermolecular interactions and rearrangements in the membrane by monitoring the intensities of the monomer and excimer emissions. The *bis*-labeled conjugates, on the other hand, are useful for conformational studies of the individual lipid moieties and for some imaging experiments in cell biology.

Domain formation and lateral distribution of the individual lipids in a cell membrane play important role in many of its vital properties, e.g., signal transduction, fluidity, surface binding. By monitoring the excimer pyrene emission of bilayers comprised of lipids that are singly derivatized with pyrenyl probes, it is easy to determine if the labeled

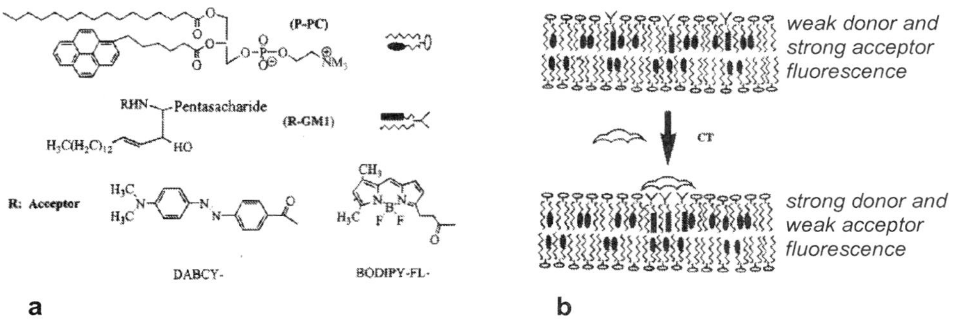

Figure 7.4. (a) Structures and acronyms of selected probes. **(b)** Schematic illustration of the signal transduction mechanism: (top) acceptor-tagged receptors distributed homogeneously in the outer leaflet of the bilayers of natural and donor-derivatized phospholipids: energy transfer results in weak donor (i.e., pyrene) fluorescence; (bottom) aggregation of the acceptor-tagged receptors resulting from CT binding. (Reproduced with permission from Reference 33.)

lipids tend to aggregate and form domains or they randomly diffuse throughout the membrane. The lateral distribution of the lipids is strongly dependent on the properties of the hydrophilic heads. For example, pyrenyl-labeled lipids with positively charged heads will tend to distribute randomly in a phospholipid bilayer. Under proper dilution of the fluorescently labeled lipid, only monomer emission will be observed. When a multivalent macromolecule (e.g., a protein) that has affinity for the positively charged heads binds to the surface of the membrane, it will cause aggregation of the pyrene-labeled lipids and thus, induce excimer emission. Furthermore, if the protein carries a quencher (e.g., a FRET energy acceptor) the excimer emission will not be observed, rather, the emission of the quencher will be recorded if the quencher is fluorescent. Alternatively, if the protein contains an FRET energy donor, e.g., tryptophan, the observed excimer emission of the pyrene probe will be enhanced when the protein-bound chromophore is excited.[40]

Based on such an approach, Song and Swanson designed a system for sensing multivalent proteins.[33] The authors prepared a membrane that contains a mixture of pyrenyl-labeled lipid with a zwitterionic head and a lipid functionalized with an energy acceptor (pyrromethene or azobenzene derivative). The head of the latter was modified with carbohydrate GM1 receptor for the cholera toxin, CT (Figure 7.4a). When the toxin was bound to the surface of the membrane, it sequestered the energy accepting lipids away from the pyrene derivatives, hence, causing regeneration of the pyrene excimer emission (Figure 7.4b).

Pyrene-labeled lipids proved to be convenient accessory for study of membrane fusion of biological importance. For example, viruses whose coats are biosynthetically labeled with pyrene-containing phospholipids have been used for studying mechanisms of viral infection. When the coat of the fluorescently labeled virus is fused with the wall of a cell (or a vesicle), the lateral dilution of the pyrene lipid conjugates results in a decrease in the observed excimer emission.[35,36]

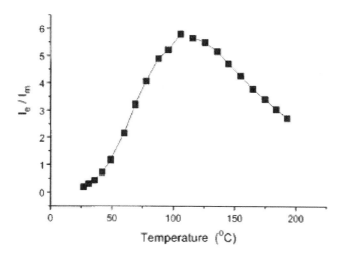

Figure 7.5. Calibration curve showing the ratio of excimer to monomer fluorescence as a function of temperature for BPP in polyethylene. (Reproduced with permission from Reference 42.)

7.3. PYRENE EXCIMER EMISSION IN ENVIRONMENTAL AND CHEMICAL SENSING

The excimer fluorescence of the parent pyrene molecule or pyrene conjugates has been used to sense environmental parameters such as temperature, pressure or pH. The change of excimer fluorescence intensity reflects the change of the environment. In addition, it can also be used to detect guest molecules such as gases (O_2 or NH_3), organic molecules, metals, or other miscellaneous analytes.

7.3.1. Sensing of Temperature, Pressure and pH

Pyrene fluorescence is moderately sensitive to medium temperature. Its emission characteristics in several different solvents over a wide temperature range from −100 to 90 °C have been studied by Birks *et al.*[41] The observations were made of the temperature dependence of the parameter I_D/I_M, where I_D and I_M were the quantum yields of excimer and monomer fluorescence, respectively. The plot of $lg(I_D/I_M)$ versus $1/T$ shows that as temperature is increased, I_D/I_M increases to a maximum at $T = T_{max}$ in the region of room temperature and then decreases. The behavior was concluded to be due to an excimer formation process which was diffusion controlled and reversible.[41] Bur *et al*[42,43] developed a fluorescence based method for non-invasive temperature measurements during polymer processing. They doped the excimer-forming bis-pyrene propane (BPP) into the polymer resin at a low mass fraction ($c < 1 \times 10^{-5}$) so as not to disturb other properties of the medium. The dependence of I_e/I_m, the ratio of excimer to monomer fluorescence intensity, on temperature was first calibrated for a specific polymer-dye system in the quiescent state. From this calibration curve (shown in Figure 7.5 is an example for BPP in polyethylene), a simple functional form can be generated to describe

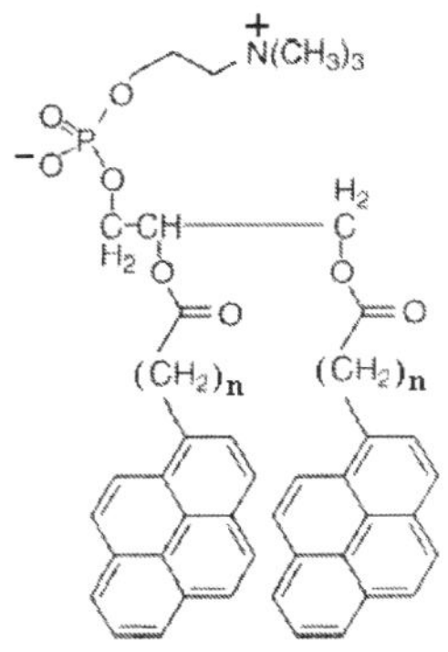

(1): **a** (n = 3); **b** (n = 5); **c** (n = 7); **d** (n = 9)

Scheme 7.3. The dipyPC fluorescent probe molecules

the temperature dependence, which then can be used to estimate the temperature during on-line measurements. Based on this concept, a device has been developed and tested for on-line temperature measurements during polymer processing. On comparison with thermocouple monitoring, this device gave satisfactory results.[42,43]

The effect of pressure on pyrene excimer fluorescence has been studied extensively for different systems; and several reports have been focused on pressure sensing. Templer *et al*[44,45] reported a homologous series of di-pyrenyl phosphatidylcholine (dipyPC) (Scheme 7.3) that could sense lateral pressure variations in the chain region of the

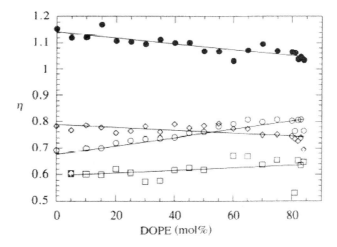

Figure 7.6. Variation in η at 25 °C as a function of DOPC/DOPE composition in the L_α phase in excess water. The excimer to monomer ratio has been measured using **1a - 1d** (●, ◊, □ and ○, respectively). Linear fits to the data are shown. (Reproduced with permission from Reference 45.)

amphiphilic membrane. At low dilutions of **1** in model membranes, the excimer signal is entirely intra-molecular. Since it depends on the frequency with which the pyrene moieties are brought into close proximity, the relative intensity of the excimer to monomer signal, η, is a measure of the pressure. As shown in Figure 7.6, for fully hydrated bilayers composed of dioleoylphosphatidylcholine and dioleoylphosphatidyl-ethanol-amine (DOPC and DOPE respectively), a dip in the excimer signal is observed (**1b** and **1d** have bigger η values than **1c**), which should be due to the dip in the lateral pressure in the region of the DOPC/DOPE cis double bond.[45]

Some polymers such as poly(acrylic acid) (PAA) and poly(methacrylic acid) (PMAA) exhibit a well-documented pH-dependent conformational change.[15] Several pyrene labeled PAA derivatives have been developed by Bossmann *et al*[46] for pH sensing. As shown in Scheme 7.4, these derivatives were achieved by randomly attaching perfluorinated ($-O-C_8F_{17}$) functions and chemically linked pyrene fluorescent units to the PAA backbone. In the notation PAA-pyrene/l-C_8F_{17}/n, l and n stand for the ratios of the total number of carboxylate groups to pyrene and perfluorinate labeled carboxylate groups, respectively. The pH-induced changes of the polymer conformation and dynamics trigger the changes in I_M/I_E, where I_M and I_E are the integrated pyrene monomer ($\lambda = 360 - 450$ nm) and excimer ($\lambda = 450 - 650$ nm) emission intensities, respectively, making detection of pH possible. For the most strongly responding modified polymer PAA-pyrene/151-C_8F_{17}/2.85, in order to eliminate other interfering influences, the I_M/I_E values obtained at high polymer concentration (1.0 g·L^{-1}) were subtracted from the values measured at low concentration (0.01 g·L^{-1}). The dependence of the resulting y values on pH could then be fitted to the fifth order polynomial equation [R = 0.996] as shown in Figure 7.7:

$$y = 34.56 - 29.68 \, pH + 15.113 \, pH^2 - 3.529 \, pH^3 + 0.349 \, pH^4 - 0.012262 \, pH^5$$

Therefore, if the value of y is known, the pH can be calculated easily.[46]

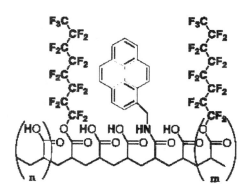

Scheme 7.4. The PAA-pyrene/l-C_8F_{17}/n polymers.

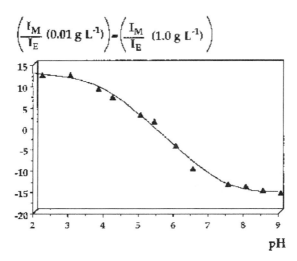

Figure 7.7. Dependence of the normalized quotient I_M/I_E versus pH for PAA-pyrene/151-C_8F_{17}/2.85 (see Scheme 4). (Reproduced with permission from Reference 46.)

7.3.2. Sensing of Oxygen

Pyrene monomer or excimer fluorescence has been used to detect oxygen, either by fluorescence quenching[47-49] or by phase modulation.[50,51] Sharma[48] immobilized pyrene in silicone rubber and observed higher oxygen quenching efficiency for excimer than for monomer fluorescence. The respective Stern-Volmer constants are K_{SV} (monomer) = 0.05 %$^{-1}$ and K_{SV} (excimer) = 0.14 %$^{-1}$ within the 24.0% oxygen concentration range in N_2. The process of quenching is diffusion controlled and fully reversible, and the sensor response time is ~3 sec. Amao *et al*[49] prepared an optical oxygen sensor using a chemisorption technique. The 4-(1-pyrenyl)butanoic acid (PBA) was chemisorbed onto nano-porous oxidized aluminium plate. The ratio of I_0/I_{100} was used as a sensitivity of the sensor, where I_0 and I_{100} represented the detected fluorescence intensities from a substrate exposed to 100% argon and 100% oxygen, respectively. In this case, the I_0/I_{100} value of monomer and excimer emissions of PBA chemisorbed layer were estimated to be 27.9 and 73.4, respectively.

On the other hand, Nagel *et al*[50,51] developed a method to prepare sensor systems for monitoring blood oxygen concentrations. One of the sensors applies bis-(1-pyrene) derivatives in highly oxygen permeable polymer matrixes, either by physical mixing or dispersing or by covalent bonding. The excitation light is intensity modulated (sine wave). Therefore, the resulting monomer and excimer emissions are also modulated. The processor then determines the extent of the phase shift between and/or the ratio of demodulation factors of these two emitted signals. Alternately, the extent of the phase shift and/or the magnitude of relative demodulation between the excimer emission signal and the excitation signal can be determined. Either way, the extent of this phase shift and/or the ratio of demodulation factors (or the magnitude of relative demodulation) are dependent on the concentration of oxygen in blood.

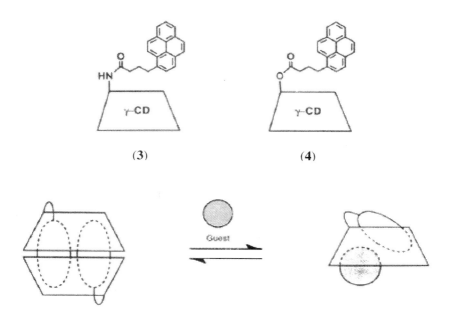

Scheme 7.5. Chemical structures of pyrene modified γ-cyclodextrins and schematic representation of the guest-induced dissociation. (Reproduced with permission from Reference 53.)

7.3.3. Sensing of Organic Guests by Modified γ-Cyclodextrins

Pyrene-modified γ-cyclodextrins (γ-CD) have been designed to sense organic molecules that are involved in host-guest inclusion.[52-55] The Ueno group has published several informative reviews.[56,57]

In 1990, Ueno et al[53] developed the first example of a host-guest sensory system, in which γ-CD was modified with pyrene moiety via amide or ester link (Scheme 7.5). These derivatives can form a complex by dimerization of the 1:1 intra-molecular CD-pyrene complexes in a 10% DMSO aqueous solution. However, upon addition of a guest, the dimers can be converted into 1:1 host-guest complexes, which exhibit only monomer fluorescence (Scheme 7.5). Consequently, in the presence of guest molecules, the pyrene excimer fluorescence decreases, while that of monomer increases. For quantification purpose, the excimer emission intensities in the absence and presence of guest species are measured at 470 nm (abbreviated as I_{ex}° and I_{ex}, respectively), while the monomer emission intensities are measured in the same manner at 378 nm (abbreviated as I_m° and I_m). The $\Delta I_{ex}/I_{ex}^{\circ}$ and $\Delta I_m/I_m^{\circ}$ values at certain guest concentrations, where $\Delta I_{ex} = I_{ex}^{\circ} - I_{ex}$, and $\Delta I_m = I_m^{\circ} - I_m$, can then be used as a measure of sensitivity. The authors tested 42 different organic compounds, including biologically important steroids such as bile acids, and concluded that the sensitivity depended on the polarity, shape and the size of guest molecules.

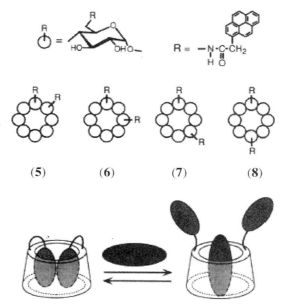

Scheme 7.6. Chemical structures of pyrene modified γ-cyclodextrins and schematic representation of the guest molecule inclusion: intra- and inter-molecular inclusion complexes. (Reproduced with permission from Reference 55.)

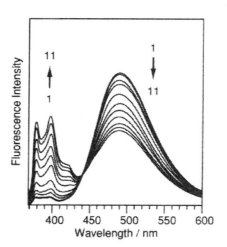

Figure 7.8. Fluorescence spectra of **7** (0.15 μ M) in a 20% DMSO aqueous solution at various lithocholic acid concentrations (1, 0; 2, 1; 3, 3; 4, 5; 5, 10; 6, 15; 7, 20; 8, 25; 9, 30; 10, 35; 11, 40 μ M). Excitation wavelength was 344 nm. (Reproduced with permission from Reference 55)

Ueno *et al*[55] also prepared several γ-CD derivatives bearing two pyrene moieties (Scheme 7.6). For 20% DMSO aqueous solutions, these derivatives show predominant excimer emission. This finding indicates that two attached pyrene rings can be included in one γ-CD cavity, irrespective of the point of relative attachment. In the presence of guest molecules, however, the two pyrene rings are shown to be excluded from the cavity and separated from each other (Scheme 7.6). Again, this host-guest interaction leads to a decrease in pyrene excimer emission with increased monomer emission. Shown in Figure 7.8 are the results for the example of sensing lithocholic acid by **7**. Similar to one pyrene moiety modified γ-CDs, these isomers show subtle but different sensitivity and selectivity for different organic compounds.

Recently, Ueno *et al*[54] also synthesized γ-CD-peptide hybrids, which have pyrene and γ-CD side chains placed in proximity along the peptide backbone (Scheme 7.7). In these compounds, peptides maintain relatively rich α-helix content and the inclusion of two pyrene units in a single γ-CD cavity causes inter-molecular peptide dimerization. Therefore, all these hybrids show excimer fluorescence at high concentrations. Upon the addition of guest molecules, the intensity of pyrene excimer emission decreases, while that of monomer emission increases. Again, this is believed to be due to the dissociation of γ-CD-peptide dimers.

A γ-CD dimer modified with two pyrene moieties has been designed and prepared by Hamada *et al*[52] for sensing steroids and endocrine disruptors (Scheme 7.8). In dilute 10% ethylene glycol aqueous solution, **12** shows both pyrene monomer and excimer fluorescence. This finding indicates that the two pyrene moieties can achieve a face-to-face orientation. In the presence of guest molecules, the two pyrene units should be displaced and far away from the γ-CD cavities, causing decreased excimer emission and increased monomer emission. A quantification method similar to that used by Ueno[53-55] was applied for assays for various steroids and endocrine disruptors.

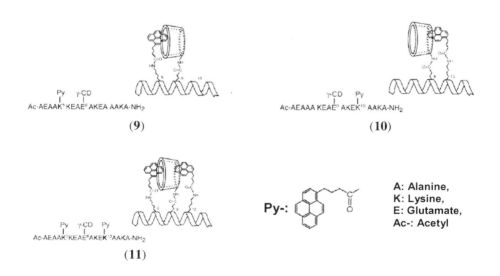

Scheme 7.7. Chemical structures of γ-CD-peptide.

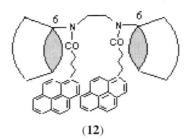

(**12**)

Scheme 7.8. Chemical structure of γ-CD dimer modified with two pyrene moieties.

7.3.4. Sensing of Metal Cations

The detection of metal cations is of great interest to many scientists, and it is always an intensive research field. Different design principles of fluorescent molecular sensors have been reviewed by Lelay *et al*[58]. According to the structure of the complexing moiety, the sensor systems can be roughly divided into 5 classes: chelators, podands, coronands (crown ethers), cryptands, and calixarenes.[58] Examples that utilize pyrene as chromophore can be found for each category, except for cryptands.

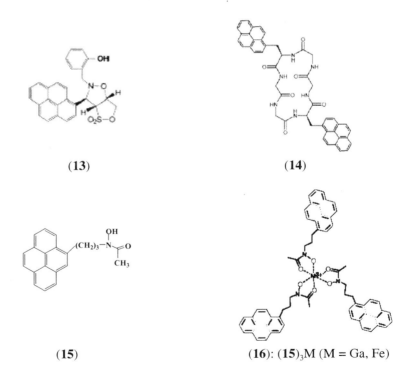

(**13**)   (**14**)

(**15**)   (**16**): (**15**)$_3$M (M = Ga, Fe)

Scheme 7.9. Chemical structures of chelator-based sensors.

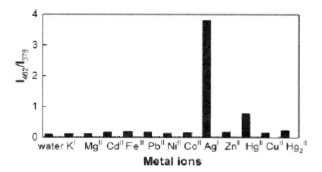

Figure 7.9. The responses of compound **13** to metal solutions (150 μM, *x*-axis markers). Excitation was at 344 nm, and emission was at 462 and 378 nm. (Reproduced with permission from Reference 60.)

7.3.4.1. Chelator-Based Sensors

The deprotonated form of the simple 4-(1-pyrenyl)butanoic acid (PBA) was found to form 1:2 (metal : ligand) complexes with Zn^{2+} or earth alkali metal ions in acetonitrile solution.[59] This complexation process brings two pyrene moieties to close contact to form dimers, leading to the progressive disappearance of monomer fluorescence and the appearance of excimer fluorescence. Using the change of the ratio of the two emission intensities, a detection limit lower than 10^{-7} M (4 ppb) was claimed for Ca^{2+} ions.[59]

Recently, a pyrene-functionalized heterocycle receptor (**13** in Scheme 7.9) was developed as a ratiometric sensor for Ag^{+} with high selectivity and sensitivity.[60] Sensor **13** can form a 1:2 (metal : ligand) sandwich complex via Ag^{+} induced self-assembly in 50% ethanol aqueous solution, resulting in a dramatic increase in excimer fluorescence at the expense of monomer fluorescence. As shown in Figure 7.9, the dependence of the intensity ratios (I_{462}/I_{378}) on the cations indicates exceptionally high selectivity for Ag^{+}

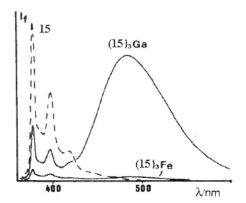

Figure 7.10. Corrected fluorescence emission spectra of **15** and **16** [(**15**)$_3$Ga and (**15**)$_3$Fe] in acetonitrile (25 °C, λ_{exc} = 340 nm, <10^{-6} M). (Reproduced with permission from Reference 62.)

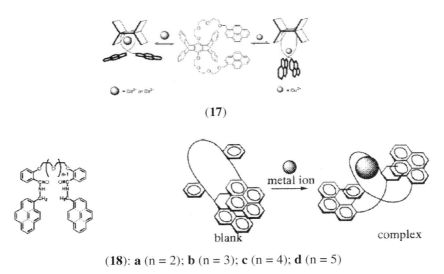

(17)

(18): **a** (n = 2); **b** (n = 3); **c** (n = 4); **d** (n = 5)

Scheme 7.10. Chemical structures of podand-based sensors and the schematic repre-sentation of their metal ion complexes. (Reproduced with permission from Reference 63 and 64.)

ions. It was believed that the phenolate hydroxyl group and the nitrogen atom of the heterocycle played an important role in ion binding.

Daub *et al*[61] reported a pyrene modified cyclic hexapeptide (**14** in Scheme 7.9) for alkaline earth metal recognition. For 1,1,1,3,3,3-hexafluoro-2-propanol/methanol (4/1) solutions, **14** exhibits both monomer and excimer fluorescence. Addition of Ca^{2+} or Ba^{2+} causes a considerable increase in the excimer to monomer emission ratio (I_E/I_M). The result was considered to be due to pyrene ground state dimerization after **14** formed 2:1 (peptide : metal ion) sandwich complexes with Ca^{2+} or Ba^{2+} cation.

An interesting self-assembly system involves pyrene labeled hydroxamate ligand (**15**

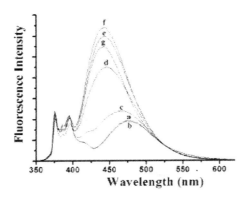

Figure 7.11. Fluorescence of **17** in CH_2Cl_2 (1×10^{-5} M, excitation at 335 nm) in the presence of (a) 0, (b) 1.0, (c) 5.0, (d) 7.0, (e) 9.0, (f) 11.0, and (g) 15 equiv. of $Cu(ClO_4)_2 \cdot 6H_2O$ predissolved in MeCN (0.005 M). (Reproduced with permission from Reference 63.)

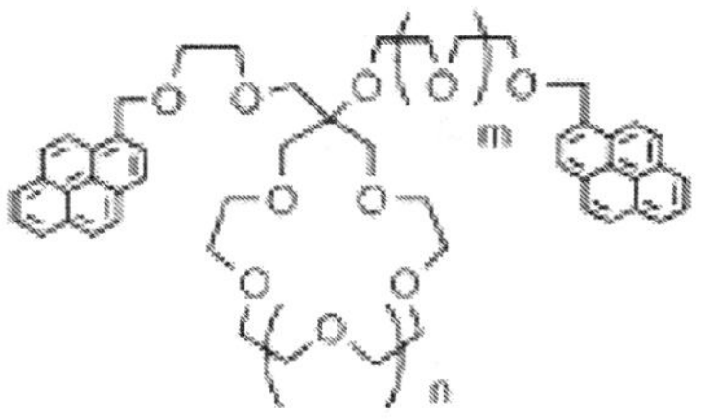

(**19**): **a** (m = 0, n = 1); **b** (m = 1, n = 1); **c** (m = 1, n = 2)

Scheme 7.11. Chemical structure of coronand-based sensor.

in Scheme 7.9).[62] In acetonitrile, the ligand alone shows pyrene monomer emission only. However, the trichromophoric gallium(III) chelate, **16** in Scheme 7.9, shows intense excimer fluorescence; while the iron(III) analogue exhibits much lower monomer and excimer emissions (Figure 7.10). For both chelates, ^{1}H NMR and other evidence support a prevailing ground state intra-molecular interaction between pyrene chromophores, leading to excimer at the expense of monomer emission. However, in the case of iron, the energy transfer can occur readily from the pyrene excimer to the metal, leading to effective quenching of fluorescence.

3.4.2. Podand-Based Sensors

A pentiptycene with linked pyrenes (**17** in Scheme 7.10) has been investigated as a sensor for metal ions.[63] In CH_2Cl_2, **17** displays both monomer and excimer fluorescence, in which excimer formation is dynamic (no ground state dimerization). In the presence of Ca^{2+} or Cd^{2+}, the I_E/I_M ratio decreases without changing the λ_E, which suggests that in the ion-bound complexes the pyrene groups cannot adopt an overlapping geometry. However, the addition of Cu^{2+} ion shifts the λ_E toward the blue ($\lambda_E = 440$ nm), along with an enhanced I_E/I_M (Figure 7.11). This blue shifted excimer was suggested to be due to more fully overlapping pyrene dimers. In this compound, the flanking benzene rings of the pentiptycene group should play a role in the recognition of Ca^{2+}, Cd^{2+} and Cu^{2+}.

Nakamura et al[64] prepared a series of noncyclic polyethers modified with two pyrene units (**18** in Scheme 10) for sensing alkali and alkaline earth metal cations. **18b** -

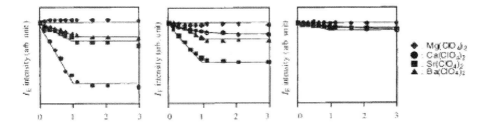

Figure 7.12. Changes of I_E of **19a** (left), **19b** (center), and **19c** (right) upon addition of alkaline earth metal cations in 99:1 v/v $CH_3CN/CHCl_3$. [probe] = 1×10^{-6} M. Excitation wavelength: 340 nm. (Reproduced with permission from Reference 65.)

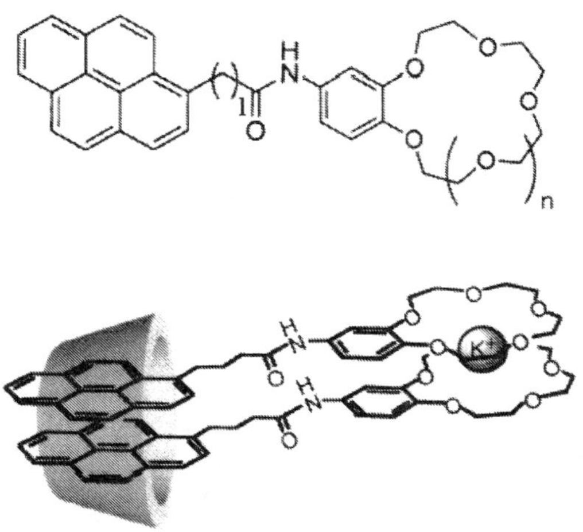

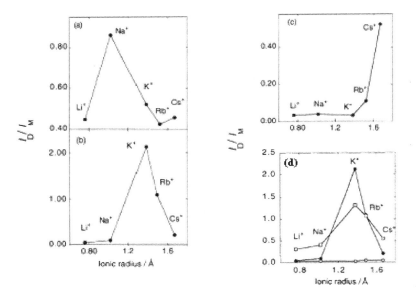

(20): **a** (n = 0, l = 3), **b** (n = 1, l = 3), **c** (n = 2, l = 3), **d** (n = 1, l = 1), **e** (n = 1, l = 5)

Scheme 7.12. Chemical structure of coronand-based sensor and schematic representation of this crown ether probe/γ-cyclodetrin/metal ion complex. (Reproduced with permission from Reference 66 and 67.)

Figure 7.13. Dependence of fluorescence intensity ratio, I_D/I_M, of (a) **20a**, (b) **20b**, (c) **20c**, and (d) **20d** (○), **20b** (●), and **20e** (□) on the ionic radius of alkali metal ions. [probe] = 0.50 μM in 99% water/1% MeCN (v/v). [γ-CD] = 5.0 mM. [MCl] = 0.10 M (M⁺ = Li⁺, Na⁺, K⁺, Rb⁺, Cs⁺). The emission wavelengths for the dimer/monomer fluorescence are (a) 480/377 nm, (b) and (d) 470/378 nm, and (c) 470/377 nm, respectively. (Revised and reproduced with permission from Reference 66 and 67.)

18d show strong intra-molecular excimer emission for acetonitrile solutions. In the presence of particular metal ions, **18b** - **18d** can form 1:1 metal : ligand complexes, causing increased monomer emission accompanied by decreased excimer fluorescence. **18a**, however, does not interact with any alkali or alkaline earth metal ions. **18b** shows preference for Li^+ and Mg^{2+}, while **18c** and **18d** are favorably associated with Mg^{2+}, Ca^{2+}, Sr^{2+} and Ba^{2+}. On the basis of ^1H NMR and fluorescence spectral data, a structural change before and after the addition of metal cation was anticipated. Without the interaction with metal cation, the ligand forms an intra-molecular pseudocyclic structure similar to a crown ether with the aid of a π-π interaction of two pyrene units. On the other hand, in the presence of metal cations, the complex forms a helical structure, separating two pendant pyrenes. The match of the sizes between the cavity in the rolled-up conformation and cation should be important for the observed metal ion selectivity.

7.3.4.3. Coronand-Based Sensors

New lariat ethers having plural pyrenylmethyl groups attached to sidearms (**19** in Scheme 11) have been prepared for alkaline earth metal sensing by Nakatsuji et al.[65] In 99:1 (v/v) $CH_3CN/CHCl_3$, compound **19a** and **19b** show strong excimer fluorescence, which is attributed to the intra-molecular π-π stacking of the two pyrene rings. Upon complexation with selected metal cations, the excimer emission decreases, accompanied with an increase in monomer emission. This effect can be ascribed to the conformational change of the ligand caused by the cooperative coordination of the crown ring and one of the sidearms to cation. The metal selectivity shown in Figure 7.12 (**19a** for Ca^{2+}, 19b for Sr^{2+}, **19c** almost no interaction) suggests the importance of the ligand cavity size (**19b** is

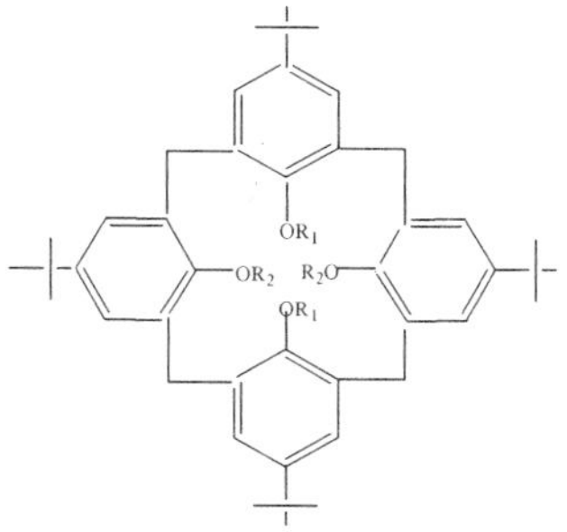

(**21**): $R_1 = CH_2CO_2Et$, $R_2 = CH_3CO_2CH_2$

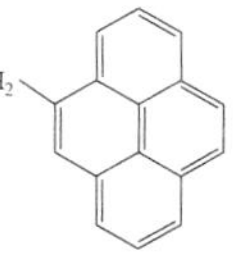

(**23**): $R_1 = CH_3$, $R_2 =$

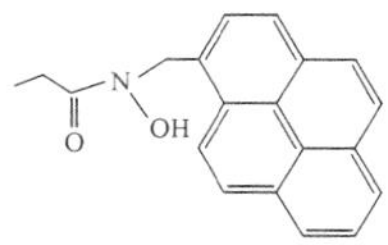

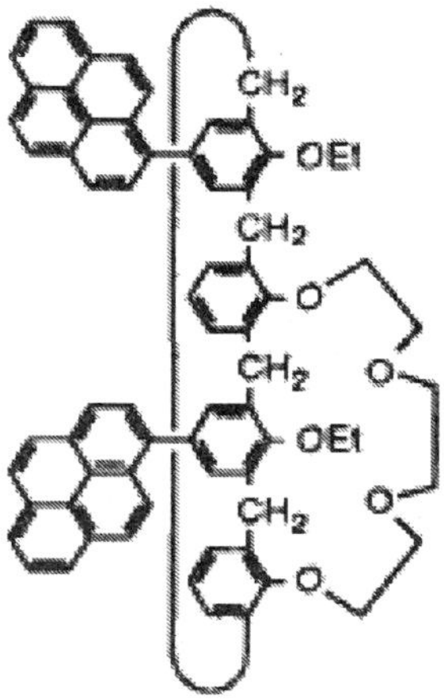

(**22**)

Scheme 7.13. Chemical structures of calixarene-based sensors.

bigger than **19a**) and the cooperative participation in metal complexation of the electron-donating sidearm (only the oxygen atoms of the crown ring coordinated to metal cation in **19c**).

Teramae *et al*[66,67] tested the selectivity for alkali metal ions of a series of crown ether probes (Scheme 7.12) in the presence of γ-cyclodextrin in 1% MeCN aqueous solution. Except for **20a**, the probes show monomer emission in the absence of metal cation. In the presence of particular metal cations, except for **20d**, the probes can form 2:1:1 ligand : metal : γ-cyclodextrin complexes, which brings two pyrene units in close contact (dimer) inside the γ-cyclodextrin cavity, exhibiting excimer fluorescence (Scheme 7.12). The dependence of fluorescence intensity ratio, I_D/I_M, on the ionic radius of the metal ions (Figure 7.13) indicates the respective selectivity of the probes. The results also suggest the importance of crown ether ring size (**20 a, b,** and **c**) and length of the linker (**20 b, d** and **e**), which will affect the flexibility and hydrophobicity, for proper metal recognition.

7.3.4.4. *Calixarene-Based Sensors*

Modified calixarene compounds have been used as sensors and a recent review is available.[68] In 1992, the first example of a pyrene-labeled calixarene Na^+ sensor (**21** in Scheme 7.13) was reported by Jin *et al.*[69] Later, Shinkai *et al*[70] prepared another compound (**22** in Scheme 7.13). For both of them (**21** in 15:1 (v/v) MeOH-THF, and **22** in 1000:1 (v/v) MeCN-THF), fluorescence spectra indicate dominant excimer emission. In the presence of metal cations, the complexation can induce the reorientation of the two pyrene units, changing their relative configuration and distance. Consequently, the excimer fluorescence decreases, while monomer emission increases. These two sensors show extremely high selectivity toward Na^+.

As noted before, when pyrene labeled hydroxamate ligand (**15** in Scheme 7.9) complexes with Fe^{3+}, excimer fluorescence can be quenched due to energy transfer.[62] This was further elucidated by Fages *et al*[71] in their new calixarene sensor system (**23** in

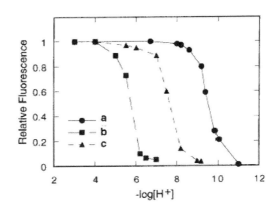

Figure 7.14. -log[H⁺]-fluorescence profiles of the pyrene labeled bishydroxamate ligand **23**: (a) free ligand, (b) in the presence of one molar equivalent of Cu^{2+}, (c) in the presence of one molar equivalent of Ni^{2+} (methanol / water 80/20 (v/v), -log[H⁺] = 7.4 (HEPES, 10 mM), μ = 0.1 (KCl), 25 °C). (Reproduced with permission from Reference 71.)

Scheme 7.13). In 80% MeOH aqueous solution, **23** exhibits both monomer and excimer (dominant) emissions. However, as shown in Figure 7.14, the fluorescence intensity is sensitive to pH due to the dissociable hydrogen atoms in the hydroxyamino groups. The deprotonated hydroxamato species can quench the pyrene fluorescence. In the presence of Cu^{2+} or Ni^{2+}, the metal binding causes the fluorescence to be quenched at a lower range of pH. However, the addition of other divalent metal ions does not display this effect. The effective quenching of Cu^{2+} or Ni^{2+} was believed to be due to energy and/or electron transfer to the metal cations.[71]

7.3.5. Other Miscellaneous Sensing Systems

A *cis*-cyclohexane-1,3-dicarboxylate (Scheme 7.14) was used as an allosteric switching sensor for metal cations as well as strong and weak acids.[72] In MeCN solution, in the conformation B, the two pyrenyl groups are in close proximity to each other and are able to show the excimer fluorescence. With conformations A, C and D, the pyrenyl groups are separated and the favor for excimer formation is decreased. The transition from conformation B to D is induced by strong or weak acids such as HNO_3 or NH_4NO_3; while formation of C is favored by metal cations.[72]

Recently, Jones and Jiang[73] developed a spermine-pyrene conjugate (Sp-Py, Scheme 7.15) for sensing polyelectrolytes in aqueous solution. Sp-Py shows only monomer fluorescence near neutral pH. In the presence of negatively charged polyelectrolyte such as poly-L-glutamic acid (PLGA), Sp-Py will bind to the polymer due to electrostatic interaction. Then pyrene moiety will dimerize, driven by hydrophobic interaction, leading to increased excimer fluorescence along with decreased monomer emission. Taking the change of the ratio of excimer to monomer intensity, the detection limit is estimated to be at about the ppb (10^{-9}) level.

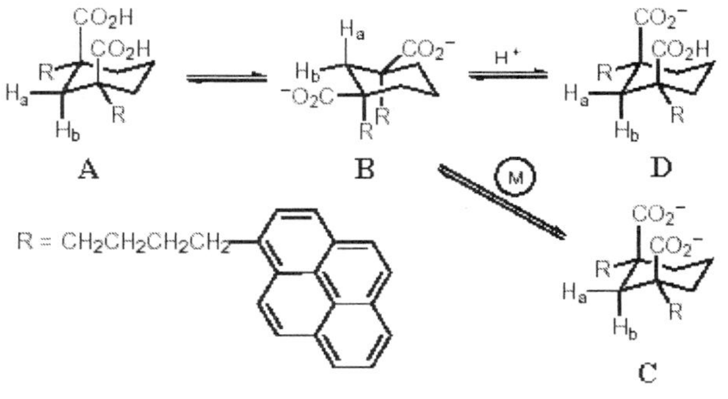

(24)

Scheme 7.14. Chemical structure of *cis*-cyclohexane-1,3-dicarboxylate sensor and its different conformations. (Reproduced with permission from Reference 72.)

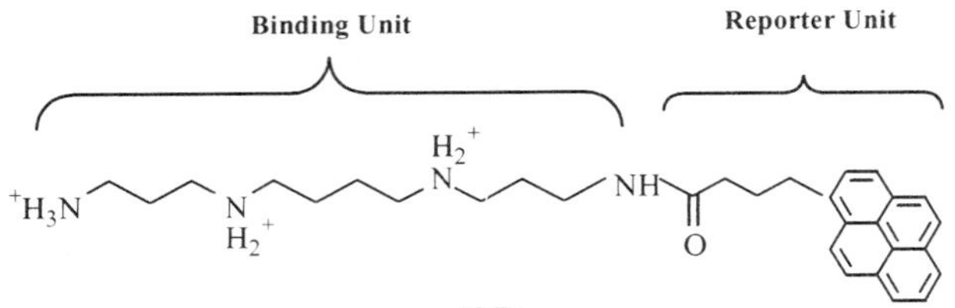

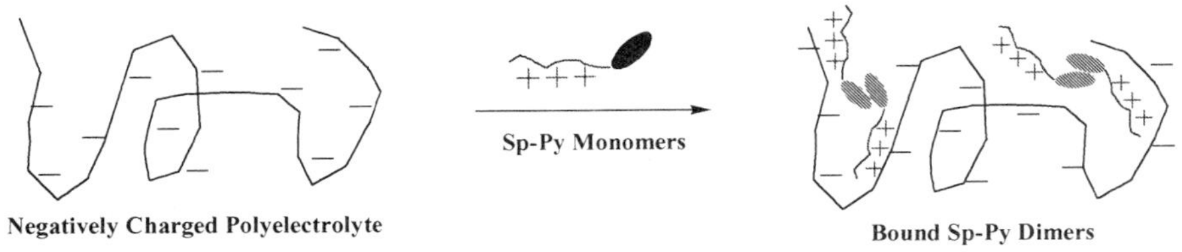

Scheme 7.15. Chemical structure of Sp-Py conjugate and schematic representation of detection of negatively charged polyelectrolytes.

7.4. CONCLUDING REMARKS

Pyrene derivatives are broadly used for photoprobes in excimer sensing experiments. The popularity of these photoprobes results from the spectral distinction between the monomer and excimer emissions and the availability of easy ways for derivatization of macromolecules with pyrene fluorophores. Pyrene taggants, however, do not always serve as good probes. The strong hydrophobicity of the pyrene containing molecules sets limitations on their use in aqueous media for biological systems, under circumstances in which pyrene π-stacking may be perturbing. Pyrene excimer analysis still can be performed in buffered water solutions if the photoprobes are attached to very hydrophilic conjugates that are used in low concentrations, e.g., for DNA sequence analysis. Alternatively, pyrene derivatives have been an invaluable tool for steady-state and dynamic structural studies of biomolecules when the probe is placed in lipophilic media (e.g., in the interior of proteins or in lipid membranes).

The excimer fluorescence of pyrene itself or its series of derivatives in various architectures can be utilized for sensitive monitoring of environmental parameters, such as temperature, pressure or pH, and chemical analytes in a variety of systems. The interchange of monomer and excimer fluorescence intensities indicates the change of the microenvironment around the chromophores, or the presence of different guest molecules.

7.5. ACKNOWLEGEMENT

The authors acknowledge with thanks support of this work by the U. S. Army Research Laboratory.

7.6. REFERENCES

1. J. B. Birks, *Photophysics of Aromatic Molecules* (John Wiley & Sons, New York, 1970).
2. T. Forster and K. Kasper, Concentration reversal of the fluorescence of pyrene, *Zeitschrift fuer Elektrochemie und Angewandte Physikalische Chemie*, **59**, 976-980 (1955).
3. G. Jones, II and V. I. Vullev, Ground- and excited-state aggregation properties of a pyrene derivative in aqueous media, *J. Phys. Chem. A* **105**(26), 6402-6406 (2001).
4. O. Shoji, D. Nakajima, M. Ohkawa *et al.*, Temperature dependence of circular dichroism and fluorescence decay of pyrene appended to the side chains of poly-L-glutamine, *Macromolecules,* **36**(12), 4557-4566 (2003).
5. G. Jones, II, V. Vullev, E. H. Braswell *et al.*, Multistep photoinduced electron transfer in a *de novo* helix bundle: multimer self-assembly of peptide chains including a chromophore special pair, *J. Am. Chem. Soc.* **122**(2), 388-389 (2000).
6. V. I. Vullev and G. Jones, II, Photoinduced charge transfer in helical polypeptides, *Res. Chem. Intermed.* **28**(7-9), 795-815 (2002).
7. V. I. Vullev, *Towards Artificial Photosynthesis: Photoinduced Multiple-step Electron Transfer in Supramolecular Structures Based on Synthetic Polypeptides* (Boston University, Boston, MA, USA, 2001).
8. C. Armbruster, M. Knapp, K. Rechthaler *et al.*, Fluorescence properties of 1-heptanoylpyrene: a probe for hydrogen bonding in microaggregates and biological membranes, *J. Photochem. Photobiol. A* **125**(1-3), 29-38 (1999).
9. V. I. Vullev and G. Jones, Photoinduced electron transfer in alkanoylpyrene aggregates in conjugated polypeptides, *Tetrahedron Lett.* **43**(47), 8611-8615 (2002).
10. G. Jones, II and V. I. Vullev, Photoinduced electron transfer between non-native donor-acceptor moieties incorporated in synthetic polypeptide aggregates, *Org. Lett.* **4**(23), 4001-4004 (2002).
11. P. L. G. Chong and T. E. Thompson, Oxygen quenching of pyrene-lipid fluorescence in phosphatidylcholine vesicles. A probe for membrane organization, *Biophys. J.* **47**(5), 613-621 (1985).
12. D. L. Daugherty and S. H. Gellman, A fluorescence assay for leucine zipper dimerization: avoiding unintended consequences of fluorophore attachment, *J. Am. Chem. Soc.* **121**(18), 4325-4333 (1999).
13. G. Jones, II and V. I. Vullev, Contribution of a pyrene fluorescence probe to the aggregation propensity of polypeptides, *Org. Lett.* **3**(16), 2457-2460 (2001).
14. C. Garcia-Echeverria, Unequivocal determination by fluorescence spectroscopy of the formation of a parallel leucine-zipper homodimer, *J. Am. Chem. Soc.* **116**(13), 6031-6032 (1994).
15. F. M. Winnik, Photophysics of preassociated pyrenes in aqueous polymer solutions and in other organized media, *Chem. Rev.* **93**, 587-614 (1993).
16. M. F. Blackwell, K. Gounaris, and J. Barber, Evidence that pyrene excimer formation in membranes is not diffusion-controlled, *Biochim. Biophys. Acta*, **858**(2), 221-234 (1986).
17. M. Goedeweeck, M. Van der Auweraer, and F. C. De Schryver, Molecular dynamics of a peptide chain, studied by intramolecular excimer formation, *J. Am. Chem. Soc.* **107**(8), 2334-2341 (1985).
18. P. Hammarstrom, B. Kalman, B.-H. Jonsson *et al.*, Pyrene excimer fluorescence as a proximity probe for investigation of residual structure in the unfolded state of human carbonic anhydrase II, *FEBS Lett.* **420**(1), 63-68 (1997).
19. D. Sahoo, P. M. M. Weers, R. O. Ryan *et al.*, Lipid-triggered conformational switch of Apolipophorin III helix bundle to an extended helix organization, *J. Mol. Biol.* **321**(2), 201-214 (2002).
20. D. Sahoo, V. Narayanaswami, C. M. Kay *et al.*, Pyrene excimer fluorescence: a spatially sensitive probe to monitor lipid-Induced helical rearrangement of Apolipophorin III, *Biochemistry,* **39**(22), 6594-6601 (2000).
21. H. Mihara, Y. Tanaka, T. Fujimoto *et al.*, A pair of pyrene groups as a conformational probe for designed four-α-helix bundle polypeptides, *J. C. S. Perkin 2*, (10), 1915-1921 (1995).
22. M. J. Dabrowski, M. L. Schrag, L. C. Wienkers *et al.*, Pyrene•pyrene complexes at the active site of Cytochrome P450 3A4: evidence for a multiple substrate binding site, *J. Am. Chem. Soc.* **124**(40), 11866-11867 (2002).
23. T. Ahn, J.-S. Kim, H.-I. Choi *et al.*, Development of peptide substrates for trypsin based on monomer/excimer fluorescence of pyrene, *Anal. Biochem.* **306**(2), 247-251 (2002).
24. P. L. Paris, J. M. Langenhan, and E. T. Kool, Probing DNA sequences in solution with a monomer-excimer fluorescence color change, *Nucleic Acids Res.* **26**(16), 3789-3793 (1998).
25. F. D. Lewis, Y. Zhang, and R. L. Letsinger, Bispyrenyl excimer fluorescence: a sensitive oligonucleotide probe, *J. Am. Chem. Soc.* **119**(23), 5451-5452 (1997).

26. K. Yamana, T. Iwai, Y. Ohtani *et al.*, Bis-pyrene-labeled oligonucleotides: sequence specificity of excimer and monomer fluorescence changes upon hybridization with DNA, *Bioconjugate Chem.* **13**(6), 1266-1273 (2002).

27. K. Yamana, M. Takei, and H. Nakano, Synthesis of oligodeoxyribonucleotide derivatives containing pyrene labeled glycerol linkers: enhanced excimer fluorescence on binding to a complementary DNA sequence, *Tetrahedron Lett.* **38**(34), 6051-6054 (1997).

28. G. Tong, J. M. Lawlor, G. W. Tregear *et al.*, Oligonucleotide-polyamide hybrid molecules containing multiple pyrene residues exhibit significant excimer fluorescence, *J. Am. Chem. Soc.* **117**(49), 12151-12158 (1995).

29. M. Ollmann, G. Schwarzmann, K. Sandhoff *et al.*, Pyrene-labeled gangliosides: micelle formation in aqueous solution, lateral diffusion, and thermotropic behavior in phosphatidylcholine bilayers, *Biochemistry,* **26**(18), 5943-5952 (1987).

30. E. H. W. Pap, A. Hanicak, A. van Hoek *et al.*, Quantitative analysis of lipid-lipid and lipid-protein interactions in membranes by use of pyrene-labeled phosphoinositides, *Biochemistry,* **34**(28), 9118-9125 (1995).

31. T. V. Kurzchalia and R. G. Parton, Membrane microdomains and caveolae, *Curr. Opin. Cell Biol.* **11**(4), 424-431 (1999).

32. M. Pitto, P. Palestini, A. Ferraretto *et al.*, Dynamics of glycolipid domains in the plasma membrane of living cultured neurons, following protein kinase C activation: a study performed by excimer-formation imaging, *Biochem. J.* **344**(1), 177-184 (1999).

33. X. Song and B. I. Swanson, Rational design of an optical sensing system for multivalent proteins, *Langmuir,* **15**(14), 4710-4712 (1999).

34. K. K. Eklund, J. A. Virtanen, P. K. J. Kinnunen *et al.*, Conformation of phosphatidylcholine in neat and cholesterol-containing liquid-crystalline bilayers. Application of a novel method, *Biochemistry,* **31**(36), 8560-8565 (1992).

35. J. M. Smit, R. Bittman, and J. Wilschut, Low-pH-dependent fusion of Sindbis virus with receptor-free cholesterol- and sphingolipid-containing liposomes, *J. Virol.* **73**(10), 8476-8484 (1999).

36. A. Irurzun, J.-L. Nieva, and L. Carrasco, Entry of Semliki Forest virus into cells: effects of concanamycin A and nigericin on viral membrane fusion and infection, *Virology,* **227**(2), 488-492 (1997).

37. T. Pillot, M. Goethals, B. Vanloo *et al.*, Fusogenic properties of the c-terminal domain of the Alzheimer β-amyloid peptide, *J. Biol. Chem.* **271**(46), 28757-28765 (1996).

38. M. O. Pentikainen, E. M. P. Lehtonen, K. Oorni *et al.*, Human arterial proteoglycans increase the rate of proteolytic fusion of low density lipoprotein particles, *J. Biol. Chem.* **272**(40), 25283-25288 (1997).

39. P. Somerharju, Pyrene-labeled lipids as tools in membrane biophysics and cell biology, *Chem. Phys. Lipids,* **116**(1-2), 57-74 (2002).

40. J. A. McCann, J. A. Mertz, J. Czworkowski *et al.*, Conformational changes in Cholera Toxin B subunit-Ganglioside GM1 complexes are elicited by environmental pH and evoke changes in membrane structure, *Biochemistry,* **36**(30), 9169-9178 (1997).

41. J. B. Birks, M. D. Lumb, and I. H. Munro, 'Excimer' fluorescence V. Influence of solvent viscosity and temperature, *Proc. Roy. Soc. Ser. A* **280**(1381), 289-297 (1964).

42. K. B. Migler and A. J. Bur, Fluorescence based measurement of temperature profiles during polymer processing, *Polym. Eng. Sci.* **38**(1), 213-221 (1998).

43. A. J. Bur, M. G. Vangel, and S. C. Roth, Fluorescence based temperature measurements and applications to real-time polymer processing, *Polym. Eng. Sci.* **41**(8), 1380-1389 (2001).

44. S. J. Cowsley, R. H. Templer, and D. R. Klug, Dipyrenylphosphatidylcholine as a probe of bilayer pressures, *J. Fluoresc.* **3**(3), 149-152 (1993).

45. R. H. Templer, S. J. Castle, A. Rachael Curran *et al.*, Sensing isothermal changes in the lateral pressure in model membranes using di-pyrenyl phosphatidylcholine, *Faraday Discuss.* **111**, 41-53 (1998).

46. M. R. Pokhrel and S. H. Bossmann, Synthesis, characterization, and first application of high molecular weight polyacrylic acid derivatives possessing perfluorinated side chains and chemically linked pyrene labels, *J. Phys. Chem. B* **104**, 2215-2223 (2000).

47. E. D. Lee, T. C. Werner, and W. R. Seitz, Luminescence ratio indicators for oxygen, *Anal. Chem.* **59**(2), 279-283 (1987).

48. A. Sharma, Excimer fluorescence quenching based oxygen sensor, *Proc. SPIE,* **2131**, 598-602 (1994).

49. U. Fujiwara and Y. Amao, Optical oxygen sensor based on controlling the excimer formation of pyrene-1-butylic acid chemisorption layer onto nano-porous anodic oxidized aluminium plate by myristic acid, *Sens. Act. B* **89**(1-2), 58-61 (2003).

50. C. C. Nagel, J. G. Bentsen, M. Yafuso *et al.*, *Sensors and methods for sensing*, US 5,498,549, 1996.

51. C. C. Nagel, J. G. Bentsen, J. L. Dektar *et al.*, *Sensors and methods for sensing*, US 5,409,666, 1995.

52. M. Narita, S. Mima, N. Ogawa *et al.*, Fluorescent molecular sensory system based on bis pyrene-modified γ-cyclodextrin dimer for steroids and endocrine disruptors, *Anal. Sci.* **17**(3), 379-385 (2001).
53. A. Ueno, I. Suzuki, and T. Osa, Host-guest sensory systems for detecting organic compounds by pyrene excimer fluorescence, *Anal. Chem.* **62**(22), 2461-2466 (1990).
54. M. A. Hossain, S. Matsumura, T. Kanai *et al.*, Association of α-helix peptides that have γ-cyclodextrin and pyrene units in their side chain, and induction of dissociation of the association dimer by external stimulant molecules, *Perkin 2*, (7), 1527-1533 (2000).
55. T. Aoyagi, H. Ikeda, and A. Ueno, Fluorescence properties, induced-fit guest binding and molecular recognition abilities of modified γ-cyclodextrins bearing two pyrene moieties, *Bull. Chem. Soc. Jpn.* **74**, 157-164 (2001).
56. A. Ueno, Review: fluorescent cyclodextrins for molecule sensing, *Supramol. Sci.* **3**(1-3), 31-36 (1996).
57. A. Ueno, H. Ikeda, and J. Wang, in: *Chemosensors of Ion and Molecule Recognition*, edited by J. P. Desvergne and A. W. Czarnik (Kluwer Academic Publishers, Netherlands, 1997), pp. 105-119.
58. B. Valeur and I. Leray, Design principles of fluorescent molecular sensors for cation recognition, *Coord. Chem. Rev.* **205**, 3-40 (2000).
59. L. Prodi, R. Ballardini, M. T. Gandolfi *et al.*, A simple fluorescent chemosensor for alkaline-earth metal ions, *J. Photochem. Photobiol. A* **136**(1-2), 49-52 (2000).
60. R.-H. Yang, W.-H. Chan, A. W. M. Lee *et al.*, A ratiometric fluorescent sensor for AgI with high selectivity and sensitivity, *J. Am. Chem. Soc.* **125**(10), 2884-2885 (2003).
61. J. Strauss and J. Daub, Optically active cyclic hexapeptides with covalently attached pyrene probes: selective alkaline earth metal ion recognition using excimer emission, *Org. Lett.* **4**(5), 683-686 (2002).
62. B. Bodenant, F. Fages, and M.-H. Delville, Metal-induced self-assembly of a pyrene-tethered hydroxamate ligand for the generation of multichromophoric supramolecular systems. The pyrene excimer as switch for iron(III)-driven intramolecular fluorescence quenching, *J. Am. Chem. Soc.* **120**, 7511-7519 (1998).
63. J.-S. Yang, C.-S. Lin, and C.-Y. Hwang, Cu^{2+}-induced blue shift of the pyrene excimer emission: a new signal transduction mode of pyrene probes, *Org. Lett.* **3**(6), 889-892 (2001).
64. Y. Suzuki, T. Morozumi, H. Nakamura *et al.*, New fluorimetric alkali and alkaline earth metal cation sensors based on noncyclic crown ethers by means of intramolecular excimer formation of pyrene, *J. Phys. Chem. B* **102**(40), 7910-7917 (1998).
65. Y. Nakahara, Y. Matsumi, W. Zhang *et al.*, Fluorometric sensing of alkaline earth metal cations by new lariat ethers having plural pyrenylmethyl groups on the electron-donating sidearms, *Org. Lett.* **4**(16), 2641-2644 (2002).
66. A. Yamauchi, T. Hayashita, A. Kato *et al.*, Selective potassium ion recognition by benzo-15-crown-5 fluoroionophore/γ-cyclodextrin complex sensors in water, *Anal. Chem.* **72**(23), 5841-5846 (2000).
67. A. Yamauchi, T. Hayashita, A. Kato *et al.*, Supramolecular crown ether probe/γ-cyclodextrin complex sensors for alkali metal ion recognition in water, *Bull. Chem. Soc. Jpn.* **75**(7), 1527-1532 (2002).
68. R. Ludwig and N. T. K. Dzung, Calixarene-based molecules for cation recognition, *Sensors*, **2**, 397-416 (2002).
69. T. Jin, K. Ichikawa, and T. Koyama, A fluorescent calix[4]arene as an intramolecular excimer-forming Na$^+$ sensor in nonaqueous solution, *J. Chem. Soc. Chem. Commun.*, 499-501 (1992).
70. H. Matsumoto and S. Shinkai, Metal-induced conformational change in pyrene-appended calix[4]crown-4 which is useful for metal sensing and guest tweezing, *Tetrahedron Lett.* **37**(1), 77-80 (1996).
71. B. Bodenant, T. Weil, M. Businelli-Pourcel *et al.*, Synthesis and solution structure analysis of a bispyrenyl bishydroxamate calix[4]arene-based receptor, a fluorescent chemosensor for Cu^{2+} and Ni^{2+} metal ions, *J. Org. Chem.* **64**, 7034-7039 (1999).
72. C. Monahan, J. T. Bien, and B. D. Smith, Fluorescence sensing due to allosteric switching of pyrene functionalized cis-cyclohexane-1,3-dicarboxylate, *Chem. Commun.*, 431-432 (1998).
73. G. Jones II and H. Jiang, In preparation.

LIFETIME BASED SENSORS / SENSING

Sonja Draxler[*]

8.1. INTRODUCTION

The importance of fluorescence lifetime is well reflected by the frequency of occurrence in this book series: Nearly half the contributions in the volumes published so far refer more or less extensively to fluorescence lifetime. Chemical sensing by fluorescence techniques was the main topic of Volume 4 that appeared in 1994, and even then it was clear that lifetime based schemes have considerable advantages over intensity based ones. In recent years improvement of existing techniques as well as development of new ones has broadened the field significantly. The most dramatic change, however, was brought about by the progress in optoelectronics. Lifetime measurement until the eighties of the past century was the sole domain of a handful of specialists and required sophisticated and expensive equipment as well as painstaking evaluation procedures to obtain useful results, precluding any practical use outside the labs of basic research. Starting with the first use of a light-emitting diode as the light source in lifetime measurement[1], instrumentation for lifetime measurement gradually became smaller, cheaper and easier to handle, so that today it is quite customary to have lifetime measurements implemented in routine instrumentation for medical, environmental, and process technology applications. Nevertheless, among non-specialist sometimes there remain some of misconceptions about fluorescence lifetimes, leading sometimes to improper application and unreliable results.

The purpose of this contribution is evident from what has been said above: The photophysics behind lifetime-based sensors will be reviewed, various methods will be described and potential caveats in evaluating and interpreting the results will be discussed. In the first paragraph, where the indispensable prerequisites for discussing lifetime based sensors will be given shortly, the more general term 'luminescence' is used

[*] Institut für Experimentalphysik, Karl-Franzens-Universität Graz, A-8010 Graz Austria

to adequately cover also cases where the distinction between fluorescence and phosphorescence becomes problematic.

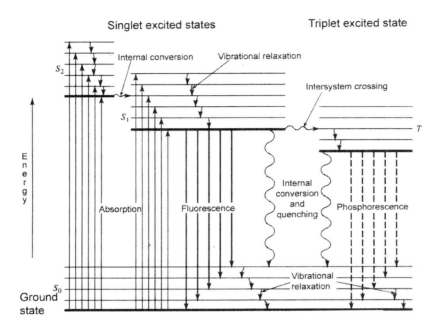

Figure 8.1. Schematics (Jablonski diagram) for a molecular system, with excitation and relaxation paths.

8.2. PHOTOPHYSICAL BACKGROUND

8.2.1. Luminescence Lifetime

8.2.1.1. Basics

The IUPAC Compendium of Chemical Terminology[2] defines the luminescence lifetime as "the time required for the luminescence intensity to decay from some initial value to 1/e of that value". This simple sentence hides a number of problems that have to be scrutinized carefully.

Luminescence is the phenomenon of non-thermal light emission. This implies that the radiation field and the emitting body are far from mutual thermal equilibrium and that the disequilibrium is caused by non-thermal energy supply. This state of non-equilibrium is called an excited state. Excitation may be achieved by light (photoluminescence), chemical reactions (chemoluminescence) or mechanical energy (sonoluminescence, triboluminescence).

The excited state of a molecular system has a higher energy than the ground state and may differ from the ground state by its electronic configuration, spin orientation, rotation of the whole system in a certain reference frame (usually defined by the environment) or parts of it (with respect to each other), and vibrating motion of the nuclei with respect to each other. They represent more or less independent degrees of freedom of the system, each of them being capable of accepting and releasing energy. The corresponding energies are quantized, that means that only states with certain energy values exist for measurable durations. The states are characterized by their quantum numbers. Going from one state to another one means changing its quantum number(s). Usually electronic states have energies an order of magnitude higher than vibrational ones, and vibrational energies are an order of magnitude higher than rotational ones. Electronic states can be categorized according to their spin-multiplicity: Those with total electron spin $s = 0$ are called singlet states S, and those with total spin $s = 1$ triplet states T.

Once excited to a higher energy state j, the system has a certain probability k_{ji} to relax to another state i by releasing the excess energy ΔE_{ji}. This probability may be constant or time-dependent. The relaxed state may be the ground state from which excitation started, or some other excited state. Transitions with high probability are called 'allowed', those with low probability 'forbidden'. The transition may be radiative (excess energy is emitted as electromagnetic radiation) or non-radiative (excess energy is dissipated to heat). Hence we can state that luminescent light is emitted as long as the system is in a state capable of performing radiative transitions to states of lower energy. A schematic representation of the energy states (a so-called Jablonski diagram) and their possibilities to relax is given in Figure 8.1.

The case of special interest in the present context is the radiative one, where the excited state relaxes by releasing energy via electromagnetic radiation. In fact this is what we call luminescence. But radiation is by no means the only way of returning to the ground state. In the case of non-radiative relaxation, the excess energy is released in a mechanical way, either by impact in direct collisions or via short-ranged forces. Since vibrational and rotational energies are so much smaller than the electronic ones, it is much easier to get rid of the excess energy. For this reason mechanical relaxation of vibrational or rotational states is much faster than relaxation of electronic states. Therefore usually an excited molecular system within less than 10^{-12} s looses all its vibrational and rotational energy and only the electronic excitation persists. Also the relaxation of electronic states with higher quantum numbers to lower-lying states of same spin multiplicity is very fast, so that in most cases the first excited electronic state with an allowed radiative transition to the ground state (called S_1) remains as the only luminescent one (Kasha's rule[3]).

While for the higher excited states vibrational relaxation and internal conversion are the main non-radiative processes, for the lowest excited state the most effective processes are intersystem crossing and quenching. Intersystem crossing is a radiationless transition between states of different spin multiplicity. If one of the two states involved in the crossing is luminescent, the other one necessarily exhibits a 'forbidden' transition, thus reducing the luminescence. While all processes discussed so far are intramolecular, i. e. involve only processes within the excited molecular system, quenching is any kind of luminescence reduction by interaction with other molecules. Most lifetime-based sensors rely on quenching of luminescence by the analyte of interest.

The term luminescence used so far is valid to radiant transitions between any kinds of states. If the transition is between states of the same spin multiplicity, especially from one singlet state to another one, then the transition is rather fast (usually in the nanosecond regime) and the emission process is called fluorescence. 'Forbidden' transitions between triplet and singlet states are slow (microseconds to hours) and are called phosphorescence. Although this distinction is not always straightforward (due to various reasons certain states may be 'halfway in between' singlet and triplet) we will use the term fluorescence for all cases of interest in the following. The molecular entity capable of emitting fluorescence will be called a fluorophore.

8.2.1.2. Fluorescence Decay

The various processes discussed in the previous paragraph collectively are responsible for the decay of the excited-state population in an ensemble of molecules. Any of these processes may be described separately by a particular relaxation probability or rate. These rates may be constant in time or, in certain cases, time-dependent. Usually an overall relaxation rate can be defined as the sums over the particular rates. This overall rate determines the time course for the decay of an excited-state population and hence is essential for the quantity called fluorescence lifetime.

The relaxation rate $k_{r(ji)}$ (where the index r stands for radiative and j and i denote the upper and lower state of the transition, respectively) of a purely radiative relaxation is constant in time and given by

$$k_{r(ji)} = \frac{64\pi^4 e^2 n^3}{3\hbar^4 c^3} \Delta E_{ji}^{3} \langle \psi_j | \vec{r} | \psi_i * \rangle^2 \qquad (1.1)$$

where n is the index of refraction of the medium surrounding the molecules, $\hbar = h/2\pi$ with h the Planck constant, c the speed of light, n the index of refraction of the surrounding medium, $\Delta E_{ji} = E_j - E_i$, $E_{j,i}$ being the energy of the respective states, $\psi_{j,i}$ the wave function of these states, and $\vec{r}$ the position operator (the term in angle brackets representing the so-called transition dipole moment). From this equation it becomes clear that the probability depends on intramolecular ($E_{j,i}, \psi_{j,i}$) as well as extramolecular (n) properties. If any of these quantities changes during the time the system is in the excited state, the probability becomes time-dependent. From the properties of the wavefunctions it can be deduced that the dipole moment for transitions from singlets to triplets and *vice versa* are small, hence these transitions are radiatively 'forbidden'.

The rates for non-radiative relaxations depend on details of the respective processes but it should be emphasized, that also these rates may be constant or time-dependent. Most lifetime-based optical sensors depend on a modification of some non-radiative rate by the presence and concentration of the respective substance to be detected. The various mechanisms by which a substance may modify a non-radiative rate will be given below (Chapter 8.3.3).

In fact these rates may influence each other and hence may be interdependent. But in many cases (and fortunately in those most interesting for sensor purposes) the rates of the various relaxation processes are independent of each other. Therefore it is possible to define an overall relaxation rate that is simply the sum of the particular rates. Of course it follows that, if at least one of the particular rates is time-dependent, this is also true for the overall rate.

Suppose that of an ensemble of molecules at a given time t a number $n(t)$ is in a fluorescent excited state. In the absence of further excitation this excited-state population relaxes according to

$$\frac{dn}{dt} = k_{tot}(t)\, n \qquad (1.2)$$

where k_{tot} is the overall relaxation rate. If all particular rates k_p are constant, then $k_{tot} = \sum_p k_p = const.$ and integration of this differential equation yields

$$n(t) = n_0 \, \exp(-t \sum_p k_p) \qquad (1.3)$$

where n_0 is the number of excited molecules at an arbitrary starting point. Thus in this case, when the overall rate is not time-dependent, the resulting decay is an exponential function, falling within a time $\tau = 1/k_{tot}$ to $1/e$ or 37 % of its initial value. τ is called the overall time constant or relaxation time of the decay. This is true irrespective of the kind and number of particular relaxation rates contributing to this overall rate. A simple exponential decay hence always represents a single population of excited molecules under the action of one or more time-independent relaxation rates.

If there exist two or more different populations $n_i(t)$ of excited molecules, differing in their respective (and constant) overall rates $(k_{tot})_i$, then every population decays independently, and the observation yields a superposition of these decays,

$$n(t) = n_0 \sum_i a_i \, \exp\left(-t/\tau_i\right) \qquad (1.4)$$

where a_i is the respective fraction of the i-th population and τ_i is the time-constant of its decay. This may occur if fluorescent molecules are present in several distinct environments, but also if two or more 'species' of fluorescent molecules exist (for example, the protonated and the unprotonated form of a fluorescent pH indicator). A special case of a multiexponential decay is that of a continuous distribution of relaxation rates, when the sum in eq. 1.4 has to be replaced by an integral

$$n(t) = n_0 \int a(\xi) \, \exp\left(-k(\xi)\, t\right) d\xi \qquad (1.5)$$

with the parameter $a(\xi)$ describing the distribution function.

Time-dependent relaxation rates may occur for various reasons. The equilibrium geometry of the fluorophore may be different in the excited state from that in the ground state, thus causing a geometry change and thus a change in transition moment after excitation. The solvent molecules surrounding the fluorophore may re-orientate upon its excitation, thus changing the local dielectric constant. Photochemical complexations or even reactions may be triggered by the excitation, influencing the fluorescence properties. And diffusion of charge, energy, or matter may be initiated, altering the excited-state population. All these phenomena cause the decay function to become non-exponential. But any mathematical function may be Laplace transformed into a distribution of exponentials. Hence from a mathematical point of view the distinction between multi-exponential (as in eq. 1.4) or non-exponential decays is not tenable. From the perspective of relaxation processes involved, however, this distinction is quite reasonable. Unfortunately a measured decay curve does not reveal the underlying processes, and it is not possible to decide from the decay curve whether it is (multi-)exponential or not. In Fig. 1.2 examples of decays with single-exponential and continuously distributed functions are shown.

8.2.2. Quantum Efficiency and Fluorescence Intensity

The relaxation rates are not only important for the fluorescence lifetime, but they also determine what fraction of the absorbed photon number I_a is re-emitted as fluorescence photons (fluorescence intensity I_f, measured here in photon numbers, but easily converted to intensity in the strict sense, in units W/m^2, by multiplication with the

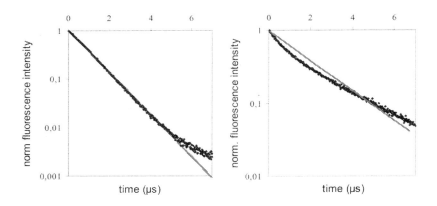

Figure 8.2. Logarithmic display of a single-exponential decay (left) and a non-exponential decay (right). The straight line represents the calculated exponential function.

photon energy and division by time). This fraction I_f/I_a is called the quantum efficiency and depends on the rates in the following way:

$$\Phi = k_r / (k_r + k_{nr}) \qquad (1.6)$$

8.3. LIFETIME BASED SENSOR DEVICES AND INSTRUMENTATION

8.3.1. Optical Chemical Sensors

In this text, following Wolfbeis[4] a sensor is defined as "a device capable of continuously and reversibly recording" some quantity of interest. This rather narrow use of the term is preferred over a broader terminology that makes no clear distinction between sensors, indicators, probes, or microanalysis systems. An 'optical sensor' then is a sensor using optical properties as the information carrier, and 'chemical' refers to the quantity to be monitored, namely the concentration of a chemical species, the so-called analyte. The complete sensor device consists of the sensor element, a light source, some structures transporting the light, a detector and the necessary electronics. The sensor element contains the indicator that responds to changing analyte concentrations by corresponding changes in at least one optical property, the indicator being incorporated in a matrix material that may be fixed to a carrier or some light-guiding structures. While the performance of the sensor device depends on all its components, the sensor element is the specific point of interaction with the analyte and hence plays the decisive role.

8.3.2. Advantages and Drawbacks of Lifetime-Based Sensors

If an optical chemical sensor uses the fluorescence intensity as the information carrier, there may arise severe problems for the practical application that are connected with the peculiarities of fluorescence intensity measurements. The signal S_f obtained from a fluorescent sensor element is the product of a multitude of instrumental parameters:

$$S_f = I_s t_e (1 - 10^{-c\varepsilon_e l_e}) \Phi \times 10^{-c\varepsilon_f l_f} t_f s_d \qquad (1.7)$$

where I_s is the intensity of the light source, t_e and t_f the light throughput in the excitation and fluorescence paths, c the concentration of the indicator in the sensor element, ε_e and ε_f the molar extinction coefficients at the excitation and fluorescence wavelength, l_e and l_f the optical path lengths within the sensor element for the exciting and fluorescent light, Φ the quantum yield of fluorescence and s_d the detector sensitivity. All these quantities with the exception of the extinction coefficients and the quantum efficiency are instrument parameters and hence depend on the elaborateness of the production process and are subject to aging. Hence the reproducibility between different sensor elements and the repeatability with a single sensor element are quite critical. In an industrial production process instrument specifications cannot be made too stringent because of expenses in manufacturing and quality control. Every parameter contributes the uncertainty of the signal, and the only possibility to overcome this problem is to calibrate every sensor element before and repeatedly during use, which is highly undesirable in a practical situation.

The fluorescence lifetime, on the other hand, does not depend on any of these instrument parameters within rather broad. Therefore it is much easier to make a reliable and stable sensor based on lifetime than on intensity. In particular, the lifetime-based sensor is independent of any change in indicator concentration, which frequently is brought about by dye bleaching or leaching from the sensor element. Hence, wherever a fluorescent sensor is necessary that has to work properly over a long time without being recalibrated, the only solution is to rely on lifetime.

One drawback of lifetime-based sensors should not be overlooked, however. The indicator dye in a sensor element may be subject to chemical modifications over the shelf-life or the operational lifetime of the sensor. As long as these modifications yield non-fluorescent products, this is no problem. Fluorescent products may occur, however, and they contribute to the observed decay without being affected by the analyte concentration. Obviously this is detrimental to the accuracy of the sensor.

8.3.3. Transduction Schemes

To be suitable for fluorescence lifetime-based sensing a transduction scheme has to be applied that constitutes an influence of the analyte on the fluorescence lifetime of the indicator molecules in the sensor element. Common schemes are dynamic and static quenching, energy transfer, photo-induced electron transfer, or ion exchange, all of them resulting in a decreasing fluorescence intensity or a shortening in the decay time of a given substance. In this chapter only mechanisms will be included that are important for lifetime-based sensing.

8.3.3.1. Dynamic Quenching of Fluorophores

Dynamic quenching[5], often also called collisional quenching, occurs when the fluorophore in its excited state interacts with a quencher molecule. Due to diffusion processes the quenching molecule comes close to the excited fluorophore. Electronic interactions during the excited state lifetime of the fluorophore result in an energy transfer from the fluorophore to the quencher. As a consequence the fluorophore returns to the ground state without radiation. The efficiency of the fluorescence quenching is, therefore, determined by the number of collisions between the fluorophore and the quencher molecules.

In its simplest case, which means with a fluorophore in a homogeneous environment and molecular fluctuations averaged out on the time scale of the excited state lifetime, the theory of dynamic quenching is described by the Stern-Volmer equation

$$\frac{\Phi^0}{\Phi} = 1 + \tau_f^0 k_q \left[Q\right] = 1 + K_{SV}\left[Q\right] \qquad (1.8)$$

where Φ and Φ^0 are the overall quantum yields with and without quencher, τ_f^0 the fluorescence decay time of the fluorophore without quencher, k_q the bimolecular quenching rate constant and $\left[Q\right]$ the concentration of the quenching molecule and

$$K_{SV} = \tau_f^0 k_q \qquad (1.9)$$

the Stern-Volmer quenching constant. Since the over-all rate constant is connected with the fluorescence decay time and the quantum efficiency of the fluorophore by the following relation

$$\Phi = k_f \tau_f \tag{1.10}$$

and therefore is also related to the fluorescence intensity (see above, 8.2.2), the Stern-Volmer equation can be written as

$$\frac{\tau_f^0}{\tau_f} = \frac{I_0}{I} = 1 + K_{SV}[Q] \tag{1.11}$$

with I_0 and I being the fluorescence intensity without and with quencher. The quenching efficiency in this simple case can be directly observed from the slope of linear Stern-Volmer plot (figure 8.3).

It should be mentioned in parentheses that a linear Stern-Volmer plot does not necessarily prove dynamic quenching. Reactions influencing the ground state of the fluorophores like static quenching (Chapter 8.3.3.2) in some cases may also yield a linear a Stern-Volmer plot.

If the environment of the fluorophore is not the same for all indicator molecules, the emission process occurs from several slightly different, independently emitting excited states and can be described, as shown above, by a sum of single-exponential decays. Each individual lifetime component fulfils a normal Stern-Volmer equation. By using this multisite model[6], however, the whole system has to be described by a more sophisticated Stern-Volmer relation

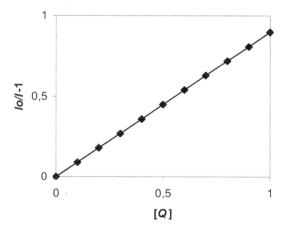

Figure 8.3. Fluorescence intensity in dependence of the quencher concentration, presented as a so-called Stern-Volmer plot.

$$\frac{I_0}{I} = \frac{\tau_{0m}}{\tau_m} = \left[\sum_i \frac{f_{0i}}{1 + K_{SVi}[Q]}\right]^{-1} \qquad 1.12)$$

with τ_{0m} being a preexponentially weighted mean lifetime, calculated according to

$$\tau_m([Q]) = \frac{\sum_i A_{i0}\tau_i([Q])}{\sum_i A_{i0}} \qquad (1.13)$$

A_{i0} is the amplitude and τ_i the decay time of the i-th component without quencher, f_{0i} is the fractional contribution to the unquenched fluorescence of the i-th component, and K_{SVi} is the quenching constant for the respective component. The different components refer to individual quenching constants for the different sites. Practical applicability of this model is rather be limited, since the number of different sites may be very high, while fitting algorithms usually do not give well-defined solutions for more than three different decay times.

The Stern-Volmer plot of multisite systems is non-linear. Non-linear Stern-Volmer plots are often observed in connection with solid state materials like polymer membranes. The micro-heterogeneity results from a spatial non-uniformity in the refractive index of the matrix, variations in the steric stabilization of the fluorophore by the polymer environment, or non-resonant energy transfer from the fluorophore to acceptor sites in the matrix.

One of the best known dynamic quenchers is molecular oxygen. If an oxygen molecule collides with a fluorophore in its excited state, the oxygen molecule will accept the excess energy of the fluorophore and quench the luminescence undergoing a triplet-to-singlet transition, while the fluorophore undergoes a non-radiative relaxation. Oxygen itself is not consumed in this process.

The first fiber-optic, lifetime-based oxygen sensor was described by Lippitsch et al.[1] in 1988. Tris (2,2'-diphenyl) ruthenium (II) dichloride hydrate in a silicone membrane was used as the sensing element, a blue light emitting diode as the excitation source and a photo-multiplier as detector. The fluorescence decay time is quenched by oxygen, thus a decrease in decay time is observed by increasing the concentration of oxygen. The Stern-Volmer plot was nonlinear showing a slightly downward curvature at oxygen levels above 20 %.

8.3.3.2. Static Quenching

Static quenching is observed when a fluorophore forms a ground-state complex with a quencher molecule. It is normally assumed that complex formation prevents the fluorophore from emitting fluorescence, while those fluorophores which are not complexed with the quencher emit normally with the same lifetime as in the absence of the quencher.

The association constant of the complex formation is determined by

$$K_S = \frac{[FQ]}{[F][Q]} \qquad (1.14)$$

where $[FQ]$ is the concentration of the complex and $[F]$, $[Q]$ the concentrations of fluorophore and quencher molecule. If the complex is non-radiant, the observed intensity is given by

$$\frac{I_0}{I} = 1 + \frac{[FQ]}{[F]} = 1 + K_S[Q] \qquad (1.15)$$

It is remarkable that equation (1.15) is identical to equation (1.11) observed for dynamic quenching if the quenching constant is replaced by the association constant. As a consequence fluorescence intensity measurements are not able to decide unambiguously between static and dynamic quenching, as the Stern-Volmer plots are identical for both processes.

To differentiate between dynamic and static quenching fluorescence decay time measurements and the temperature or viscosity dependence of quenching can be used. If there is only a static quenching the fluorescence decay time of the fluorophore should not be affected as the fluorescence results from the same molecules. One observes only a decrease in fluorescence intensity due to the formation of non-fluorescent complexes. A recent example is the change in fluorescence intensity of an aqueous solution of anthracylazacrown ethers[7] by adding paramagnetic metal cations like Mn^{2+}, Co^{2+} or Cu^{2+}. The quenching ions form the complexes in the ground state of anthracylazacrown ethers through the electron-donating nitrogen site.

On the other hand it is also possible that a fluorescing complex is formed. In this case fluorescence intensity measurements may be able to distinguish between static and dynamic quenching when the complex formation is connected with a spectral shift in the intensity. Fluorescence decay time measurements in this case will clearly deviate from single exponential behavior due to the fact that the decay time of the complex differs from that of the uncomplexed fluorophore. Assuming the simplest case with fluorophore and complex having single exponential decays, the measured signal can be fitted with a sum of two exponentials. The relative concentration of the complexes can be deduced from the pre-exponential factors.

8.3.3.3. Energy Transfer

A special type of fluorescence quenching is resonance energy transfer, a non-radiative transfer of energy from an excited donor to an acceptor. The energy transfer occurs without photon emission. The rate of transfer depends on the spectral overlap between the emission spectrum of the donor and the absorbance spectrum of the acceptor, the quantum yield of the donor, the relative orientation of the transition dipoles of donor and acceptor, and the distance between the donor and acceptor.

The theory of radiationless energy transfer was developed by Förster[8] in 1948, describing the following process:

$$D^* + A \quad D + A^* \qquad (1.16)$$

During the absorption process the fluorophore is excited to one higher vibrational level of its first electronic excitation state. From there it is converted to lower vibrational levels of the same electronic state by obtaining thermal equilibrium with the surrounding medium. In solution this thermal relaxation takes place during $10^{-13} - 10^{-12}$ seconds. If the energy difference for radiative deactivation corresponds well to that for a possible absorption transition in a nearby acceptor molecule, a transfer of excitation energy from the fluorophore to the acceptor is possible, depending on the coupling between the molecules. As a result the observed fluorescence of the acceptor molecule is moderately shifted to the red.

The electric field near an electronically excited molecule is assumed to behave like a field generated by a classical oscillating dipole. The charge oscillations cause electrostatic forces to be exerted on the electronic system of nearby molecules. According to Förster the rate of energy transfer from a donor to the acceptor is given by

$$k_{DA} \propto \frac{\kappa^2 \phi_D}{n^4 \tau_D R^6} \int_0^\infty \frac{f_D(\nu) \cdot \varepsilon_A(\nu)}{\nu^4} d\nu \qquad (1.17)$$

where κ is an orientation factor describing the relative orientation between the transition moment vectors of donor and acceptor, ϕ_D is the quantum yield of the donor in the absence of an acceptor, n is the refractive index of the medium, τ_D is the mean lifetime of the donor in absence of the acceptor, R is the distance between donor and acceptor, ν is the frequency, $f_D(\nu)$ is the spectral distribution of the donor fluorescence (normalized to unity on a frequency scale) and $\varepsilon_A(\nu)$ is the molar decadic extinction coefficient of the acceptor at the frequency ν. Thus, Förster's theory predicts that the rate constant k_{DA} for an energy transfer via a Coulombic interaction will be proportional to the inverse sixth power of the separation of D^* and A.

More conveniently, equation 1.17 can be rewritten as

$$k_{DA} \propto \frac{1}{\tau_D} \left(\frac{R_0}{R} \right)^6 \qquad (1.18)$$

with R_0 being the critical transfer distance for which the excitation transfer and the spontaneous deactivation of the donor are of equal probability. This equation is valid for any thermal equilibrium distribution over the vibrational levels of both molecules, however, in liquid or solid medium, where the energy transfer is very rapid due to strong interactions, the transfer may take place from the excited vibrational level directly and depend therefore on the exciting wavelength.

Especially in solid medium the orientation factor κ may be important. It can be calculated by

$$\kappa = \cos \vartheta_{DA} - 3 \cos \vartheta_D \cdot \cos \vartheta_A \qquad (1.19)$$

where ϑ_{DA} is the angle between the transition moment vectors of donor and acceptor molecules, while ϑ_D and ϑ_A are the angles between the transition moment vector of donor or acceptor (figure 8.4) and the direction from donor to acceptor. κ can take values between 0 and 4 (the average value for a random distribution is $^2/_3$, in the case of a two-dimensional layer this value is $^5/_4$, orientation along a straight line would give 4). There are numerous examples for sensors using energy transfer as the transducing

Figure 8.4. Orientation factor describing the relative orientation between the transition moment vovtors of donor and acceptor

mechanism. One of the first was implemented by Jordan[9] *et al.*, who described a pH sensor composed of immobilized eosin as a fluorescent donor and immobilized phenol red as the pH dependent acceptor.

A very clever energy-transfer based method has been described recently[10]. The donors (ruthenium complexes) and acceptors (cyanine dyes) are incorporated in polyacrylonitrile-derived nanospheres. Varying acceptor species and concentration spheres are prepared that can be distinguished by emission wavelength and decay time. By using suitable copolymers, nanospheres with reactive surfaces for selective covalent coupling to proteins or other biomolecules can be produced. They act as markers to quantitatively detect those biomolecules.

8.3.3.4. Photoinduced Electron Transfer (PET)

Photoinduced electron transfer (PET) means an electron transfer resulting from an electronic state produced by the resonant interaction of electromagnetic radiation with one molecular entity to another, or between two localized sites in the same molecular entity.

Figure 8.5 shows possible mechanisms for the electron transfer: The donor is excited by light. From the excited state an electron transfer to the acceptor takes place, leaving a positive charged donor and a negative charged acceptor. In this process no luminescence will be observed.

For practical sensor applications electron transfer preferably takes place within two

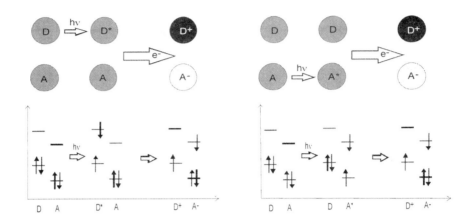

Figure 8.5. Photoinduced electron transfer

different sites of one molecule, the photon- and cation interaction sites usually are localized in spatially distinct regions by means of a spacer molecule. Very often some overlapping of a conjugated π-electron system (the photonic site) with an analyte-binding
receptor like cryptands or crown ethers are used. When the crown ether binds an analyte
the photoinduced electron transfer is blocked. As a result now the fluorescence of the
donor is observed which was quenched by the PET.

An example for a sensor using photoinduced electron transfer was published by
Bryan[11] *et al.* in 1989 using anthracen-9-yl methyl azacrown ether as a metal ion sensor.
Another example is a pH sensor using diethylaminomethyl pyrene as fluorescence indi-

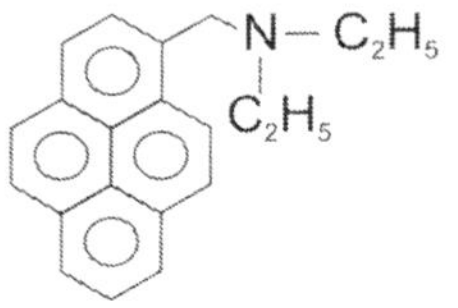

Figure 8.6: Diethylaminomethyl pyrene

cator incorporated in a hydrogel[12] (figure 8.6).

The absorption spectra of the acid and the base form of the indicator are identical,
but the fluorescence quantum efficiencies are quite different. When the chromophore is
excited there exists a certain probability for an electron transfer from the amino group to
the pyrene, even if the amino group is decoupled from the chromophoric system of the
pyrene by the methyl group. The process of electron transfer competes with fluores-

cence emission and hence influences the decay time. At acidic pH the indicator is protonated, in this case almost no electron transfer occurs and the observed decay time is
rather long. At higher pH the indicator is in its neutral form, and the electron transfer
becomes higher shortening the fluorescence decay. If the acid and base forms of the
indicator show a single exponential decay time, the decay of the fluorescence can be
described by the following equation:

$$I(t) = A_a \exp\left(-\frac{t}{\tau_a}\right) + A_b \exp\left(-\frac{t}{\tau_b}\right) \qquad (1.20)$$

with *I(t)* being the normalized time-dependent fluorescence intensity, A_a and A_b the
pre-exponential factors and τ_a and τ_b the decay times of acid and base form of the
indicator. The pre-exponential factors depend on pH according to

$$A_a = \left[1 + 10^{pK_a - pH}\right]^{-1} \qquad (1.21)$$

$$A_b = \left[1 + 10^{pH - pK_a}\right]^{-1} \qquad (1.22)$$

A closer look at the decay curve shows that the fluorescence decay of the acid or the
base form of the indicator is not really single exponential. Due to the microheterogene-
ity of the hydrogel the fluorescence decay can be assumed to be an average over a distribution of relaxation rates dependent on interactions with the neighboring regions of
the hydrogel-water system. The influence of a non-uniform microenvironment and possibilities of fitting such curvatures will be discussed later (eq. 1.27). Nevertheless it is
still possible to define a ratio of pre-exponential factors that is a function of pH.

The sensor performance is demonstrated in figure 8.7, showing the pH dependence
of the pre-exponential factor for the base form of the indicator. From the measured
values a pK_a of 7.7 is deduced.

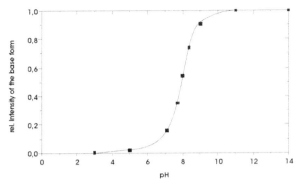

Figure 8.7. Fluorescence intensity of the base form of the pyrene compound in dependence on pH

Long-term stability tests were performed with this sensor over more than one year. The sensing membranes were stored without any precaution regarding temperature or humidity in an open glass beaker in a laboratory case. The hydrogel dried out completely between two measurements, however, after immersing it into water for some

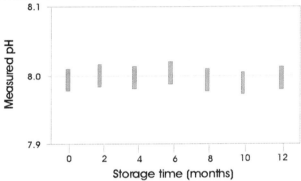

Figure 8.8. Long-term stability test for the pH sensor

hours to let the hydrogel swell again measurement was successful. The films proved to be very stable over a whole year (figure 8.8) and did not show significant changes over that period.

8.3.3.5. Ion Exchange

The scheme of ion exchange is based on the use of a (for example) cation selective ionophore together with an acidic indicator. Both are incorporated into a hydrophobic matrix. Cations usually do not penetrate into a hydrophobic membrane unless they are carried by an ionophore. When a cation is imported the acidic indicator is deprotonated, and the proton leaves the membrane to achieve electro neutrality. In this case the state of protonation becomes a function of the cation concentration.

This mechanism is used to detect cations. It is necessary that the fluorophore is protonated. To achieve electro neutrality a proton has to leave the membrane if a cation
penetrates the membrane. The ion exchange equilibrium between solution and hydrophobic membrane is given by

fluorophore
  ○ H+ ⟶
  ○ ◄──── K+
ionophore

Hydrophobic membrane   solution

Figure 8.9. Principle mechanism of ion exchange

$$k_{exch} = \frac{X_a \cdot IH \cdot C}{H_a \cdot I \cdot CX}$$

(1.23)

with X_a being the concentration of the cation in the solution, IH the concentration of the
protonated fluorophore in the membrane, C the concentration of the protonated ionophore in the membrane, H_a the concentration of the positively charged protons in the
solution, I the concentration of the neutral fluorophore in the membrane and CX the
concentration of the ionophore-cation complex in the membrane.

An example for the ion exchange mechanism is a nitrate sensor[13] used for ground
water monitoring. Pyrene butyric acid (figure 8.10) is used as the fluorophore, tridode-
cylmethylammoniumchlorid as a nitrate carrier (Fluka Chemika), both incorporated into
PVC together with a plasticizer and ammonium-tetraphenyl-borate to facilitate proton
transfer in the membrane.

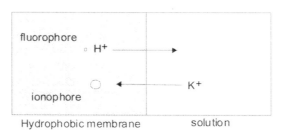

Figure 8.10, Pyrene butyric acid

Pyrene butyric acid is in its neutral form at acidic pH, with a fluorescence decay time of a few nanoseconds. In a base environment the indicator is deprotonated which

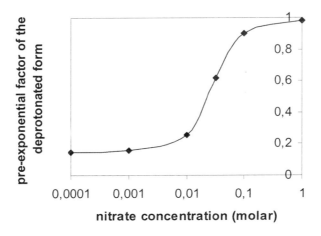

Figure 8.11. Working function of the nitrate sensor

increases its decay time to about 90 ns. As a consequence the measured fluorescence decay can be fitted with a sum of two exponentials, one decay time belonging to the neutral fluorophore, the other one to the deprotonated. The pre-exponential factors of the two decay times are related to the concentrations of neutral and deprotonated fluorophore, therefore the relative amplitudes being a function of the nitrate concentration. Different concentrations of ammonium-tetraphenyl-borate can shift the working function (figure 8.11) to higher or lower concentrations of nitrate, thus enabling the measurement of nitrate with high accuracy over a wide range of nitrate concentrations.

8.3.4. Sensor Elements

8.3.4.1. Indicators

Materials suitable for lifetime-based sensors should fulfill, in addition to the necessary properties for any sensor, some additional conditions. Indicators should be excitable preferably in the visible range, their luminescence decay should be of moderate complexity (if possible single exponential) and lifetimes should be long enough to allow measurements with reasonable efforts. Polymer matrices should exhibit no major influence on the decay properties. In reality these conditions are rarely met and compromises have to be accepted.

The number of indicators with rather long fluorescence decay time (that means longer than 20 ns) is not very large. Suitable candidates may be found among poly-cyclic hydrocarbons, hetero-cyclic compounds and especially among organo-metallic complexes. Only a few analytes directly influence the fluorescence decay time of the indicator. Usually it is necessary to attach some functional group to the fluorophore to obtain a certain analyte response.

If sensors for different analytes are required within the same instrument the concept of indicator families[14-16] has proofed to be successful. That means that a group of indicators has to be found which have similar spectroscopic properties, especially similar absorptions and emission wavelengths, and exhibit rather long fluorescence decay times of comparable magnitude with each other. Indicators belonging to one family are derived from the same luminophore. Functionality with respect to analyte recognition is added without significantly altering the spectroscopic properties of the luminophore. If the fluorescence decay time of the fluorophore is used for measuring, this interaction between the functional group and the fluorophore should be restricted to the excited state of the fluorophore without relevant interaction in the ground state.

To achieve this, various approaches could be envisaged. One is that of excited-state energy transfer between the luminophore and the functional group. Another one is the application of photoinduced excited-state electron transfer. Both principles guarantee high excited-state interactions with negligible interactions in the ground state, and both influence the fluorescence decay time of the fluorophore.

The first indicator family developed so far was based on excited-state electron transfer with pyrene as the fluorophore. The absorption of pyrene is in the near UV (λ_{abs} = 335 nm), and the monomer fluorescence can be observed between 360 and 400 nm. The fluorescence decay time of pyrene is in the range of 150 – 400 nm, the value strongly depends on the kind of solvent and the matrix used. Attaching different functional groups with specific analyte recognition to the fluorophore leads to indicators for various analytes, for example oxygen, pH, carbon dioxide or potassium. Using the principle of ion exchange the pH indicator can be used with various commercially available ionophores.

The absorption of pyrene is in the near UV, thus it is not possible so far to use commercially available and cheap light emitting diodes. Therefore, another approach was undertaken to develop an indicator family based on fluorophores excitable in the visible with fluorescence decay times in the microsecond range. This family is based on ruthenium (II) tris-chelated complexes. Such a complex consists of a central ruthenium ion surrounded by three heterocyclic ligands, bound to the ion via nitrogen atoms. An example is shown in Fig. 1.12. The photophysical properties of this class of fluorophores are well studied. They show absorption and emission in the visible range, high Stokes shift and long fluorescence lifetime, all favorable properties for optical sensing. Excitation of a ruthenium complexes results in a metal-to-ligand charge transfer (MLCT). The luminescence decay is determined by transitions to competing states. Investigations of the dependence of the decay time on solvent polarity and viscosity, on temperature, and on substitution of the ligands showed that the decay time is not only determined by crossing to the dd-state, but also by population of a further MLCT state that is much lower in energy than the dd, but has a rather low crossing probability. The position of this state is dependent on the electron distribution within the ligands. Transfer of electrons within the ligands towards the nitrogen atoms generally destabilizes the excited state and shortens the decay time, while drawing electrons off the nitrogen increases the decay time. This effect may, however, be counteracted by the different dependence of energetic position and crossing probability of the low-energy state. Also the degree of delocalization of the charge over the three ligands has to be taken into

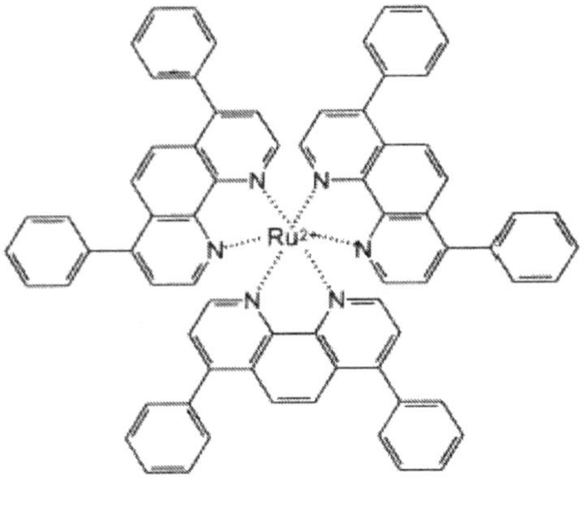

Figure 8.12. Ruthenium(II)-tris-(4,7-diphenyl-1,10-phenanthroline) complex

account, together with the dipole and higher electric moments induced in the complex by substituent(s). These findings allow deliberately tailoring the ligands to the desired properties required within the sensor family.

The ruthenium complexes are dynamically quenched by oxygen (see above) and hence are suitable for oxygen sensing. Numerous investigations[17,18] made use of preferably the ruthenium (II)-tris(4,7-diphenyl-1,10-phenanthroline) complex (figure 8.12) and its lifetime dependence on oxygen. For use with other analytes either the matrix has to be chosen as to suppress oxygen quenching, or the measured response has to be corrected for oxygen.

To measure pH or other ions the ligands have to be suitably modified by proper substitution. Photoinduced electron transfer as the transduction mechanism is problematic in this case since excited ruthenium complexes usually are electron donors, while virtually all feasible ionophores need an acceptor. Nevertheless binding of ions may influence the lifetime by other mechanisms. Ruthenium(II)-tris(4-(4-hydroxyphenyl)-1,10-phenanthroline)[19] , for example, is quite useful as a pH indicator. In deoxygenated aqueous solvents the decay function is pH dependent, clearly deviating from a single exponential at pH > 4.5. The preexponentially weighted mean lifetime varies monotonically with pH from 3.6 µs at pH 3 to 250 ns at pH 11 with an apparent pK_a of 7.47 (figure 8.13). Sensor membranes for pH were produced by immobilizing the fluorophore in a hydrogel that was cast on a solid carrier. In this polymeric surrounding the decay times were generally longer than in solution.

Interestingly, with carboxy instead of hydroxy ligands the behavior on protonation is quite different. In the hydroxy compound the acid form has a long decay time (3.6 µs, the base form has only about 250 ns). The carboxy compound, on the other hand, has 1.2 µs in the acid and about 3.7 µs in the base form. This opposite behavior of the pH dependence is due to different electron distributions in both compounds and to different changes in the crossing probability to the dd- and MLCT state. The hydroxyl compound efficiently shifts electronic charge to the nitrogen upon deprotonation. In the carboxy compound the charge remains delocalized over the two nitrogens. The longer decay time is due to increased decoupling of the substituent from the phenanthroline. Both effects can be achieved also with suitable ionophores binding sodium or potassium.

To obtain a sensor for carbon dioxide these pH sensors were equilibrated with alkaline buffer. Mixtures of nitrogen and carbon dioxide saturated with water vapor were

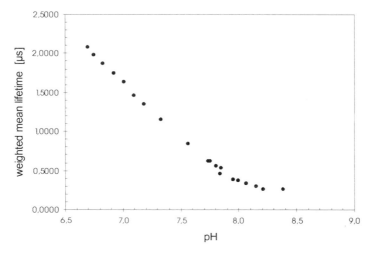

Figure 8.13. The preexponential weighted mean fluorescence decay time $_m$ of the pH sensitive Ruthenium complex in solution, plotted versus pH.

blown over the sensor membrane. Highest sensitivity could be achieved for partial pressures between 0 and 50 mm Hg. Covering the membrane with a thin Teflon film makes the sensor usable in liquid environments.

The three ligands on the ruthenium need not be equal. Using three different ligands, a so-called tris-heteroleptic complex is obtained[17]. This kind of complex may be highly advantageous for optical sensing. As described earlier, indicator molecules have to fulfill multiple requirements. They must be specific for a certain analyte, show favorable response properties, meet certain spectroscopic and kinetic specifications and must provide the possibility to be properly incorporated in a matrix. These requirements can be distributed over the three different ligands of a tris-heteroleptic complex: One ligand may be responsible for the analyte specificity, the second for the excitation and emission properties, and the third one may serve as a means to covalently anchor the indicator to the matrix polymer, as exemplified in a potassium indicator in figure 8.14. Here covalent binding to a cellulose matrix is achieved via a primary amino group on the first ligand. This matrix minimizes oxygen cross sensitivity. A calyx crown ether on the second ligand serves as the potassium-specific moiety. And the pyrene substituent on the third ligand effectively lengthens the fluorescence decay time.

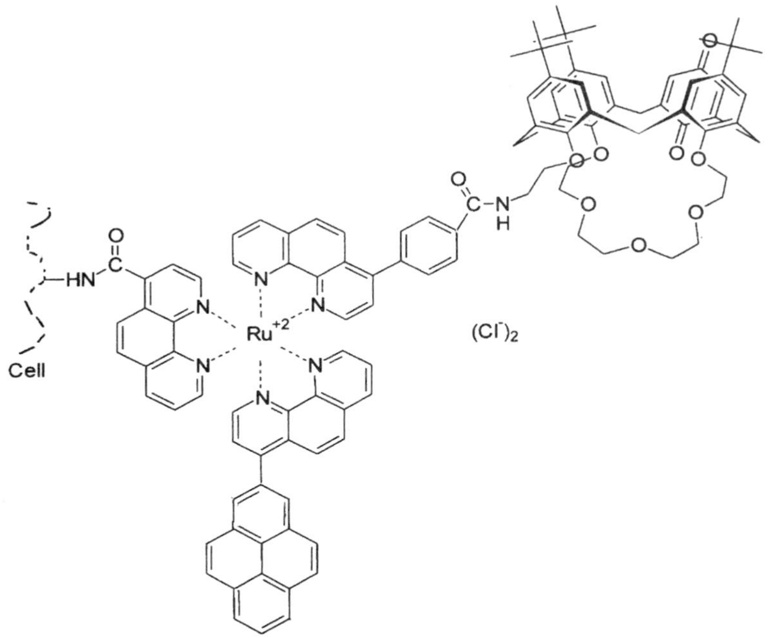

Figure 8.14. Example of a tris-heteroleptic complex.

8.3.4.2. Matrix Materials

To achieve a working sensor element indicators have to be incorporated into a suitable matrix material. Conveniently polymer materials are used like polyvinylchloride, silicone, ethylcellulose, polystyrene or polyurethanes, but there are also successful approaches to use glassy materials like sol-gel matrixes. Indicators can be incorporated into a membrane, they can be immobilized at a surface or encapsulated into special cavities.

It is well appreciated that the properties of the medium affect the luminescence decay properties of the molecules. As long as the temporal fluctuations of the environmental influences are much faster than the luminescence decay, each molecule seems to be affected by the same interactions, and the kinetics of the ensemble is uniform. However, that assumption is not valid for molecules in a polymer membrane, where different molecules have different influences from their respective microenvironment. Most polymers are non-uniform media in a sense that there is no long-range structural order. In this case the fluorescence decay profiles of molecules can be assumed as being an average over a distribution of relaxation rates, the decay function of the molecules becomes non-exponential.

As described in paragraph 8.2.1.2, the time-dependent probability $p(t)$ for an excited fluorophore to be in the excited state at a time t after excitation is found from the solution of the following differential equation

$$-\frac{dp(t)}{dt} = p(t)\left(k_r + k_{nr}\right) \qquad (1.24)$$

with k_r being the radiative and k_{nr} the non-radiative deexciation rate. Both rates are determined by internal properties of the fluorophore as well as by properties of the surrounding medium. Due to spatial non-uniformities in the medium the rates may be different for molecules situated at different sites inside the medium. As a consequence a discrete or quasi-continuous distribution of rates may be applied, leading to distributions of lifetimes[18]. A deeper understanding of the decay process suffers from the fact that the physical reason for the distribution, and hence the form of the distribution function, is generally not known. The medium may influence the radiative rate by shifting the absorption maximum or distorting the shape of the absorption band. Furthermore, in a broad fluorescence band, local variations in the refractive index may be considered. The main influence of the medium surrounding the fluorophore on its decay time should enter via the non-radiative decay rate, which is the sum of the rates of internal conversion, intersystem crossing and quenching. The intersystem crossing probability depends on the spin-orbit coupling in the excited fluorophore which could be influenced by an external heavy atom effect. Fortunately such atoms usually are not present in a polymer membrane. The internal conversion rate decreases on inclusion of the fluorophore in a rigid matrix, because of the limitations imposed on the molecular flexibility. If the stabilization effect of the matrix shows spatial variations, a distribution of rates and hence to a non-exponential decay might follow. Last but not least the matrix itself can act as a quencher. This quenching is brought about by interactions with neighboring regions of the polymer and depends on the distance between the fluorophore and the nearest interacting polymer site. The quenching rate $k_q^{(i)}$ for the ith excited molecule is the sum over the distance dependent interactions with j quenching sites

$$k_q^{(i)} = \sum_i k_{ij}\left(r_{ij}\right) \qquad (1.25)$$

with k_{ij} being the quenching rate for the ith molecule caused by the jth interaction site, r_{ij} being the distance between the interacting partners. This rate has to be included in equation 1.24. The resulting differential equation can be solved, if the distribution of distances is assumed to be homogeneous and isotropic.

The distance dependence of the quenching rate k_{ij} can have different forms, depending on the quenching mechanism, two of them being physically important: electron transfer and electromagnetic near-field interactions. Polymer matrices used for chemical sensing usually do not have strong electron acceptors, as necessary for electron transfer. Therefore, the main influence can be attributed to electromagnetic interactions.

As described in 8.3.3.3. electromagnetic interactions between close-lying molecules were first treated by Förster[8]. He considered the resonant case only, which means that the absorption band of the energy acceptor matches the emission band of the fluorophore. This is certainly not the case in transparent polymers. However, for fluorophores with a rather long excited-state lifetime non-resonant interactions may contribute significantly. After excitation of the fluorophore the equilibrium in the microenvironment of the indicator is disturbed due to a change in the dipole moment. Therefore, a reorientation of the local environment due to minimizing the local field energy leads to an inhomogeneous spectral broadening[20] of the fluorescence.

In contrast to solvents, where only a solvation shell around the indicator has to be taken into account, in a membrane the whole volume surrounding the indicator has to be considered. Assuming an electromagnetic interaction between indicator and polymer and no point charges present in the matrix, only dipole interactions have to be considered. This leads to the so-called "stretched" exponential of the form

$$I(t) = I(0) \exp\left[-\frac{t}{\tau} - \left(\frac{t}{\tau}\right)^{\Delta/n} \right]$$

(1.26)

with $I(t)$ being the time-dependent fluorescence intensity of the indicator, $I(0)$ the fluorescence intensity for $t = 0$, Δ being the dimensionality of the system and n the exponent in the power law describing the dependence of the interaction on the distance between the indicator molecule and the interaction site. For many systems, $\Delta = 3$ and $n = 6$. In this case the fluorescence decay profile is found to deviate from the usual exponential form even in the absence of an external quencher.

For optical sensors the influence of this fluorophore-polymer interactions on a dynamical quenching process is of considerable interest. By neglecting the interactions between quencher and polymer the time-dependent fluorescence intensity can be described by the following equation

$$I(t) = I(0) \exp\left[-(1+c)\frac{t}{\tau} - a\left(\frac{t}{\tau}\right)^{\Delta/n} \right]$$

(1.27)

with $c = \tau k_q$ being the dynamical quenching parameter, k_q the bimolecular dynamic quenching rate, and a a parameter describing the influence of the environment. If no matrix interactions were present ($a = 0$) the decay would remain single exponential, but shortened according to the Stern-Volmer equation 1.8. With matrix interactions the form of the decay function is concentration dependent. The relative contribution of the power term decreases with increasing quencher concentration.

These results stress the importance of carefully studying the matrix properties in optical sensors. Another aspect of matrix materials should be concerned, especially when sensing ions. Usually hydrophobic polymers are used as the matrix material. The reactions between the ion carriers and the respective ions in the sample occur at the interface between polymer and aqueous solution, probing the activities of ions near the interface. These activities are not necessarily the same as in the bulk of the sample. The optical measurement is done in the bulk of the sensor membrane, the signal depends on the concentration of reacted ions in the polymer matrix. This concentration is dependent not only on the amount of ions undergoing reaction at the interface, but also on the extraction coefficients from the sample to the polymer for the reaction products in the bulk membrane. These coefficients and constants may be influenced by processes determined by the concentration in the sample of compounds other than the ions to be measured. Thus the concentration of reacted indicator in the sensing membrane may not be a well-defined function of the ion activity in the sample[21].

The alternative is to use a hydrogel as the matrix material. Such a matrix, when soaked with water, can be considered as a mixture of fluids rather than a solid. There exist no hard interfaces between the aqueous and polymer domains. However, the indicator favourably has to be covalently immobilized to avoid washing out of the hydrogel.

8.4. LIFETIME MEASUREMENT METHODS AND INSTRUMENTATION

Fluorescence lifetimes range from picoseconds to microseconds, those of phosphorescence up to many hours. Fluorophores of interest for optical sensing usually exhibit lifetimes from nanoseconds to microseconds. Several conceptually different schemes have been developed, that can be divided (somewhat arbitrary) into time-domain and frequency-domain methods.

8.4.1. Time-Domain Methods

Under this heading all methods are united that are based on measuring the time evolution of the emitted fluorescence radiation after the fluorophore after switching off the exciting action. This means that during the measuring time no further excitation takes place and only relaxation mechanisms are at work. Therefore the decay function depends only on the relaxation rates.

Excitation is usually achieved by a short light pulse of rather high intensity. The fluorescence decay is monitored by a photo multiplier or, in favourable cases, a photodiode. Usually it is not possible to measure the whole fluorescence decay with sufficient signal-to noise ratio in a single flash. Various methods to measure the fluorescence decay with repetitive pulses. One is single-photon timing, which over the last decades has been applied in laboratory instruments by several companies. In this method very low light levels are used. For every excitation pulse the time is measured that elapses before the first fluorescence photon reaches the detector. The statistics over these arrival times gives the fluorescence decay. Since in every single measurement only one electron is involved, a high number of flashes (and hence a long collection time) is necessary to obtain a good signal-to-noise ratio. This method therefore is rather inappropriate for optical sensors.

Another method is pulse sampling, which is more or less a stroboscopic scheme, applying a successively delayed gate pulse to the modulator and thus sampling the decay. This method necessitates fast high-voltage pulses to gate the multiplier and hence also is uncommon in optical sensing.

Modern electronics provide us with fast AC-converters, making possible to digitally store the whole decay from a single flash. Averaging over multiple flashes is used to improve the signal-to-noise ratio. Evaluation of the obtained decay curve usually is performed by nonlinear fitting routines using statistical criteria of goodness for the fit. The main problem is to apply a physically meaningful fitting function. Formal fitting of non-exponential decays with a sum of exponentials, as is often reported in literature, is the frequent cause of misinterpretations. A thorough understanding of the underlying molecular properties is essential to prepare the adequate description of the decay curve.

In sensor applications the decay curve is usually well known and the only problem remains to extract the information on the analyte concentration from the decay. With strongly multi- or non-exponential decays the only reliable procedure is to directly compare the measured curve with a set of pre-determined curves for various analyte concentrations. Figure 8.15 gives an example for a CO_2 sensor[19]. The previously recorded curves are stored in the instrument. A new curve is checked for minimum difference to the stored ones. The analyte concentration is determined by interpolating between the two neighbouring reference curves.

In the case of rather simple decays it is sufficient to register a few points in order to determine the analyte concentration. In the single-exponential case (with some inevitable background) three parameters fully determine the function:

$$I(t) = a \, \exp(t/\tau) + b \qquad (1.28)$$

The function can be determined from the measured intensity at $t = 0$ (time of excitation switch-off) and two other points in time with

$$\tau = \frac{t_2 - t_1}{\ln(I_1/I_2)} \qquad (1.29)$$

and

$$b = I_0 - I_1 \left(\frac{t_1 \ln I_1 - t_2 \ln I_2}{t_1 - t2} \right) \qquad (1.30)$$

In a practical sensor device the procedure to determine the relevant quantities is to use comparators set to different thresholds, delivering pulses when the signal falls below the

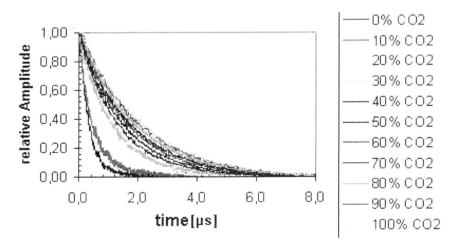

Figure 8.15. Fluorescence decay curves at different CO_2 concentrations

threshold. The intensities then are determined by the threshold values and the times are the intervals between the pulses delivered.

In a real measurement there is inevitably some noise superimposed on the signal (Fig. 8.16). This means that the pulses are delivered with some error. This error depends on the steepness of the decay at the given point. Noise will be more detrimental when the decay is steep. In an exponential decay a noise amplitude ΔI translates into a tim-

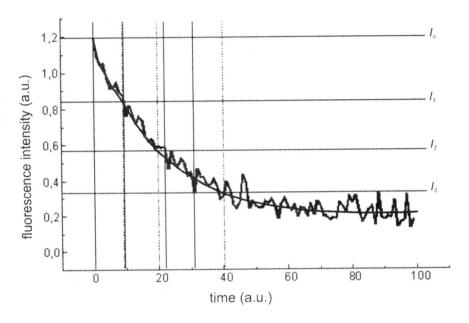

Figure 8.16. Fluorescence intensity in dependence on time

ing error Δt according to

$$\Delta t = -\Delta I \frac{\tau}{(I_0 - b)\exp(-t/\tau)} \qquad (1.31)$$

The exponent in the denominator yields large errors at times long compared with the time constant. Furthermore, the error increases with increasing background b. The comparator delivering the timing pulse is nonlinear regarding error propagation. This means that the average over the timing error is not zero. Errors in time and intensities lead to an error in the calculated time constant that is given by

$$\Delta\tau = \frac{\tau\sqrt{(I_0 - I_1)^2 + (I_0 - I_2)^2}}{(I_0 - I_1)(I_2 - b)\ln\left(\dfrac{I_0 - b}{I_2 - b}\right) - (I_0 - I_2)(I_1 - b)\ln\left(\dfrac{I_0 - b}{I_1 - b}\right)} \Delta I \qquad (1.32)$$

From this equation can be deduced how the thresholds should be positioned to give the lowest error. The result is that without background the optimum value for the first threshold is when $I_0/I_1 = -\ln(I_1/I_0)$ or $I_1 = 0.567 I_0$. With background, the optimum value is shifter slightly higher. The optimum for the second threshold is $I_2 = b$. This is trivial insofar as the error goes to zero for infinitely long measuring times. Obviously this is impracticable, but it is clear that the I_2 has to be as low as possible. For this case it can be shown that the total error is only weakly dependent on the position of the first threshold.

If the decay is single-exponential and the thresholds are positioned appropriately, statistical evaluation of the timing intervals gives reliable results for the decay parameters. Unfortunately the error estimation is not straightforward in multi- and non-exponential decays.

8.4.2. Frequency-Domain Methods

Under the term frequency-domain methods all procedures are subsumed where the excitation is not performed by switching on and off the excitation light but by modulating it by a periodic, continuously differentiable function.

8.4.2.1. Phase Shift Measurement

If the sensor membrane is excited with light whose intensity is modulated sinusoidally, then the emission is a forced response to the excitation, and therefore the emission is modulated at the same circular frequency as the excitation. Because of the finite lifetime of the excited state, the modulated emission is delayed in phase by an angle φ relative to the excitation with

$$\tan \varphi = \omega \tau_p \qquad (1.33)$$

with ω being the circular frequency of the exciting light and τ_p the phase lifetime of the emission, which is identical with the actual fluorescence lifetime in case of single exponential decay.

If the decay consists of multiple exponentials with multiple time constants τ_i and relative contributions a_i, the phase lifetime is given by

$$\tau_p = \frac{\sum_i a_i \tau_i^2 / (1 + \omega^2 \tau_i^2)}{\sum_i a_i \tau_i / (1 + \omega^2 \tau_i^2)} \qquad (1.34)$$

The sums have to be replaced by integrals in case of a continuous contribution. To find out the time constants involved the measurement has to be performed with as many modulation frequencies as there are time constants. It is clear that the multi-equation system obtained yields results with high uncertainty, as is the case with multi-exponential fitting in time-domain measurements.

If the decay consists of only two exponential components as, for example, from analyte-bound and unbound indicator molecules, the problem is made easier. The time constants for the bound and unbound indicator can be determined independently, so that the only unknown variable is the ratio of the relative contributions of the two components that is a function of the analyte concentration. This ratio can be determined from the phase lifetime at a single modulation frequency.

The advantage of the phase shift method is that excitation and detection are modulated by the same frequency. The optimum modulation circular frequency ω for a single-exponential decay with time constant τ is given by

$$\omega \approx \tau^{-1} \tag{1.35}$$

This means that modulation frequencies rarely have to be above 100 MHz, which is conveniently mastered by modern technologies. In practical sensor applications inevitable ambient light fluctuations will cause unwanted noise. This can be eliminated by a narrow-band filter centered at the modulation frequency, but it must not be forgotten that filters always induce an additional and sometimes uncontrollable phase shift, thus deteriorating the signal accuracy.

8.4.2.2. Demodulation Depth

When excitation is modulate sinusoidally with circular frequency ω, the demodulation factor m of the fluorescence signal is defined as

$$m = Ba/bA \tag{1.36}$$

(see Fig. 8.17). For a single-exponential decay with time constant τ this demodulation depends on the fluorescence lifetime as

$$m = \left(1 + \omega^2 \tau^2\right)^{-1/2} \tag{1.37}$$

In this case it is straightforward to deduce the time constant from the demodulation. For a multiexponential decay this expression converts to a lengthy expression containing sums over linear and quadratic functions of the time-constants involved. In principle measuring the demodulation at various modulation frequencies gives a set of equations from which all time constants can be deduced. In practical work this is virtually impossible if more then two time constants are present.

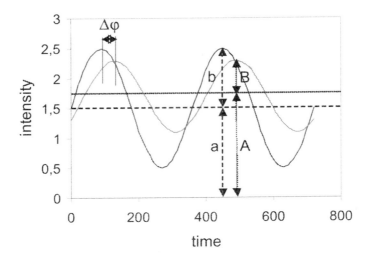

Figure 8.17. Determination of demodulation factor

8.4.2.3. Dual Lifetime Referencing

This sensing method was first suggested in a patent application by Klimant[22] to make sensors based on static quenching accessible to lifetime methods. In essence it is a phase-shift method, but it uses two different fluorescing dyes present in one sensor membrane. The first one, called the indicator, has a short fluorescence decay time in the nanosecond region and changes its fluorescence intensity upon interaction with the analyte. The second dye, named the reference, has a long luminescence decay time in the microseconds and its fluorescence intensity or decay time is not affected by the analyte. As a necessary condition both dyes need to have overlapping excitation and emission spectra so that they can be excited with the same light source and the fluorescence can be detected by using the same photodetector. The measured phase shift ϕ_m then depends an the ratio of fluorescence intensities A of the indicator and the reference dye according to

$$\cot \phi_m = \frac{1}{\sin \phi_{ref}} \left(\cos \phi_{ref} + \frac{A_{ind}}{A_{ref}} \right) \qquad (1.38)$$

When the phase shift for the reference dye is known, the measured phase shift depends only on the ratio A_{ind}/A_{ref}, which in turn depends on the analyte concentration. In this way it is possible to apply lifetime methods to analytes that exhibit static quenching of the indicator only. But the drawback of the method is obvious: The signal quantity is an intensity ratio. Therefore the advantage of lifetime methods discussed above, being independent of intensities and hence of indicator concentrations, is lost in this method. There exist several important applications where this drawback is not significant[10], but in all other cases the method has to be regarded with caution.

8.4.2.4. Optical Design

Despite the fact that we are dealing with optical sensors, in the technical literature the optics of sensor devices has not been gained much attention. For many years a considerable appreciation of fiber optics was en vogue, but few thoughts were spent on the question if using fibers instead of conventional optics could have any advantage for the respective purpose. In most fluorescent sensors the only task assigned to optical fibers was to transport light from the source to the sensor and from there to the detector, much as a garden hose delivers water to the lawn.

In a practical sensor device the importance of a well considered optical design cannot be overestimated. Analysis of optical systems as usually applied reveals a total efficiency (power of fluorescent light at the detector divided by power emitted by the light source, assuming quantum efficiency $\phi = 1$) of about 10^{-4} only, 99.99 % of the light is lost! It is clear that an improvement of this value is highly desirable.

To obtain an optimum efficiency in the excitation path the only possibility is to put the source as close as possible to the sensor. Any optical system, however sophisticated, will only induce losses.

In the emission path the conditions are a little more complicated. Fast photodiodes as needed in lifetime measuring systems have only small surface areas (a few square millimetres at most). Therefore also the sensor element has to be small or it must be imaged in reduced size onto the detector. Small sensor elements need high light intensities to yield sufficient signal, bringing about disadvantages like photochemical indicator bleaching. Imaging with reduced size means lenses with rather low f-numbers and hence loss of light. In order to collect as much fluorescent light as possible, it is advantageous to resort to the principle of luminescent concentrators. When a transparent body containing fluorescent material is illuminated from the side, a considerable portion of the fluorescent light is guided inside the body. A detector placed at the end face of the waveguiding body yields a high light-collecting efficiency. This principle can be implemented with planar as well as cylindrical light-guiding bodies.

A typical waveguide collects light in a solid angle corresponding to a lens with f-number 1.7. In addition there is no restriction to the size of the sensor spot. A careful design of the geometry and the incorporation of absorbing dyes into the waveguide to avoid the necessity of external filters may boost the optical efficiency by a factor of 150 to 6 %, as shown by ray analysis and confirmed experimentally.

A special case of light-guiding structures useful for sensor applications is the capillary waveguide. The schematic of this sensor type is shown in Fig. 8.18. A fluorescent sensor layer is applied to the inner wall of a glass capillary. Excitation is performed from outside the capillary. Fluorescence generated inside the sensor layer is radiated into the glass capillary wall and efficiently guided therein. The advantage of this structure is a double one: First, since only light generated in the sensor layer is guided, efficient suppression of excitation light is achieved, lowering the requirement for filtering considerably. And second, the capillary serves as a sample compartment. This is espe-

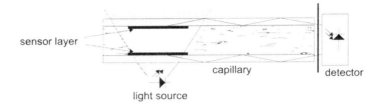

Figure 8.18. Capillary waveguide scheme

cially useful in medical applications, where standard capillaries commonly applied for collecting blood from a pierced fingertip may be used.

8.4.2.5. Sensor Optoelectronics

The main progress achieved over recent years was in electronics and optoelectronics. While 20 years ago a lifetime instrument was a large and expensive laboratory equipment, today handheld instruments are possible that allow use in near-patient testing or environmental field work.

The main breakthrough was the application of all solid-state optoelectronics, starting with the first use of a light-emitting diode (LED) for lifetime-based sensing in 1988[1]. At that time the output of blue LEDs was so weak that a photomultiplier was indispensable as a detector. Since then the brightness of the LEDs has improved dramatically. Therefore photodiodes can now be used savely for fluorescence detection. A circuit diagram for a low-cost instrument featuring blue LED excitation, photodiode detection, all solid-state processing, fully digital decay curve storage, and digital processing for capillary sensors is shown in figure 8.18.

Over the last years solid-state light sources have made great progress. Especially the choice of wavelengths has broadened considerably. Presently, LEDs and even laser diodes are available on the market down to 370 nm, and in the development laboratories wavelengths of below 300 nm have been obtained. This opens a rich field of indicator dyes that where hitherto excluded from practical sensor applications because of their absorption in the UV. On the other hand, these possibilities will only be utilized adequately if problems with the optical materials (transmission, radiation damage) can be overcome.

8.5. SUMMARY

Fluorescence decay time based sensors gained much interest in recent years due to the advantages mentioned in this chapter. As the number of sensing principles, transduction schemes and sensing devices is growing very rapidly, a detailed review of all possible aspects was impossible. Therefore, this chapter aimed at showing the power of lifetime based sensing by explaining a few subjectively chosen examples and discussing advantages and disadvantages or problems connected with the used method. What should always be kept in mind is that lifetime-based methods have their advantages only if their peculiar problems are not overlooked.

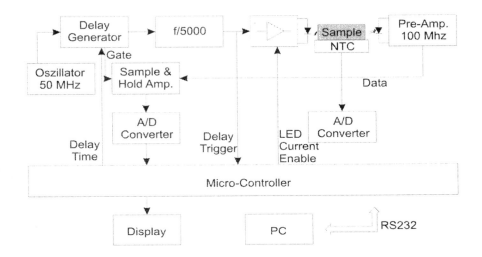

Figure 8.19. Circuit diagram for low cost sensor instrumentation

8.6. ACKNOWLEDGEMENT

The author would like to thank Max E. Lippitsch for many ideas and helpful discussions.

8.7. REFERENCES

1. Lippitsch, M. E., Pusterhofer, J., Leiner, M. J. P., and Wolfbeis, O. S.: Fibre-optic oxygen sensor with the fluorescence decay time as the information carrier. *Anal. Chim. Acta 205* (1-2), 1 (1988)
2. McNaught, A. D., and Wilkinson, A. (eds.): Compendium of chemical terminology. The gold book, 2nd edition. Blackwell Science London (1997)
3. Kasha, M.: Characterization of electron transitions in complex molecules. *Diss. Faraday Soc. 1950* (9) 14-19 (1950)
4. Wolfbeis, O. S.: Fiber Optic Chemical Sensors and Biosensors. CRC Press Boca Raton 1991
5. Lakowicz J. R.: Principles of Fluorescence Spectroscopy. Chapter 9: Quenching of Fluorescence. 1983 Plenum Press New York and London
6. Demas, J. N., DeGraff B. A., and Wenying Xu: Modeling of Luminescence Quenching-based Sensors: Comparison of Multisite and Nonlinear Gas Solubility Models. Anal. Chem. 67, 1377 – 1380 (1995)
7. Jeong Ho Chang, Hae Joong Kim, Jeung Hee Park, Young-Kook Shin, and Yongseong Chung: Fluorescence Intensity Changes for Anthracylazacrown Ethers by Paramagnetic Metal Cations. Bull. Korean Chem. Soc. *20* (7), 796 – 800 (1999)
8. Förster, Th.: Zwischenmolekulare Energiewanderung und Fluoreszenz. Annalen der Physik 55-75 (1948)

9. Jordan, D.M., Walt D. R., and Milanovich F.P.: Physiological pH fiber optic chemical sensor based on energy transfer. Anal. Chem. *59*, 437 (1987)

10. Mayr, T., Liebsch, G., Klimant, I., and Wolfbeis, O. S.: Multi-Ion Imaging Using Fluorescent Sensors in a Microtiterplate Array Format. Analyst *127* (2), 201-203 (2002)

11. Bryan A.J., De Silva A.P., De Silva S.A., Rupasinghe R.A.D.D., and Sandanayake K.R.A.S.: Photo-induced electron transfer as a general design logic for fluorescent molecular sensors for cations. Biosensors *4*, 169 – 179 (1989)

12. Draxler S. and Lippitsch M.E.: pH sensors using fluorescence decay time. Sensors and Actuators B *29*, 199 – 203 (1995)

13. Teichmann M.T.: Optimierung von Sensormembranen am Beispiel von Anionensensoren. Diploma Thesis, Karl-Franzens University Graz 1998

14. Lippitsch, M.E. and Draxler S.: A family of lifetime sensors for medical purposes. Proc. SPIE *2388*, 132 – 137 (1995)

15. Draxler, S. and Lippitsch M.E.: A family of lifetime sensors for environmental purposes. Proc. SPIE *2508*, 30 – 35 (1995)

16. Draxler, S. and Lippitsch, M.E.: Time-resolved fluorescence spectroscopy for chemical sensors. Applied Optics *35* (21), 4117 – 4123 (1996)

17. Kimberly, A.M., Sykora M., DeSimone M., and Meyer, T.J.: One-pot synthesis and characterization of a chromophore-donor-acceptor assembly. Inorg. Chem. *39*, 71 – 75 (2000)

18. Draxler, S. and Lippitsch, M.E.: Lifetime-based sensing: Influence of the microenvironment. Anal. Chem. *68*, 753 – 757 (1996)

19. Oechs, K.: Optische CO_2-Sensoren basierend auf Fluoerszenz-Abklingzeit. Diploma Thesis, Karl-Franzens University Graz 1999

20. Nemkovich, N.A., Rubinov, A.N., and Tomin, V.I.: In *Topics in Fluorescence Spectroscopy*, Lakowicz, J.R., Ed., Plenum Press: New York and London, 1991, Vol. 2

21. Janata, J.: Do optical sensors really measure pH? Anal. Chem. *59*, 1351 – 1356 (1987)

22. Klimant, I. German Patent Application DE 198.29.657 (1997)

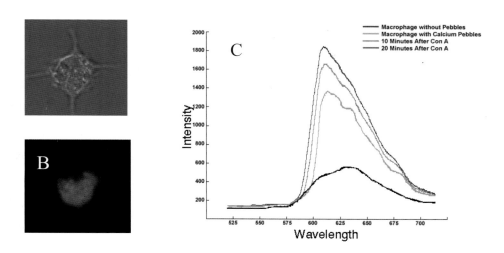

Figure 4.19. (Page 101, H. Xu *et al.*) A rat alveolar macrophage with calcium PEBBLE sensors (60x). A) Nomarski illumination and B) Fluorescence illumination, C) The increasing intracellular calcium, monitored by calcium PEBBLEs in alveolar macrophage following stimulation with 30 µg/ml concanavalin A.

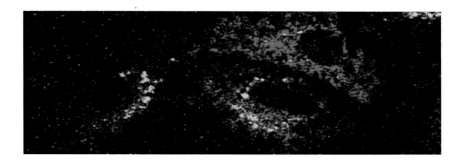

Figure 4.20. (Page 102, H. Xu *et al.*) Confocal microscope image of human C6 glioma cells containing Calcium Green/sulfarhodamine PEBBLEs (DNB moving left to right).

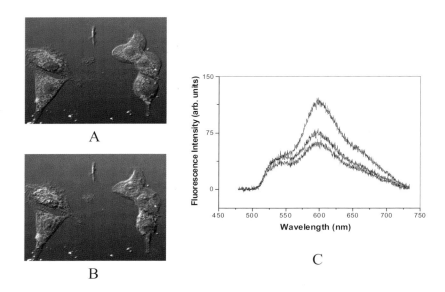

Figure 4.22. (Page 104, H. Xu *et al.*) Confocal images of rat C6 glioma cells loaded with sol-gel oxygen PEBBLEs by gene-gun injection. (A) Nomarski illumination overlaid with Oregon Green fluorescence of PEBBLEs inside cell. (B) Nomarski illumination overlaid with [Ru(dpp)3]2+ fluorescence of PEBBLEs inside cells. (C) Fluorescence spectra of a typical ratiometric sensor measurement of molecular oxygen inside rat C6-glioma cells; bottom line: cells (loaded with sol-gel PEBBLEs) in air-saturated DPBS; middle line: cells in N2-saturated DPBS, 25 seconds after replacing the air-saturated DPBS; top line: cells in N2-saturated DPBS, after 2 minutes.

Proteins: *see also* Green fluorescent protein
  (GFP)-based sensors
  molecularly imprinted polymers sensors for,
    159
  pyrene probe-based structural studies of,
    216–218
Protein transducers, in fluorescence anisotropy
  accuracy of anisotropy measurement in, 15
  advantages of, 3–4
  anisotropy calculation for, 1–3
  energy transfer in, 13, 15
  fluorescence polarization immunoassay
    applications of, 4–8
  metal ion-based fluorophore quenching-
    based, 12–15
  for metal ions, 9–15, 16
  with metal-to-ligand charge transfer (MLCT)
    probes, 6
  multiphoton excited, 4
  for nonmetal analytes, 15–16
  polarization calculation for, 1–4
  principles of operation, 1–3
  as radiometric measurement, 2, 3
Proteomes, aptamer array-based screening for,
  143–145
Pulse sampling, 265
Pyrene
  energy transfer with phenanthracene, 52–53
  fluorescence decay time of, 259
  as lifetime-based sensor fluorophore, 259
  molecularly imprinted polymer sensors for,
    175, 177, 178
  as saccharide sensor fluorophore, 50, 51
Pyrene butyric acid, as nitrate sensor
  fluorophore, 257–258
Pyrene derivatives, as excimer sensing
  photoprobes, 211–239
  advantages of, 215
  applications of, 211
  disadvantages of, 215–216
  for DNA assays, 218–219
  as emission probes for macromolecules,
    215–216
  environmental and chemical sensing
    applications of, 221–236
    calixarene-based sensing, 234–235
    chelator-based sensing, 229–231
    cis-cyclohexane-1,3-dicarboxylates-based
      sensing, 235
    coronand-based sensing, 233–234

Pyrene derivatives, as excimer sensing
  photoprobes (*cont.*)
  environmental and chemical sensing
    applications of (*cont.*)
    host-guest organic molecule sensing, 225–
      228
    metal cations sensing, 228–235
    oxygen sensing, 224
    pH sensing, 223–224
    podand-based sensing, 231–233
    pressure sensing, 222–223
    spermine-pyrene conjugate sensors, 235,
      236
    temperature sensing, 221–222
  excimer photophysics of, 212–215
  fluorescence quantum yield of, 215
  ground-state aggregation in, 212–213
  hydrophobic interactions of, 216
  on macromolecular templates, 212–215
  partial ground-state aggregation of, 212, 216
  π-stacking geometry of, 215
  pyrenyl-containing lipid membranes, 219–
    220
  for structural studies of proteins and
    peptides, 216–218
  use in aqueous media, 216
N-(1-Pyrene)maleimide, 217
N-(1-Pyrenemethyl)iodoacetamide, 217
4-(1-Pyrenyl) butanoic acid (PBA), 216, 224,
  229
di-Pyrenyl phosphatidylcholine, 222–223
Pyrocatechol violet, 56, 57

Quantum dots, 200
Quartz crystal microbalance, 178
8-Quinoline boronic acid, as saccharide sensor
  fluorophore, 46

Rac 1, green fluorescent protein-based probe
  for, 33
Ran, green fluorescent protein-based probe for,
  33
Rap 1, green fluorescent protein-based probe
  for, 32–33
Ras, green fluorescent protein-based probe for,
  32–33
Rational design, of molecular sensing probe
  aptamers, 133–134
Reactive oxygen species, use in photodynamic
  therapy, 121

Red fluorescent protein (DsRed), 24

Reduction-oxidation (redox) potential, yellow
fluorescent protein-based sensors for,
33–34

Resonance energy transfer, as lifetime-based
sensing transduction scheme, 248, 251–
253, 263–264

Rhodamine B, as L-phenylalaninamide
fluorescent competitor, 181, 183

Rhodamine Green-5, as adenosine triphosphate
signaling aptamer component, 133

Ribonucleoside 5′-triphosphate sensor, 60

Ricin
aptamer array-based sensor for, 146–150
as molecularly imprinted polymer target
molecule, 175, 179

Ruthenium (II)-tris-chelated complexes, as
lifetime-based sensor fluorophores,
259–262
metal-to-ligand charge transfer (MLCT) of,
259

Ruthenium (II)-tris (4,7-diphenyl-1,10-
phenanthroline), as PEBBLE
nanosensor indicator dye, 94, 96, 104,
108, 111

Ruthenium metal ligand complex/boronic acid-
based glucose sensor, 58

Saccharide fluorescent sensors, boronic acids-
based, 41–67
as competitive assays, 56–60
dye molecules for, 56, 59
as selective assays, 59
ternary complex formation in, 60
diboronic, 47–55
allosteric boronic acid-based, 49, 50
anthracene-based, 52, 53
calixarene-based, 49, 50
dendritic boronic acid-based, 49, 50
as disaccharide sensors, 54–55
fluorescence energy transfer (ET)-based,
52–53
fluorophore selectivity of, 52
with guanidinium receptor units, 53–54
with hexamethylene spacer, 50, 51, 52
ortho aminomethyl boronic acid-based, 50
saccharide selectivity of, 50–52
with small bile angle, 49, 50
stability constants of, 51
diol-boronate binding equilibria of, 42

Saccharide fluorescent sensors, boronic acids-
based (cont.)
with guanidinium receptors, 53–54
historical background of, 42
with hydrogen bonding receptors, 62
importance of, 41
industrial applications of, 41
internal charge transfer (ICT)-based, 47, 48
Lewis bases contraindication for, 42–43
medical applications of, 41
with metal chelates, 53
monoboronic, 42–47
in competitive assays, 56
selectivity of, 42
photoinduced electron transfer (PET)-based,
48–54
chiral recognition property of, 48, 49, 54
fluorophore selectivity of, 52
OFF/ON, 45–46, 48
stability constants of, 53–54
polymer-supported, 60–61
read-out units, 43–47
β-cyclodextrins-based, 46, 47
internal charge transfer (ICT)-based, 43–
45
N-phenyl boronic acid derivatives-based,
46, 47
photoinduced electron transfer (PET)-
based, 45–46
8-quinoline boronic acid-based, 46, 47
rigidification of, 54–55
selective interfaces of, 62
signal read-out ensembles of, 62
ternary complex formation in, 42–43
Schiff bases, 163
Second messengers, fluorescent protein-based
sensors for, 28–30, 35
Sensors, definition of, 211
Shaker potassium channel-based membrane
potential sensor, 34–35
Sialic acid receptors, 60
Sialyl Lewis X, 52
Signaling transduction agents, aptamers as,
129–143
anti-adenosine, 130–132
definition of, 130
design of, 134–139
DNA-signaling, 131, 132
evolutionary engineering of, 133, 134
fluorescence anisotropy-based, 142–143